GENETICS, GENOMICS
AND BREEDING OF BANANAS

香蕉遗传学基因组学与育种

[南非]Michael Pillay
[美]George Ude　Chittaranjan Kole　主编
冯慧敏　徐小雄　译

中国农业出版社

图书在版编目（CIP）数据

香蕉遗传学基因组学与育种/（南非）迈克尔·皮莱（Michael Pillay），（美）乔治·乌德（George Ude），（美）吉德伦金·科莱（Chittaranjan Kole）主编；冯慧敏，徐小雄译．—北京：中国农业出版社，2017.5
ISBN 978-7-109-23098-9

Ⅰ.①香… Ⅱ.①迈… ②乔… ③吉… ④冯… ⑤徐… Ⅲ.①香蕉—遗传学②香蕉—基因组③香蕉—育种 Ⅳ.①S668.103.2

中国版本图书馆 CIP 数据核字（2017）第 127446 号

中国农业出版社出版
（北京市朝阳区麦子店街 18 号楼）
（邮政编码 100125）
责任编辑　郭　科　孟令洋

北京万友印刷有限公司印刷　新华书店北京发行所发行
2017 年 5 月第 1 版　2017 年 5 月北京第 1 次印刷

开本：700mm×1000mm 1/16　印张：19.5
字数：400 千字
定价：60.00 元

Genetics，Genomics and Breeding of Bananas
Edited by Michael Pillay，George Ude，Chittaranjan Kole
ISBN 978－1－4665－0516－2

北京市版权局著作权合同登记号：图字 01-2017-8699 号

译者序

香蕉和大蕉（*Musa* spp.）是全球非常重要的水果，其出口贸易总量可能仅次于柑橘。香蕉栽培品种起源于东南亚，由 *M. acuminata*（贡献 A 基因组）和 *M. balbisiana*（贡献 B 基因组）种内或种间自然杂交而来。

目前，在热带和亚热带地区广泛种植的香蕉（AAA、AAB）、煮食蕉（ABB）和大蕉（AAB）仅有少数几个优质的品种，由于种质单一，在面对逆境、病虫害等不利因素时抗性较弱，因此对香蕉进行遗传改良刻不容缓。但是由于大多数的香蕉主栽品种为三倍体，不育，行无性繁殖，因此香蕉的常规育种进展缓慢。随着分子标记技术的发展，各种组学的研究等为香蕉的遗传改良和育种提供了新的思路。

《香蕉遗传学 基因组学与育种》由美国克莱姆森大学出版，全书共 16 章，涉及香蕉的经典遗传学与传统育种、香蕉品种起源与多样性、分子标记技术、分子连锁图谱、分子作图、图位克隆、功能基因组学、转录组学、蛋白质组学、代谢组学、主要病原菌基因组学、抗病相关基因和分子育种案例等内容。本书综述了近年来在上述领域的主要成就和今后的发展方向，可加深香蕉研究领域的科学家和育种工作者对相关领域的认识。

本书中文版本由海南热带海洋学院冯慧敏和徐小雄翻译，其中冯慧敏负责第 1 章至第 8 章及第 15 章的翻译，合计字数 21 万字，徐小雄负责第 9 章至第 14 章及第 16 章的翻译，合计字数 19 万字，全书由海南大学武耀廷研究员最终审核。

译著在完成的过程中，得到了很多同学的帮忙，感谢四川外国语学院余珊珊同学，重庆大学黄大千同学，海南热带海洋学院熊字雯、郑晓凤、杨凡、周展华、李洁、肖梦缘、陈惠、刘文侃、何伟浚、张舒婷、李莛烨、王涵、曾凡苏、于紫娟同学的帮助！

感谢国家自然科学基金（31440075）和农业部热带作物种质资源保护项目（15RZZY-46）对译著出版的资金资助。

最后需要声明的是，尽管笔者有拆句意识并且得到很多同学的鼎力相助，但限于水平、时间、精力，译文中难免存在不少问题，读者或多或少会遭遇一些阅读不适，所以，诚恳地欢迎读者批评指正，并提出宝贵意见。

译　者

2017年4月2日

序言

《香蕉遗传学 基因组学与育种》

香蕉和大蕉（*Musa* spp.）是世界上非常重要的水果。香蕉出口在贸易量上可能仅超过柑橘。*Musa* 栽培种由两个种 *M. acuminata*（提供 A 基因组）和 *M. balbisiana*（提供 B 基因组）天然杂交而来，其起源于东南亚。

甜蕉（AAA，AAB）、煮食蕉（ABB）和大蕉（AAB）的生产依赖于在世界范围广泛传播的几个优越的地方品种和栽培品种，它们被种植在热带和亚热带地区，并且在各自的地点都以本地名而熟知。

遗传改良通过杂交育种而实现，通常以三倍体地方品种为母本、二倍体为父本。高级的倍性操作可能导致二级三倍体杂交种，其为经过挑选的四倍体杂交种和优越的二倍体纯种杂交而来，这二者都能产生 *n* 配子。如果其中一个亲本产生 2*n* 配子，则由于在选择的二倍体种子中的单倍性多倍化，也可能产生芭蕉属三倍体杂种。已故的 Dirk Vuylsteke，Norman Simmonds，Ken Shepherd，Phil Rowe 和 Ren Golsalves 是世界上最受尊敬的芭蕉育种者。他们于 20 世纪在热带美洲和撒哈拉以南非洲地区进行的香蕉和大蕉遗传改良的开创性工作为改善这些小农作物和了解芭蕉属基因组铺平了道路。

分子生物学和细胞生物学研究，以及随之而来的生物技术工具可以帮助育种者进行转基因研究，从而满足 21 世纪对香蕉和大蕉的需求。现有的选择和转基因技术可以显著缩短育种过程，克服一些传统方法无法实现的农艺和环境问题。同样，分子标记辅助育种有可能显著提高芭蕉属植物遗传改良的速度和效率。

正如 Dirk Vuylsteke 博士论文的最后一段所指出的那样，“具有广泛基础的害虫/病害抗性，改良的芭蕉属种质将成为实现这种无性繁殖的多年生作物可持续生产的主要组成部分。这种种质可以通过常规的杂交育种来产生，通过利用创新方法来增加额外的遗传变异。另外，分子标记使用的增加将加速改良芭蕉属种质反复选择的进程，并因此促进新的杂交种的发展。香蕉和大蕉育种前景无限，增加的努力将立即启动一个芭蕉属进化的新阶段。”我完全同意他的观点，并且相信芭蕉属植物基因组学和转基因技术的新突破将使其能够用于培育世界上香蕉和芭蕉农民迫切需要的新品种，并可持续地生产这种作物，特别是在气候变化和高物价的时期。

Rodomiro Ortiz，博士（威斯康星大学麦迪逊分校，1991 年）

Chaclacayo（秘鲁），2011 年 2 月

目录

第 1 章　引　　言

Michael Pillay[①,*]　**Abdou Tenkouano**[②]　**Rodomiro Ortiz**[③]

摘　要： 香蕉起源于东南亚和西太平洋地区，随后传播至世界其他地区。虽然香蕉的起源与演化仍未有定论，但从植物硅酸体和分子研究中提出了一些新的观点。本章介绍了芭蕉属（*Musa*）的分类地位，重点指出新的分子数据的争议性；介绍了芭蕉属植株一般形态、不同的基因组及其大小和倍性；突出介绍了芭蕉属植物作为粮食作物的作用、主要香蕉生产国和种植面积及产量；概述了香蕉的营养信息，以及香蕉在基因组学研究、多倍体、作为模式植物在蛋白质组学的研究、育种和细胞遗传学领域的学术价值。虽然香蕉育种受到许多因素的制约，但该作物为独具慧眼的科学家提供了很多遗传改良的新机遇。香蕉育种家在很高程度上依赖于近缘野生种质的渗入基因来增加栽培种的遗传变异。另外，本书讨论了野生种质作为香蕉主要病虫害抗性资源的价值。在撰写本章时，芭蕉属植物的研究很明显地仍充满生机，新的研究小组正源源不断地添加该作物的信息。

关键词： 芭蕉属，起源，经济价值，学术价值，种质，育种

1.1　作物简史

虽然香蕉的历史悠久，但其起源和演变仍处于推测阶段，没有定论。主要原因或许是香蕉无法形成化石。随着新数据的出现，特别是来自植物硅酸体这种形成于植物细胞的、非常持久的二氧化硅的新数据和分子研究的发展，不断

① Vaal University of Technology，Private Bag X021，Vanderbijlpark 1900 South Africa；e-mail：mpillay@vut. ac. za.

② Regional Center for Africa，AVRDC-The World Vegetable Center，P. O. Box 10，Duluti，Arusha，Tanzania；e-mail：Abdou. Tenkouano@worldveg. org.

③ Department of Plant Breeding and Biotechnology，Swedish University of Agricultural Sciences，Box 101，SE-230 53，Alnarp，Sweden；e-mail：rodomiroortiz@gmail. com.

*　通信作者。

刷新对香蕉起源的看法。Lejju 等（2006）和 De Langhe 等（2010）分别描述了对香蕉的早期历史和起源的新看法，其中也附有参考文献。然而，人们普遍接受的观点是香蕉起源于东南亚的原始森林和太平洋西部，这些地区现在仍分布野生、不能食用、有硬籽的二倍体种（*Musa acuminata*）（Robinson，1996）。基于 Vavilov's（1935）的假设，作物起源于野外有最多该物种种类的地区，因而有多种芭蕉属植物的东南亚、太平洋西部被认为是芭蕉属植物的原生多样性中心。最近的遗传学研究已经证实，野生、有硬籽的 *Musa acuminata* ssp. *banksii* F. Muell. 在新几内亚被驯化，然后传播至东南亚（Lebot et al.，1993；Lebot，1999）。除了一小部分"Fei"香蕉，可食用香蕉被认为源自基因组分别是 AA 和 BB 的两个野生二倍体种（$2n=2x=22$）*M. acuminata* Colla 和 *M. balbisiana* Colla 的种间或种内杂交（Simmonds，1962）。许多被认为源自 A 和 B 基因组供体杂交的香蕉品种，其形态学显示出与 A 或 B 表型的偏差，也与 Simmonds 和 Shepherd（1955）提出的基因组简单公式不相吻合。这表明香蕉栽培种的起源可能并非一蹴而就，而是经过一系列回交、人工选择才驯化成现代作物（De Langhe et al.，2010）。通过观察大多数无性繁殖的香蕉品种的剩余繁殖力，科学家产生了一个想法，早期人类种植的那些被选中的植物通过回交后产生了种子后代（De Langhe et al.，2009）。Boonruangrod 等（2008，2009）阐述了现今关于杂交香蕉起源的新思路。*M. acuminata* 种群存在许多亚种。

下面关于香蕉早期历史的描述来自 Robinson（1996）。不可食用二倍体 *M. acuminata* 的亚种经过多年自然杂交产生了大量种间杂交种。一些杂交种可能单性结实、雌性不育，被当地人发现后成为食用水果。这些杂交种通过挑选、栽培，在当地成为食物而流通。而部分二倍体×二倍体的杂交产生了三倍体品种，这或许是由于二倍体产生了 $2n$ 配子。AAA 和 AAAA 品种或许就是这样来的。自从发现芭蕉属植物有 $2n$ 配子后这一猜测似乎合理了（Dodds，1943；Dodds and Simmonds，1946；Sathiamoorthy and Balamohan，1993；Ortiz，1997）。可食用的三倍体品种在生活力、果实大小和适应性方面通过进一步的选择而得到提升，而劣质的二倍体品种则被淘汰。*M. balbisiana* 被认为自然起源于印度的干燥季风区和菲律宾。现今在印度、斯里兰卡、缅甸和中国西南部分地区的野外，仍能觅到 *M. balbisiana* 的踪迹（Simmonds and Shepherd，1995）。在斯里兰卡一处全新世早期遗址中发现的野生香蕉种子就被认为可能是 *M. balbisiana* 的种子（Kajale，1989）。当 *M. acuminata* 的二倍体和三倍体的选种被人类带到这些干燥地区时，发生了种间杂交，从而产生了 *M. acuminata* ×*M. balbisiana* 的二倍体和三倍体品系。证明 AB 杂交的最早考古学证据来自巴基斯坦信德省果德迪吉的哈拉帕遗址发现的芭蕉属植物岩，距今已有 2 000～2 500 年历史（Kajale，1989；Fuller and Madella，2001）。另一种可能是参与了栽培种起源，特别是巴布亚新几内亚的二倍体起源的种质，如 *M. schizocarpa*（Sharrock，1989）。分子研究显示 *M. schizocarpa* 和 *M. acuminata* 关系密切（Ude et al.，2002a）。还有报道指出自然界存在 *M. schizocarpa* 和 *M. acuminata* 的杂交种（Argent，1976；Shepherd and Ferreira，1984；Tezenas du Montcel et al.，1995）。*M. schizocarpa* 和 *M. acuminata* 种质的关系，暗示着 *M. schizocarpa* 特有的 S 基因组可以被用来育种。

香蕉如何从发源地传播到世界其他地区仍无明确答案，但航海国家对这一传播可能功不可没。原马来-波利尼西亚人被认为开辟了早期横跨印度洋的远航（Lejju et al.，2006）。有迹象表明约公元前 500 年，香蕉从印度尼西亚横跨印度洋被传播到马达加斯加，然后又被传播到东非、扎伊尔（现刚果民主共和国）和西非（Robinson，1996）。而大蕉的传播则晚很多。15 世纪，葡萄牙人抵达西非，他们发现该地区已有了香蕉和大蕉。一种观点认为是葡萄牙人将香蕉从西非传播到加那利群岛，再于 16 世纪传播到海地。之后，香蕉又传入加勒比海其他地区和美洲其他热带地区。一种假设认为香蕉可能多次被引入非洲。AA 和 AAA 品种可能直接引自东南亚，而 AAB 和 ABB 杂交种则更可能从印度或斯里兰卡引入非洲（Lejju et al.，2006）。最近在乌干达 Munsa 发现的香蕉植物硅酸体证明了在公元前 4000 年非洲就有香蕉的存在（Lejju et al.，2006）。此外，该文献还报道香蕉可能远在 5 000 年前就在非洲被种植（Lejju et al.，2006），这表明香蕉的传播离不开他们的栽培历史。公元前 600—前 500 年，印度的古代经文记载了香蕉种植，这是香蕉种植最早的文字记载（Reynolds，1927），但现今人们认为巴布亚新几内亚的香蕉种植得更早，在7 000～10 000 年前就有了（Denham et al.，2003）。香蕉确切在什么时候、通过什么途径传播至热带非洲和南美洲仍然没有确切答案（Harris，2006）。然而，似乎伊斯兰征服者和阿拉伯商人在传播过程中起到关键作用。香蕉这个词源自西非，但它衍生自阿拉伯语的 *banaan*（手指），然后从西班牙语或葡萄牙语逐渐变为英语。

芭蕉属部分亚组植物的次生多样性中心被确认。至少存在 115 种已知大蕉品种的西非和中非低洼湿地被认为是大蕉的次生多样性中心（Rossel，1991；Swennen and Rosales，1994），而东非高地被认为是东非高地煮食蕉和啤酒蕉的多样性次生中心（De Langhe，1961；Stover and Simmonds，1987）。在这两个区域内，芭蕉属种植和试验的悠久历史加强了体细胞突变，这些突变不断累积增强了这些香蕉亚组的多样性（De Langhe，1964）。

1.2 植物学描述

香蕉隶属姜目，芭蕉科。芭蕉科只有两个属，芭蕉属和象腿蕉属（Simmonds，1966）。据估计，芭蕉属约有 70 个种（Hakkinen and De Langhe，2001）和 500 个栽培品种（Simmonds，1966）。由于不断有新的物种被鉴定，这些数字也在不断变化（Hakkinen，2004；Hakkinen and Meekiong，2004；Valmayor et al.，2004；Hakkinen and Hong，2007；Hakkinen，2009；Argent，2010；Hakkinen et al.，2010）。相信随着对芭蕉属物种多样性中心的不断探索，更多的芭蕉属种质会得到鉴定（Hakkinen and Wong，2007）。基于染色体数目和花序形态，芭蕉属被分为 5 组：*Eumusa*、*Rhodochlamys*、*Australimusa*、*Callimusa* 和 *Ingentimusa*。*Eumusa* 和 *Rhodochlamys* 的染色体数为 $2n=22$，而 *Australimusa* 和 *Callimusa* 的染色体数为 $2n=20$。*M. ingens* 的染色体数为 $2n=14$，这一单一物种构成了 *Ingentimusa*。*Eumusa* 包含 13～25 个种，被认为是最古老且多样性最丰富的组（Purseglove，1972），包括广泛分布在热带、亚热带国家的甜蕉、煮食蕉和大蕉。*Callimusa* 和 *Rhodochlamys* 包括 9 个非单性结实的种，它们没有营养价值，仅具有观赏价值。*Australimusa* 包括单性结实可食用的种，就是常说的 Fei 香蕉品种。Fei 香蕉的显著特点是直立果穗和红色汁液，这使它有别于其他栽培香蕉。该组香蕉是重要的食物和纤维来源，从它的假茎能提取有价值的红色染料。目前对 Fei 香蕉的起源学术界仍未达成共识。Cheesman（1950）认为 Fei 香蕉与 *M. lolodensis* 密切相关，而 Simmonds（1956）推测 *M. maclayi* 最有可能是 Fei 香蕉的祖先。Jarret 等（1992）进行的限制性片段长度多态性（RFLP）分析揭示了 Fei 香蕉确实与 *M. lolodensis* 关系最为密切。

一些学者利用分子手段检验芭蕉属下各组的分类关系后，对芭蕉属下组的分类的有效性和实用性提出了质疑（Wong et al.，2002；Nwakanma et al.，2003a）。扩增片段长度多态性（AFLP）和叶绿体分析表明 *Rhodochlamys* 和 *Eumusa* 可以合并成一组（Wong et al.，2002；Nwakanma et al.，2003a）。这一观点已得到形态数据和交叉杂交试验（Simmonds，1954）、细胞核基因组的 RFLP（Gawel et al.，1992）和 AFLP 数据（Ude et al.，2002b）的支持。Wong 等（2002）还认为 *Callimusa* 和 *Australimusa* 也可合并为一组。叶绿体 DNA（Nwakanma et al.，2003a）和 AFLP 数据（Ude et al.，2002b）显示这两个组紧密联系。

大多数栽培香蕉为二倍体或三倍体，它们被认为源自有种子的 *M. acuminata* 亚种（A 基因组供体）和 *M. balbisiana*（B 基因组供体）的种内和种间杂交（Cheesman，1948；Simmonds and Shepherd，1955）。根据形态特征和杂交研究，*M. acuminata* 的亚种包括 *banksii*，*burmannica*，*burmannicoides*，*errans*，*malaccensis*，*microcarpa*，*siamea*，*truncata* 和 *zebrine*。但是，基于分子数据，它们的分类仍然存在争议（Carreel et al.，2002；Ude et al.，2002b）。

基于细胞核和细胞质基因组的分子数据将用于阐释现今各亚种间的亲缘关

系，然而，要得出精确的和结论性的分类对更多的基因型进行分析则十分必要（Boonruangrod et al.，2008，2009）。

Musa balbisiana 广泛分布在亚洲的热带和亚热带地区，包括中国、印度、印度尼西亚、马来西亚、缅甸、尼泊尔、巴布亚新几内亚、菲律宾、斯里兰卡、泰国和越南（Ge et al.，2005）。尽管 *M. balbisiana* 种内有大量分子和形态分化，但没有任何关于 *M. balbisiana* 的亚种的描述（Ude et al.，2002a；Ge et al.，2005）。*M. balbisiana* 对香蕉的主要病虫害具有抗性，并能在干燥和寒冷的环境中茁壮成长。但 *M. balbisiana* 没有自然存在的可食用的单性结实二倍体和三倍体（Simmonds，1995）。

1.2.1 一般形态

香蕉是大型单子叶多年生草本植物，高 1.5～3 m，某些种可高达 9 m。香蕉有一个部分或全部在地下的真茎，称为球茎或根状茎。Robinson（1996）认为香蕉茎是根状茎，而非球茎。根从内层（中柱）和外层（皮层）之间的根茎发出。一条健康的根茎可产生 200～500 条主根，每条主根产生二级、三级根，从而形成根系。根茎是重要的储藏器官，供应果穗的生长和吸芽的发育。顶端分生组织位于根茎的茎尖。在开花、长出能承载果实的花序轴之前，分生组织一直长在地下。叶由根茎分生组织发育而成，由叶鞘、叶柄和叶片构成。连续的叶鞘层层堆叠、紧紧环绕形成假茎。当新叶在分生组织发育时，老叶被推向外面，变黄、干枯（Simmonds，1962）。大多数香蕉品种一生能产生30～40 片叶。固定数量的叶片发育完成后，分生组织停止分化幼叶，开始发育花序。花序通过假茎的中部向上生长。花茎从叶冠中间抽出，然后发育成穗状花序。雌花先长出，有大的子房能发育成果实。花序不断发育，最终形成一枚卵球状的、内含小花的雄蕾。然而，大多数栽培香蕉品种是单性结实，避免形成种子，以免不适合人类食用。香蕉花序形成 3 种类型的花：雌花能发育成果实；雄蕾中的雄花能生成可育或不育的花粉；第三种类型的花为雌雄同体或中性花，位于雌花和雄蕾之间的花序轴上，通常不育。大多数栽培香蕉的雌花高度不育，果实为单性结实。所有的香蕉一旦挂果，之后植株就会枯死（Simmonds，1962），它们通过根茎上的不定芽发育成的吸芽繁衍延续下去。吸芽是无性繁殖的主要材料，繁殖出无性后代。母株结果死亡后，头芽（体积大却不结果、有叶片的宿根苗）继续其生长周期。虽然野生香蕉用种子繁殖，但香蕉通过吸芽进行无性繁殖（Stover and Simmonds，1987）。吸芽的生长分为 3 个明显的阶段：褛衣芽（仅披着鳞叶的幼芽）、剑芽（叶窄如剑的吸芽）、大叶芽（体积大却不结果、有叶片的宿根苗）（Simmonds，1966；Swennen et al.，1984）。由母株和周边吸芽所形成的集群被称为“丛”。

1.2.2 基因组与基因组大小

栽培香蕉含有 A、B、S 和 T 4 种基因组（Simmonds and Shepherd，1955；Tezenas du Montcel，1988；Sharrock，1989）。其中 3 种（A、B 和 S）只存在于 *Eumusa* 中。大部分栽培香蕉只有 A 和 B 基因组，而 S 基因组仅存在于 *M. schizocarpa* 中。T 基因组为 *Australimusa* 所特有。芭蕉属植物的基因组构成在其分类中起到了关键作用。芭蕉属植物主要的基因组包括二倍体（AA，AB，BB）、三倍体（AAA，AAB，ABB）和四倍体（AAAA，AAAB，AABB，ABBB)。含有 S 基因组的包括 AS、AAS 和 ABBS，而 T 基因组出现在以 AAT、AAAT 和 ABBT 为代表的 A、B 基因组的组合中。A、B 基因组已得到分子标记的确认（Howell et al.，1994；Pillay et al.，2000；Nwakanma et al.，2003b)，其他基因组还未得到确认。

芭蕉属植物的细胞核基因组大小为 552～697 Mb，A 基因组比 B 基因组大（Dolezel，2004)。二倍体芭蕉属物种的 2C 值在 B 基因组物种的 1.16 pg 与 A 基因组物种的 1.33 pg 之间（Kamate et al.，2001)。B 基因组比 A 基因组小 12%（Lysak et al.，1999)。芭蕉属植物基因组大小为 500～600 Mb，虽然比水稻的基因组（490 Mb）约大 25%，但要分给 11 条染色体，则相对较小。因此，芭蕉属植物基因组适于全基因组测序，在本章的后面也会提到芭蕉属植物可作为开展基因组学研究的多倍体模式植物。序列可以为广泛的基因功能鉴定提供机会。

1.2.3 倍性水平

芭蕉属有 4 种基本染色体数和 3 个主要倍性水平。对于将芭蕉属划分成不同的组，基本染色体数作用很大（Cheesman，1947)。大多数食用香蕉为三倍体（$2n=3x=33$)。其他主要的倍性水平是二倍体（$2n=2x=22$）和四倍体（$2n=4x=44$)。直到 20 世纪 90 年代中期前，传统计算方法已计算了大量的香蕉染色体数目及倍性。其他表型性状包括气孔大小、气孔密度和花粉大小也被用来估算芭蕉属的倍性（Simmonds，1948；Hamill et al.，1992；Tenkouano et al.，1998)。这些方法后来发现彼此相悖，这主要由于基因型的巨大影响（Vandenhout et al.，1995；Van Duren et al.，1996)。快捷可靠的流式细胞技术在测定芭蕉属倍性上几乎取代了染色体计数（Dolezel et al.，1994；Dolezel，2004)。流式细胞技术在香蕉遗传改良项目中特别有用，它们从倍性水平相同或不同的杂交中重建了大量的群体。了解每个后代的染色体数目对育种项目至关重要。芭蕉属植物育种过程中常会出现倍性与基因组分离的情况（Bakry and Horry，1992；Osuji et al.，1997a；Pillay et al.，2004)，这对选种产生巨大阻碍，因为这些特征从形态学层面上不易察觉（Oselebe et al.，

2006)。流式细胞术可用来快速检测产生 $2n$ 配子的芭蕉属植物基因型。通过 $2x \times 2x$ 杂交得到 $3x$ 个体的数量优势，建议至少有一个亲本能产生 $2n$ 配子。

1.3 经济价值

对世界大多数不生活在热带地区的人来说，“香蕉”一词让人联想到的是甜蕉这种在世界市场上最畅销的水果之一。这可能是由于它的口感、质地和易于剥皮食用（Robinson，1996）。但是香蕉还有许多别的种类。香蕉这个名称还包括淀粉含量很高的大蕉。一般情况，香蕉（*Musa* spp.）主要分为两类：甜蕉与煮食蕉（Jones，2000）。甜蕉占世界香蕉产量的 43%，成熟后（变黄后）直接生食，特点为果实口感香甜。甜蕉在某些热带国家也被用来煮食，特别是在没有已知或可用的煮食类型的地区。由于即使在成熟阶段淀粉含量也很高，煮食蕉占到了世界香蕉产量的 57%。通过烹、煮、炸，使它们变得可口。煮食蕉中最有名的当属大蕉，它们占世界香蕉总产量的 23%。其他煮食蕉类型在果实绿色时蒸或煮，在营养成分上是提供类似马铃薯的淀粉主食（Simmonds，1966）。它们包括独特的东非高地香蕉，这种香蕉只生长在东非的中高纬度地区。

香蕉和大蕉种植在热带和亚热带地区的 130 多个国家，这些地区的气候条件似乎是香蕉生产的理想条件。就生产总值而言，香蕉属于最重要的热带食物来源之一，仅次于水稻、小麦、玉米这类谷物主食。就贸易量而言，香蕉的出口量第一，出口额仅次于柑橘，排名第二。香蕉全球生产总量现今超过 1 亿 t，然而仅有 10%进行出口贸易（FAOSTAT，2005）。2004 年，香蕉的全球出口总量接近 0.159 亿 t。表 1-1 列举了世界香蕉主产国。

表 1-1 2004 年世界香蕉主产国

国 家	产量（t）	国 家	产量（t）
印度	16 820 000	尼日利亚	2 103 000
乌干达	10 515 000	墨西哥	2 026 610
巴西	6 602 750	泰国	1 900 000
厄瓜多尔	6 552 000	喀麦隆	1 830 000
中国	6 420 000	秘鲁	1 660 310
菲律宾	5 638 060	科特迪瓦	1 602 423
哥伦比亚	4 400 000	布隆迪	1 600 000
印度尼西亚	4 393 685	刚果民主共和国	1 412 000
卢旺达	2 469 741	越南	1 353 800
加纳	2 390 858	危地马拉	1 268 000
哥斯达黎加	2 230 000	洪都拉斯	1 225 066

数据来源：FAOSTAT，2004。

2004 年，十大香蕉主产国的香蕉产量约占香蕉全球总产量的 75%。全球约 98%的香蕉产自发展中国家，超过 50%的香蕉产自印度、乌干达、巴西和厄瓜多尔。香蕉的最大出口国是厄瓜多尔、菲律宾、哥斯达黎加、哥伦比亚和危地马拉。尽管印度、乌干达和巴西是排名前 3 的香蕉生产大国，但这些国家的香蕉或大蕉的出口量很少。据估计，产自印度、乌干达、巴西和中国的香蕉，超过 90%以多种方式被本国消耗，因为这几个国家都有各自的传统菜肴和加工方式（Frison and Sharrock，1999）。新几内亚岛和东非大湖带周边国家的香蕉消耗率也相当高，平均每人每年的香蕉消耗量为 200～250 kg。除了作为主食，香蕉和大蕉贸易为世界数以百万计的人们提供重要收入来源。香蕉产业也是拉丁美洲、加勒比海、亚洲和非洲地区发展中国家收入、就业和出口创汇的重要来源，全球每年香蕉出口额超过 47 亿美元（数据来自联合国粮农组织贸易与发展会议秘书处，网址：www.unctad.org，时间：2011 年 1 月 28 日）。对于其中一些国家，香蕉出口是其主要的外汇收入来源。联合国粮农组织认为："在地方和区域层面上，香蕉带来的收入和家庭粮食安全建立了牢固的联系（FAO，1999）。"表 1-2 列举了 2004 年部分国家的香蕉生产面积。

表 1-2　部分国家的香蕉生产面积

国　　家	面积（hm^2）
印度	680 000
巴西	491 042
菲律宾	415 427
印度尼西亚	314 708
布隆迪	300 000
中国	269 150
厄瓜多尔	226 521

数据来源：FAO，2004。

1.4　营养价值

公共机构和农业产业正使用营养信息来推广鲜活农产品。富含抗氧化的维生素（维生素 A、维生素 C、维生素 E）及钙、镁与钾的食品受到关注健康的消费者追捧（Wall，2006）。食用更多水果、蔬菜意味着降低患心血管疾病、中风及口腔癌、咽喉癌、食管癌、肺癌、胃癌、结肠癌的风险（引自 Wall，2006）。香蕉富含碳水化合物、纤维、蛋白质、脂肪和维生素 A、维生素 C、

维生素 B_6（Marriott and Lancaster，1983；Robinson，1996）。卡文迪什香蕉被证实含有抗氧化剂儿茶素（Someya et al.，2002），并且果皮中的含量（1.58 mg/kg，以干重计）比果实中的含量（0.296 mg/kg，以干重计）更高，表明常被丢弃的香蕉皮有成为食物来源的潜在可能。

香蕉的矿物浓度已被报道（Wenkam，1990；Hardisson et al.，2001；USDA-ARS，2004）。表 1－3 列举出美国农业部农业科学研究院（USDA-ARS）得出的香蕉营养含量。

表 1－3 每 100g 生香蕉所含营养值

	含 量	美国推荐的成人摄入量
碳水化合物	22.84 g	
糖	12.23g	
膳食纤维	2.6g	
脂肪	0.33 g	
蛋白质	1.09 g	
硫胺素（维生素 B_1）	0.031 mg	2%
核黄素（维生素 B_2）	0.073 mg	5%
烟酸（维生素 B_3）	0.665 mg	4%
泛酸（维生素 B_5）	0.334 mg	7%
维生素 B_6	0.367 mg	28%
叶酸（维生素 B_9）	20μg	5%
维生素 C	8.7 mg	15%
钙	5 mg	1%
铁	0.26 mg	2%
镁	27 mg	7%
磷	22 mg	3%
钾	358 mg	8%
锌	0.15 mg	1%

数据来源：2011 年 4 月 21 日，美国农业部营养数据库（www.nal.usda.gov）。

虽然不清楚这组营养信息来自哪个香蕉品种，但最有可能的是卡文迪什亚组（AAA）。最近的一项综合研究发现，相比威廉姆斯香蕉（AAA），矮巴西香蕉（AAB）更有营养，含有更多的维生素 A、维生素 C 和矿物质磷、钙、镁、锰、锌。矮巴西香蕉的维生素 C 含量几乎是威廉姆斯香蕉的 3 倍（Wall，2006）。之前的研究还发现地理位置会影响香蕉的矿物质和维生素含量水平。

水果营养分析往往不考虑水果品种、成熟度、环境对该水果的维生素和矿物质含量的影响。水果收获后的营养构成在很大程度上取决于水果品种、成熟度、气候、土壤类型和肥力（Lee and Kader，2000）。香蕉中的类胡萝卜素含量随着香蕉的成熟而增加，而抗坏血酸含量随着香蕉的成熟而减少（Lee and Kader，2000）。

香蕉作为食物来源的优点被 Robinson（1996）充分列举。甜香蕉低脂、低胆固醇和低盐，并有许多特殊饮食治疗的价值。因为易消化，成熟的香蕉是婴儿以及老人极好的食物。而低钠盐、无胆固醇脂肪使得香蕉非常适合用于高血压和心脏病患者的饮食调控。由于低钠盐、低蛋白，香蕉被用于肾脏疾病患者的特殊饮食。香蕉也适于有消化性溃疡和胃炎、肠胃炎等胃部问题人群食用。香蕉集低脂肪、口感好为一体，非常适合肥胖者食用。

在许多国家，香蕉在减轻微量营养素缺乏症上发挥了重要作用。食品中的微量营养素缺乏症已受到世界卫生组织和世界银行的关注（World Bank，1994）。微量营养素营养不良症主要发生在发展中国家贫困家庭的妇女、婴儿和儿童身上（WHO，1992；Mason and Garcia，1993）。撒哈拉以南非洲、南亚和东南亚、拉丁美洲和加勒比地区的妇女、儿童正处于疾病、过早死亡，以及因为饮食差而缺少关键营养素，特别是铁、维生素 A、碘和锌而导致认知能力损伤的风险。然而，这些国家都是世界香蕉主产国。Engelberger（2003）的研究表明，密克罗尼西亚联邦的香蕉品种中含有丰富的维生素 A 原类胡萝卜素。未发表的研究也表明，巴布亚新几内亚的香蕉品种含有丰富的维生素 A、铁和锌（Fungo，2007）。鉴于这些数据，似乎应致力于鉴定并传播具有高营养价值的香蕉到这些微量元素严重缺乏的国家。然而，香蕉产品本身仅能满足人类一部分的饮食需求，因而香蕉消费应被视作饮食多样化的一部分（Honfo et al.，2007）。

1.5 学术价值

目前植物基因组学使用两种模式植物，拟南芥（*Arabidopsis*）和水稻（*Oryza sativa*）。它们作为基因组学研究模式植物的优势包括基因组相对较小（拟南芥 125 Mb，水稻 490 Mb）、便于操作、生命周期短。它们也很容易转化，并且自然分布广泛。由于具有合理的推理，模式植物被其他相关群组推荐作为它们的选择。例如，杨属（*Populus*）被推荐作为树木遗传学研究的模型（Taylor，2002），而豆类植物，如苜蓿（*Medicago truncatula*）和大豆已成为植物基因组学研究的模式植物（Barker et al.，1990；Gepts et al.，2005）。芭蕉属植物也被推荐为多倍体模式植物，用以分析其他多倍体物种的基因组

(Frison et al., 2004)。为什么将芭蕉属植物作为植物基因组学研究的模式植物？这些作者给出了理由。不仅因为芭蕉属植物重要的农业地位，还因为它在结构和功能基因组学的几个基本方面都是很好的模型物种，在这几个方面拟南芥和水稻无法取代。从形态学、细胞器到细胞核 DNA 标记，学者对芭蕉属植物的结构有丰富了解。芭蕉属植物极为丰富的多样性为检验基因调控提供了良好模型。芭蕉属植物是良好的基因组学研究模型，因为它是少数双亲细胞质遗传的植物之一，双亲细胞质遗传线粒体中为父系遗传，而叶绿体中为母系遗传。作为无性繁殖作物，芭蕉属植物为研究体细胞无性系变异的作用和“印迹”等现象提供了无限可能。基因组印迹是一种表观遗传现象，在大多数情况下被认为发生在配子发育时期。基因组印迹发生在母系和父系等位基因同时存在，一个等位基因表达而另一个等位基因沉默时（Albertson et al., 1996）。芭蕉属植物的基因组（500～600 Mb）比水稻约大 25%，对整个基因组的测序可以得到关于基因特件和功能的信息。芭蕉属植物也被认为是在基因组水平研究植物进化的理想模型。芭蕉属已在相同的环境中进化了几千年，许多不育克隆的基因组通过营养繁殖而固定下来。在同一进化时期出现的有部分生育能力和生育能力强的二倍体为基因组进化提供了重要线索。模式植物应该是容易转化的。芭蕉属已开发出了许多转化系统。源自胚性细胞悬浮的原生质体电穿孔（Sagi et al., 1994）、胚性细胞的粒子轰击（Sagi et al., 1995）、分生组织的共培养（May et al., 1995；Tripathi et al., 2005）或与根癌农杆菌（*Agrobacterium tumefaciens*）的细胞悬浮共培养（Ganapathi et al., 2001；Khanna et al., 2004）等方案在香蕉和大蕉中都是可行的。

1.5.1 多倍性

拥有两个或以上单倍体染色体组的生物称为多倍体。人们早已认识到多倍体在植物进化中起重要作用。科技进步使研究基因组进化和基因复制的影响成为可能，因而重燃了人们对多倍体的研究兴趣（Mable, 2003）。除 Tate 等（2005）对植物多倍体的修订外，最近很多关于多倍体的论文浮现，涵盖了一系列主题，包括基因的复制和进化（Lynch, 2002）、基因组进化（Wendel, 2000）、新基因的表达机制（Osborn et al., 2003）、复制出的基因对转录组的影响（Adams et al., 2003）、光合速率（Wullschleger et al., 1996）、影响植物高度和开花的数量性状位点（QTL）（Ming et al., 2002）、多倍体水分亏缺时的结构和生理特性（Li et al., 1996）及多倍体的优势和不足（Comai, 2005）。由于香蕉既是同源多倍体，又是异源多倍体，因而可作为模式植物来进行多方面的研究，如多倍体的功能遗传学、基因沉默和复制基因的组织或发育特异性表达。Mable（2003）描述了在不同栖息地二倍体和四倍体细胞型替代的适应、植物进化和植物中多倍体的形成和作用。Mable（2003）所著一篇论文提出了香蕉研究新领域的有趣的想法，为多倍体研究提供了新思路。这些想法中的一部分描述如下：虽然产生 $2n$ 配子一直被认为是多倍体的主要机制，体细胞加倍也被认为发挥着重要作用；植物研究的当前目标是细胞学机制和参与 $2n$ 配子产生的基因的数目与位置；细胞遗传学技术的很多进展，如染色体涂染技术、染色体分选技术、高分辨显带技术、显微解剖和原位杂交技术能更精准地剖析多倍体减数分裂时控制同源和部分同源配对的、可能导致影响三倍体生育力的因素。认为相比二倍体，多倍体额外的基因组拷贝使得它们能够忍耐、抵御更严苛的环境。多倍体可能更耐旱（Wullschleger et al.，1996），或具有更高水平的光合作用（Li et al.，1996）。植物生态适应的研究现在正处于分子水平，其方法为观察调控基因的变化（Lawton-Rauh et al.，2003）和 QTL（Ming et al.，2002）。

二倍体植物中大量基因复制的证据表明，多倍体促成了进化的多样性。了解这些变异的机制和影响是植物多倍体研究的重要组成部分。

复制的基因可能有多种命运——沉默、发育新功能、在不同组织和发育阶段功能分化（Mable，2003）。

植物中的中间三倍体阶段被认为是形成四倍体的最有可能的路线。因为三倍体通常不育，所以这个想法受到质疑。但三倍体中存在低频率的可育配子，这极有可能能穿越这个屏障。食用香蕉大多数是三倍体，但目前已知可育性确实能发生在三倍体水平（Pillay et al.，2002）。

与二倍体相比，芭蕉属多倍体植株更大，生态适应性更强，并且存在生殖差异。但仍不清楚这种引起多倍体在新环境中适应潜力的发育差异的成因和造成的后果（Mable，2003）。

Tate 等（2005）充分描述了不同类型的多倍体。从严格意义上来说，同源多倍体是通过一个物种的种内杂交而成，这种杂交要么来自单一个体的基因组加倍，要么来自基因相似个体的 $2n$ 配子融合。种间同源多倍体由基因不同但染色体结构相似的同一物种的不同种杂交而成。这两种类型的同源多倍体都有多价染色体和多体性遗传的特点。纯的或基因组的异源多倍体由远缘物种之间的种间杂交而成。异源多倍体与同源异源多倍体有一个显著区别。对后者而言，异源多倍体遵循的是基因组倍增。异源多倍体的特点是二价染色体和双体遗传，而同源异源多倍体中染色体配对后导致多价染色体或二价体，并且一些基因座是二体遗传，另一些则是多体遗传。部分异源多倍体被视为种内植物，但彼此的基因组却不同，有大量不同的基因或染色体片段。它们是朝着同源异源多倍体或纯的异源多倍体进化的不稳定多倍体，其特点是二价染色体和二体遗传或者多价染色体和多体遗传。关于芭蕉属植物基因遗传方式的信息几乎没有。但是，对香蕉基因型细胞学分析观察到了各种各样的染色体配对，从单价染色体到多价染色体（Adeleke，2001）。原位杂交研究报告了香蕉不同基因组之间的交叉杂交（Osuji et al.，1997b；D'Hont et al.，2000）。发现 A 和 B 基因组交叉杂交的数量比 A 和 S 基因组的多。最少出现的交叉杂交是 T 基因组与 A、B 基因组之间的。交叉杂交的强度反映了基因组之间的序列同源性和亲缘关系。上述信息表明，香蕉不仅仅是纯种的同源或异源多倍体。因此，香蕉可以作为模式植物用于多倍体研究。

1.5.2 蛋白质组学

1994 年，科技文献首次介绍了"蛋白质组学"。自此，这一领域的研究特别是在植物方面取得了巨大进展。蛋白质组学被描述为基因组表达的蛋白质系统分析。它包括鉴定其一级氨基酸序列、确定它们的相对数量、比活、修饰情况、与其他蛋白质或不同类型分子的互作、亚细胞定位和 3D 结构（Barbier-Brygoo and Joyard，2004；Kwon et al.，2006）。

蛋白质组学研究似乎在全基因组测序后的物种上有更大进展。香蕉的蛋白质组学研究仍处于起步阶段，目前仅有少量的成果发表（Carpentier et al.，2007；Samyn et al.，2007），尽管香蕉被推荐为研究分生组织蛋白质组的完美模型（Carpentier et al.，2007）。超过 130 个蛋白质在香蕉 1～2 mm 的分生组织顶层被鉴定。香蕉有多个芽分生组织，可通过成熟的芽分生组织培养技术进行繁殖（Strosse et al.，2006）。这使得用单一分生组织做多个试验成为可能。例如，Carpentier 等（2007）研究了香蕉中蔗糖介导的渗透胁迫的影响。在这个方法中，鉴定出蛋白质显著的上调或下调。使用类似的方法，香蕉的分生组织蛋白质组可用于研究不同胁迫下蛋白质的数量差异。

科学家们对植物蛋白质组学的兴趣正在增加，并且并不局限于当前作为基因组模型的物种。功能蛋白质组学被用作分析植物的特异过程和对各种生物和非生物胁迫反应的工具。模式基因组测序后，植物中成千上万的蛋白质仍有待发现。功能蛋白质组学和蛋白质表达谱的谱写将是植物蛋白质组学面临的下一挑战（Rossignol et al.，2006）。

1.5.3 育种

虽然香蕉育种受限于三倍体和雌雄生育力低，但是其为独具慧眼的科学家提供了很多遗传改良的新机遇。大多数可食用三倍体品种不携带病虫害抗性基因，因此，这些基因必须从可育的二倍体种质渗入。对于香蕉育种，已制订了各种方案（Rowe and Rosales，1996；Vuylsteke et al.，1997；Tenkouano，2005；Pillay and Tripathi，2007）。其中一个方案是二倍体作为父本与三倍体品种杂交。在这个方案中，选择四倍体是出于兴趣。但却发现，四倍体可育，并且如果暴露在可育花粉下能结籽。因此四倍体和改良后的二倍体杂交后生成次级三倍体。另外，四倍体也可与四倍体杂交。$4x$ 个体的主要花粉类别是亚倍体，由于倍减效应，该亚倍体含有一半单倍体染色体物质（Tenkouano，2005）。所以 $4x \times 4x$ 杂交的后代基本上是三倍体，该三倍体积累来自其三倍体外祖母的基因。因此这些 $4x \times 4x$ 杂交后代在形态和特征上表现出与其外祖母高度相似也不足为奇了（Pillay，未发表）。

消费者对这类杂交种的接受程度比预计要高，因为它们与农夫偏爱的原始品种类似。$4x \times 4x$ 杂交也可以作为遗传媒介，连接因不育而无法相互杂交的三倍体。通过这种方式，$4x \times 4x$ 育种方案，可新增更多重组案例，将之前的不可能变为可能。香蕉 $4x \times 4x$ 育种方案可以作为其他作物育种的模式。

1.5.4 细胞遗传学

早期，芭蕉属植物细胞遗传学的研究重点为确定染色体数目（Cheesman，1932a，b；Cheesman and Larter，1935）。随后，研究的重点放在人工杂交种的染色体配对上（Dodds，1943；Wilson，1946）。Shepherd（1996）进行了芭蕉属植物减数分裂的综合研究。Agarwal（1987，1988a，b）发表了一系列关于芭蕉属植物减数分裂的论文。而分子生物学的出现似乎减退了生物学家们对生物体细胞遗传分析的热情，因为生物体的 DNA 检验现已成为可能。香蕉染色体体积小、染色能力差，这对芭蕉属植物细胞遗传学研究又是一大阻碍。

由于香蕉染色体的体积小，缺少细胞学标记，部分香蕉独特的染色体和数目尚未确定。因此，任何关于芭蕉属植物染色体核型分析的研究一经发表都被视为首创。芭蕉属植物中唯一容易确认的染色体是随体染色体，它在 *M. acuminata* 和 *M. balbisiana* 中都是最长的染色体。对于许多染色体小的种质，减数分裂粗线期为发展中的核型分析提供了一种替代方法。更先进的染色技术使得首次观察到芭蕉属植物的粗线期染色体变为可能（Adeleke et al.，2002）。一些芭蕉属研究应用分子细胞遗传学技术，得出了关于基因组鉴定、杂交种基因组构成、NOR（核仁组织区）区域位点（Osuji et al.，1998；Dolezelova et al.，1998；D'Hont et al.，2000）、移动遗传因子的存在（Baurens et al.，1997）和转座子（Balint-Kurti et al.，2000）的有趣结果。基因组的这些方面对香蕉育种项目，特别是有多个基因组和复杂遗传系统的作物来说意义重大。

1.5.5 非整倍体

非整倍体出现在一些三倍体芭蕉属无性系的根尖分生组织中（Shepherd，

1996；Shepherd and Da Silva，1996）。

在田间试验中，非整倍体表现出不同的特征：有些相当正常和强健；有些衰弱，还未开花就死亡；有些不能存活，还有很多不能长到成熟期（Shepherd，1996）。香蕉突变体的一系列非整倍体是很有价值的遗传学工具，可用来鉴定在不同香蕉染色体中发现的基因。使用流式细胞术能实现快速检测芭蕉属植物的非整倍体（Roux et al.，2003）。尽管它非常具有研究价值，但似乎生物学家对这类关于芭蕉属的基础研究缺乏兴趣。

1.6 种质、基因库和野生种质的利用

基因资源指的是具有遗传多样性的材料，包括农家种、现代栽培品种、作物的野生近缘种和其他野生种，它们能为农民和植物育种家开发新作物和高产作物提供可能。野生种和许多作物的原始品种中含有有用的抗性基因，地方品种含有优质特性，杂交种则有优良的农艺性状。香蕉育种家很大程度上依靠野生近缘种的基因渗入来加强栽培品种的遗传变异。地方品种、可杂交的野生近缘种和杂交种构成了香蕉育种的初级基因库。在以下章节，笔者检测了香蕉一些主要的病虫害，并讨论了野生种质作物抗性资源的价值。

1.6.1 病虫害

1.6.1.1 黑叶斑病

因斐济球腔菌（*Mycospharella fijiensis*）而导致的香蕉叶斑病是对全球香蕉和大蕉经济危害最大的一种疾病（Carlier et al.，2000）。黑叶斑病最重要的抗性材料是一种野生二倍体香蕉（*M. acuminata* ssp. *burmannicoides*），俗称“Calcutta 4”（Rowe and Rosales，1990）。许多育种项目都将 Calcutta 4 用作黑叶斑病的抗性资源。Pisang Lilin 是 *M. acuminata* ssp. *malaccensis* 的单性结实二倍体（AA）种，和 Tuu Gia 一起被用作黑叶斑病的抗性资源。洪都拉斯农业研究所（FHIA）的项目通过将 Calcutta 4 和其他可繁殖的二倍体杂交，培育出了一些特殊果串大小并能抗病虫害的二倍体。能抗黑叶斑病的重要二倍体包括 SH2829、SH3437、SH3142 和 SH3217。FHIA 育种项目的成功似乎很高程度上来自这些抗性二倍体的最初的开发。同样，国际热带农业研究所（IITA）使用 Calcutta 4 和 Pisang Lilin 作为主要来源，将宿主植物的抗性引入了非洲大蕉（Vuylsteke et al.，1997）。

1.6.1.2 镰孢枯萎病

由土传真菌尖孢镰孢（*Fusarium oxysporum*）引起的镰孢枯萎病是香蕉最具毁灭性的病害之一。尖孢镰孢分为 4 个生理小种（1、2、3、4 号生理小

种）。体细胞的相容性被用来确定镰孢种群的遗传分离。营养体亲和群（VCGs）的测定是通过不利用硝酸盐营养体缺陷型突变体形成异核体进行的。*F. oxysporum* f. sp. *cubense* 有 16 个 VCGs 或 VCG 复合物（Ploetz，1990；Ploetz et al.，1997）。

据报道，野生二倍体 *M. acuminata* ssp. *malaccensis*、Calcutta 4 和可食用二倍体 Pisang Lilin 是镰孢 1 号和 2 号生理小种良好的抗性资源（Shepherd et al.，1994；Rowe and Rosales，1996）。Pisang Lilin 是培育抗枯萎病品种 Gros Michel 的第一种抗性资源作物。Gros Michel×Pisang Lilin 杂交产生的 Bodles Altafort 或 1847，Highgate×Pisang Lilin 杂交产生的 2390-2，是两个抗枯萎病品种（Jones，2000）。随后，从 *M. acuminata* ssp. *banksii* 和非洲桑给巴尔的 AA 二倍体 Paka 的杂交中选出父本。20 世纪 60 年代，*M. acuminata* ssp. *malaccensis* 和 *M. acuminata* ssp. *banksii* 的杂交种作为父本用于育种。随后，通过将 *M. acuminata* ssp. *malaccensis* ×ssp. *banksii* 得到的二倍体与其他种，包括 *M. acuminata* ssp. *banksii*×Paka 的杂种进行杂交，得到了改良的二倍体。

据报道，FHIA 的二倍体 SH3142、SH3362 和 SH3437 能抗镰孢 1 号和 2 号生理小种，而 SH3362 还能抗 4 号生理小种（Rowe and Rosales，1996）。利用这些二倍体，FHIA 已经培育出能抗 1 号生理小种的两种四倍体杂交种，FHIA17 和 FHIA23。此外，FHIA01 已显示出对 1 号和 4 号生理小种的抗性。巴西农牧研究院（EMBRAPA）的育种计划项目已培育出了许多抗镰孢的无性繁殖系（Shepherd et al.，1994）。它们包括分别来自 Prata Ana、Pacovan 和 Prata 的 PA03-22（AAAB）、PV03-44（AAAB）和 PC12-05（AAAB）。这些无性繁殖系对 1 号生理小种的抗性已在多个国家检测。枯萎病是威胁东非、南非 Pisang Awak（ABB）和 Sukali Ndizi（AAB）生长的主要病害之一。8 种基因型种质，Pisang Lilin、Fougamou、Pisang Awak、SH3217、SH3142、SH3362、Yangambi Km5 和 Kikundi，在乌干达受镰孢侵染的农田里进行了抗性检测。其中，Pisang Awak 和 Fougamou 在 6 个月内出现典型的枯萎病症状，叶片变色、枯萎，假茎分裂，维管组织褐变。这些种系被认为易感染镰孢 1 号生理小种。而 Pisang Lilin、SH3217、SH3142、SH3362、Yangambi Km5 和 Kikundi 在乌干达感染镰孢的田地里生长了 60 个月，未出现任何枯萎病症状。这后 6 种基因型种质被认为具有抗性（Pillay，未发表）。

1.6.1.3 细菌性枯萎病

细菌性枯萎病是影响东非香蕉的相对较新的病害。引发这种病害的细菌之前称为 *Xanthomonas campestris* p. *musacearum*，现在称为 *X. vasicola*（Carter et al.，2009）。这种病害会导致总产量减产，因为病株的果实无法食用。该病现流行于乌干达、卢旺达、刚果民主共和国东部和坦桑尼亚，最近肯尼亚也确认出现该病（Mbaka et al.，2007，2009）。该病的典型症状包括植株枯萎、过早变黄和果实腐烂（Biruma et al.，2007）。据估计，该病每年对乌干达的经济造成3.6亿美元的损失。如果对此病不加以抑制，它将对乌干达的香蕉主产区和大规模种植香蕉的毗邻国家产生巨大损失。栽培控制措施能遏制该病害的蔓延，但为了有效抑制需要所有的农民采纳（Eden-Green，2004，2007；Blomme et al.，2005a，b）。

迄今为止，未发现有抗细菌性枯萎病的香蕉基因型。

研究人员尝试筛选一系列基因型来鉴定乌干达黄单胞菌的抗性资源。该试验的初始观察表明，即使周围环绕着被感染的植物，基因型Yangambi Km5、Kikundi、Kisubi、TMB×660K1、TMB×201K、Yalim、Nakasabira、TMB×1438K-1、TMB×7197-2、SH3217、SH3362、SH3640-9和Nakitengu没有显示出被感染。尽管现有文献还不清楚该病菌是如何传播的，但Pillay先前的观察表明该病菌可通过媒介物、污染的农具、感染的种植材料和在已受感染的土地种植而传播。媒介物传播可能是该疾病传播最重要的手段。疾病的传播被认为是媒介物（昆虫、鸟类或蝙蝠）将接种物（细菌）从病株转移到健康植株上。该疾病的切入点似乎是花蕾苞片的“缢痕”，因为该疾病症状在这个地方最先被发现。植物如果没有苞片，就不会为疾病敞开入口，就可能“躲过”疾病。在东非高地香蕉中，基因型Nakitembe的无性克隆系不脱落它们的花序苞片。乌干达指导的一项试验表明，不脱落花序苞片的香蕉品种，如Nakitembe和Mbwazirume，不会感染黄单胞菌（Pillay，未发表）。

1.6.1.4 病毒病

目前已知有4种典型的病毒威胁着香蕉和其他芭蕉属物种。它们是香蕉束顶病毒（BBTV）、香蕉条纹病毒（BSV）、香蕉花叶心腐病毒（CMV）和香蕉苞片花叶病毒（Pietersen and Thomas，2001）。

随着病毒观察技术日益成熟，病毒学家能够发现危害香蕉的新病毒。从非洲、美洲、东南亚和澳大利亚的香蕉中观察到类似病毒的粒子，从而发现了一种新病毒，被称为温和的香蕉花叶病毒。另一种新病毒，香蕉枝叶枯死病毒，在尼日利亚已有记载（Hughes et al.，1998）。所有的病毒性疾病中，香蕉束顶病最严重、破坏性最大（Rishi，2009）。早期识别和根除被感染植株是控制病毒病的最佳手段。

尽管病毒诊断技术最近已大为改善，但病毒筛选程序仍不发达。抗病毒种质的大规模筛选受到限制。

因此，侵染香蕉的不同病毒的抗性资源尚不清楚。由于缺乏合适的亲本材料，抗病毒育种没有任何进展。对此，Dahal等（2000）基于病毒症状、病毒指数，以及通过酶联免疫吸附试验（ELISA）测定叶片组织中 BSV 抗原的相对浓度、生长势和产量性状，评估了 BSV 的天然发生率。在种植周期中，病毒的发生、症状的表现和 BSV 抗原的相对浓度随着季节变化剧烈波动，雨季最高，在炎热干燥的季节几乎可以忽略。叶组织样品中 BSV 抗原的相对浓度比大多数多倍体杂交种高，但比地方品种低。虽然有高的自然发病率和相对较高的叶组织抗原浓度，一些 IITA 和 FHIA 杂交种却减产不多，或根本不减产，因而可被视作“BSV 田间耐受型”。

1.6.1.5 象鼻虫

象鼻虫（*Cosmopolites sordidus*）是对香蕉和大蕉危害最大的害虫。成年象鼻虫在植物底部产卵，孵化后幼虫钻入球茎，以球茎为食进行生长。幼虫在球茎中钻出的密集通道会影响植株水分和矿物质的吸收，导致植株衰弱和减产（果穗重量）。象鼻虫危害严重时，受害植株普遍倾倒和过早死亡。象鼻虫严重危害的症状还有植株假茎基部开裂和幼叶扭曲变形。

评估抗象鼻虫的香蕉种质的研究很少（Fogain and Price，1994；Kiggundu et al.，2006）。大蕉（AAB）被认为比其他香蕉基因型更易感染象鼻虫（Ittyeipe，1986；Fogain and Price，1994）。Pavis（1991）证实，虽然 Pisang Awak 品种被象鼻虫钻出很多通道，但表现出高水平的耐受性，而 Yangambi Km5 几乎没有受到象鼻虫攻击。Pisang Awak 和 Bluggoe（AABs）通常被认为比大蕉（AAB）和东非高地蕉更耐象鼻虫（Abera et al.，1997）。大蕉和 Calcutta 4 的杂交表明，大部分的二倍体杂交种能抗象鼻虫，而大多数的多倍体则易感象鼻虫（Ortiz et al.，1995）。因此，Calcutta 4 被认为是象鼻虫的抗性资源。其他抗象鼻虫的基因型包括 Sannachenkadali（AA）、Sakkaki（ABB）、Senkadali（AAA）、Elacazha（BB）、Njalipovan（AB）、FHIA03、TMB×612-74、TMB2×6142-1、TMB2×8075-7、TMB2×7197-1、Long Tavoy、Njeru、Muraru、Bluggoe 和 *M. balbisiana*（Kiggundu et al.，2003）。现已发展出一种新型筛选香蕉基因型对象鼻虫抗性的方法（Kassim et al.，2010）。该方法可以在筛选室进行象鼻虫抗性和易感性的筛选，且仅需要 7 个月的时间，而田间检测则需要 2～3 年。

1.6.1.6 线虫

线虫被认为是危害香蕉的主要害虫，波及范围遍布世界大部分地区。据估计，线虫导致全球香蕉年均减产约 20%（Sasser and Freckman，1987）。最重要和最流行的香蕉线虫有 *Radopholus similis*（穿孔线虫）、*Helicotylenchus multicinctus*（螺旋线虫）、*Pratylenchus coffeae* 与 *P. goodeyi*（根腐线虫）、*Meliodogyne incognita*（根结线虫）。此外，其他小范围发生的线虫，如 *Heterodera oryzicola*（印度）、*Radopholus reniformis* 和 *Helicotylenchus dihystera*（马来西

亚）及 *Rotylenchus*（印度尼西亚）也在一些国家的香蕉中发现。最具破坏性的线虫是内寄生迁移性腐烂线虫 *R. similis* 和 *O. coffeae*（Sarah et al.，1996）。

线虫的抗性鉴定已在野生品种、地方品种和育种计划培育出的新合成植物中进行。野生种 *M. acuminata* Calcutta 4 在田间试验中显示抗线虫，尤其是抗 *R. similis*。盆栽试验也证实它能抗 *R. similis*（Viaene et al.，2003）。*M. acuminata* 的其他亚种（*malaccensis*、*microcarpa* 和 *zebrina*）也被发现对 *R. similis* 有中等至好的抗性（Wehunt et al.，1978；Fogain，1996）。Marau、Pora Pora、Kokopo、Pisang Mas、Saba、Gia Hiu 和 Vudu papua 香蕉能抗 *R. similes*，而 Pitu、Yalim、Yanun Yefan 和 *M. balbisiana* 显示出部分抗 *R. similes*（Dochez et al.，2006）。

FHIA 在育种项目中将 *Musa acuminata* ssp. *malaccensis* 作为线虫的抗性资源。一些二倍体栽培种，如 Pisang Mas 和 Pisang Lidi 对 *R. similes* 有中度抗性（Wehunt et al.，1978；Fogain，1996）。二倍体 Pisang Jari Buaya（AA）品系和 Pisang Batuau 被认为能抗 *R. similes*（Wehunt et al.，1978；Fogain，1996；Elsen et al.，2002）。FHIA 将 Pisang Jari Buaya 作为母本培育出抗穴居线虫的合成二倍体 SH3142。AB 二倍体的无性克隆 Kunnan 一直表现出高抗 *R. similes* 和 *P. coffeae*（Collingborn and Gowen，1997）。据报道，Yangambi Km5（AAA）能抗 *R. similis*（Pinochet et al.，1998）。东非高地蕉中，除 cv. Muvubo 外，其他和甜蕉 Valery 一样都易感 *R. similis*（Dochez et al.，2005）。经证实，13 种四倍体杂交体中的 4 种，19 种二倍体中的 13 种和 18 种次级三倍体中的 5 种，能抗 *R. similis*。因此，IITA 的香蕉育种项目培育出了一系列抗 *R. similis* 的二倍体和四倍体杂交种，它们包括 TMB2×9128-3、TMB2×5265S-1、TMB2×3107S-4、TMB2×2582S-1、TMB2×4443S-1、TMB2×2569S-1 和 TMB2×5105-1。Stoffelen 等（1999）证实，在 Fei 香蕉品种中，Rimina 和 Menei 两个品种能抗 *R. similis*。*M. balbisiana* 的大部分种质能抗 *R. similis*（Fogain，1996），ABB 品种 Pelipita 被报道对 *R. similis* 有中等抗性（Price and McLaren，1996）。

1.7 突变体

据推测，食用香蕉是自史前时代从自发突变体中选出的（Buddenhagen，1987）。自发突变的发生频率非常低，因而突变并不是影响芭蕉属植物遗传改良的显著因子。诱变剂如辐射和某些化学物质可用于诱发更高频率的突变，并产生遗传变异，从而选出期望的突变体（Roux，2004）。早期的突变研究被认为无法诱发芭蕉属植物突变，因为被用作原始材料的是芭蕉属植物种子和吸芽。随后，当植物离体材料，特别是包含分生组织的茎尖与γ射线一同应用时极成功地获得了突变体。Roux（2004）提供了关于芭蕉属植物诱变突变体的详细方案。然而，芭蕉属植物诱变突变体的难点之一是在突变体中观察到了高度嵌合状态（Hautea et al.，2004；Roux，2004）。减少嵌合现象发生的方法是在芭蕉属植物中利用胚性细胞悬浮来诱发突变体（Roux et al.，2004）。诱变育种也很耗时，需要高额的劳动力成本和实地成本，因为需要大面积的农田来种植植株直至成熟。到目前为止，Novaria 和 Klue Hom Thong KU1 这两种香蕉种质是唯一注册的改良的突变品种。Novaria 是源自 Grande Nain 的早花突变体，它比其无性繁殖系亲本早开花约 10 周。Klue Hom Thong KU1 是将组织培养物在γ射线照射下得到的。相比其亲本，它有更大的圆柱形果穗。Roux（2004）提供了体细胞无性系变异产生的一系列变异体和在γ射线照射下产生的假定的突变体。选择这些突变体的特征包括矮化、大果、早熟、抗 *Mycosphaerella fijiensis* 毒素和抗尖孢镰孢。

1.8 结论

香蕉和大蕉作为粮食作物的重要性，特别是在热带地区的重要性是公认的。关于该作物的研究趋势似乎反映了其他类似作物的研究方向。最初，研究重点是在作物分类中发挥重要作用的形态学和细胞遗传学。

分子生物学技术的发展使得大批关于作物分子方面的论文涌现，并提升了基因组学研究的学术地位。与此同时，细胞遗传学和分子细胞遗传学研究工具的革新为芭蕉属植物的基因组研究提供了新思路。流式细胞技术的发展为芭蕉属植物育种开辟了新道路。在撰写本章时，芭蕉属研究明显仍充满活力。新的物种正源源不断从偏远地区采集。这些种质对破译芭蕉属植物的复杂分类和育种具有巨大价值，蕴含着巨大的研究潜力。香蕉的类胡萝卜素含量及其在健康中的价值和营养价值正在积极调查中。同样有趣的是，芭蕉属植物已被建议作为蛋白质组学研究的模式植物。尽管要获得该作物的研究经费困难重重，但是芭蕉属植物研究的未来充满前景。

缩略语

AFLP：扩增片段长度多态性
DNA：脱氧核糖核酸
EMBRAPA：巴西农牧研究院
FHIA：洪都拉斯农业研究所
RFLP：限制性片段长度多态性

参考文献

Abera A, Gold C S, Kyamanywa S, 1997. Banana weevil oviposition and damage in Uganda [J]. African Crop Science Conference Proceedings, 3: 1199-1205.

Adams K L, Cronn R, Percifield R, et al, 2003. Genes duplicated by polyploidy show unequal contributions to the transcriptome and organ-specific reciprocal silencing [J]. Proc Natl Acad Sci USA, 100: 4649-4654.

Adeleke M T V, 2001. Meiotic studies in the genus *Musa* L. [D]. Port Harcourt, Nigeria: University of Port Harcourt.

Adeleke M T V, Pillay M, Okoli B E, 2002. An improved procedure for examining meiotic chromosomes in *Musa* [J]. HortScience, 37: 959-961.

Agarwal P K, 1987. Cytogenetical investigations in Musaceae. Ⅱ. Meiotic studies in eight male sterile triploid banana varieties of India [J]. Cytologia, 52: 451-454.

Agarwal P K, 1988a. Cytogenetical investigations in Musaceae. Ⅲ. Meiotic studies in diploid *Musa* species and banana varieties of India [J]. Cytologia, 53: 359-363.

Agarwal P K, 1988b. Cytogenetical investigations in Musaceae. Ⅳ. Cytomorphology of an interspecific triploid hybrid of *Musa acuminata* Colla. X M. *rubra* Wall [J]. Cytologia, 53: 717-721.

Albertson K, Frederickson L, Palmer S, 1996. DNA methylation and genomic imprinting [M]. Browder L W Developmental Biology. http://www.ucalgary.ca/~browder.

Argent G C G, 1976. The wild bananas of Papua New-Guinea [J]. Notes Roy Bot Gard Edinb, 35: 77-114.

Argent G, 2010. A new species of wild banana *Musa arfakiana* (Musaceae) from Papua (formerly Irian Jaya) of Indonesia [J]. Gardens' Bulletin, 62 (2): 243-248.

Bakry F, Horry J P, 1992. Tetraploid hybrids from interploid $3x/2x$ crosses in cooking banana [J]. Fruits, 47: 641-647.

Balint-Kurti P J, Clendennen S K, Dolezelovca M, et al, 2000. Identification and chromosomal localization of the monkey retrotransposon in *Musa* sp. [J]. Mol Gen Genet, 263: 908-915.

Barbier-Brygoo H, Joyard J, 2004. Focus on plant proteomics [J]. Plant Physiology and Biochemistry, 42: 913-917.

Barker D G, Bianchi S, Blondon F, et al, 1990. Medicago truncatula, a model plant for studying the molecular genetics of *Rhizobium*-legume symbiosis [J]. Plant Mol Biol Rep, 8: 40-49.

Baurens F C, Noyer J L, Lanaud C, et al, 1997. Sequence tagged site markets to draft the genomic structure of the banana chloroplast [J]. Fruits, 52: 247-259.

Biruma M, Pillay M, Tripathi L, et al, 2007. Banana Xanthomonas wilt: a review of the disease, management strategies and future research directions [J]. Afri J Biotechnol, 6: 953 - 962.

Blomme G, Mpiira S, Ssemakadde R, et al, 2005a. Controlling banana Xanthomonas wilt through de-budding [J]. InfoMusa, 14 (1): 46.

Blomme G, Mukasa H, Ssekiwoko F, 2005b. On-farm assessment of banana bacterial wilt control options [C] // African Crop Science Society Conference Proceedings. Entebbe, Uganda: 317 - 320.

Boonruangrod R, Desai D, Fluch F, et al, 2008. Identification of cytoplasmic ancestor genepools of *Musa acuminata* Colla and *Musa balbisiana* Colla and their hybrids by chloroplast and mitochondrial haplotyping [J]. Theor Appl Genet, 118: 43 - 55.

Boonruangrod R, Fluch S, Burg K, 2009. Elucidation of origin of the present day hybrid banana cultivars using the 5'ETS rDNA sequence information [J]. Molecular Breeding, 24: 77 - 91.

Buddenhagen I W, 1987. Disease susceptibility and genetics in relation to breeding bananas and plantains [M] // Persley G J, De Langhe E A. Banana and plantain breeding strategies. Proceedings No. 21, Australian Centre for International Agricultural Research, Australia: 95 - 109.

Carlier J, Foure E, Gauhl F, et al, 2000. Fungal diseases of the foliage [M]//Jones D R. Diseases of banana, abaca and enset. UK: CABI Publishing Wallingford: 37 - 141.

Carpentier S C, Witters E, Laukens K, et al, 2007. Banana (*Musa* spp.) as a model to study the meristem proteome: acclimation to osmotic stress [J]. Proteomics, 7: 92 - 105.

Carreel F, de Leon Gonzalez D, Lagoda P J L, et al, 2002. Ascertaining maternal and paternal lineage within *Musa* by chloroplast and mitochondrial DNA RFLP analyses [J]. Genome, 45: 679 - 692.

Carter B A, Reeder R, Mgenzi S R, et al, 2009. Identification of *Xanthomonas vasicola* (formerly *X. campestris* pv. *musacearum*), causative organism of banana Xanthomonas wilt, in Tanzania, Kenya and Burundi [J]. New Disease Reports, 19: 25.

Cheesman E E, 1932a. Genetic and cytological studies of *Musa*. Ⅰ. Certain hybrids of the Gros Michel banana [J]. J Genetics, 26: 291 - 312.

Cheesman E E, 1932b. Genetic and cytological studies of *Musa*. Ⅱ. Hybrids of the Mysore banana [J]. J Genetics, 26: 313 - 316.

Cheesman E E, 1947. Classification of the bananas [J]. Kew Bull, 2: 97 - 117.

Cheesman E E, 1948. The classification of the bananas [J]. Kew Bull, 3: 11 - 28, 145 - 157, 323 - 328.

Cheesman E E, 1950. The classification of the bananas [J]. Kew Bull, 5: 27 - 31, 151 - 155.

Cheesman E E, Larter L N H, 1935. Genetic and cytological studies of *Musa*. Ⅲ. Chromosome numbers in the Musaceae [J]. J Genetics, 30: 31 - 50.

Collingborn F M B, Gowen S R, 1997. Screening of banana cultivars for resistance to

Radopholus similis and *Pratylenchus coffeae* [J]. Infomusa，6：3.

Comai L，2005. The advantages and disadvantages of being polyploidy [J]. Nature Reviews，6：836 - 845.

D'Hont A，Paget-Goy A，Escoute J，et al，2000. The interspecific genome structure of cultivated bananas，*Musa* spp. revealed by genomic DNA in situ hybridization [J]. Theor Appl Genet，100：177 - 183.

Dahal G，Ortiz R，d'A. Hughes J，et al，2000. Relationship between natural occurrence，symptom expression，relative concentration of banana streak virus antigens，growth and yield characteristics of some micropropagated *Musa* accessions [J]. Plant Pathology，49：68 - 79.

De Langhe E，1961. La taxonomic du bananier plantain en Afrique Equatoriale [J]. J Agri Trop Bot Appl (Brussels)，8：419 - 449.

De Langhe E，1964. The origin of variation in the plantain banana [J]. Mededelingen van de Landbouwhogeschool en de Opzoekingsstations van de Staat te Gent，29：45 - 80.

De Langhe E，Vrydaghs L，de Maret P，et al，2009. Why bananas matter：an introduction to the history of banana domestication [J]. Ethnobot Res Appl，7：165 - 177.

De Langhe E，Hřibová E，Carpentier S，et al，2010. Did backcrossing contribute to the origin of hybrid edible bananas [J]? Ann Bot，106：849 - 857.

Denham T P，Haberle S G，Lentfer C，et al，2003. Multi-disciplinary evidence for the origins of agriculture from 6950 - 6440 cal BP at Kuk Swamp in the Highlands of New Guinea [J]. Science，301：189 - 193.

Dochez C，Whyte J，Tenkouano A，et al，2005. Response of East African highland bananas and hybrids to *Radopholus similis* [J]. Nematology，7：655 - 666.

Dochez C，Tenkouano A，Ortiz R，et al，2006. New sources of resistance to *Radopholus similis* in *Musa* germplasm from Asia [J]. Aust J Plant Pathol，35：481 - 485.

Dodds K S，1943. Genetical and cytological studies of *Musa*. Ⅴ. Certain edible diploids [J]. J Genet，45：113 - 138.

Dodds K S，Simmonds N W，1946. Genetical and cytological studies of *Musa*. Ⅷ. The formation of polyploid spores [J]. J Genet，47：223 - 241.

Dolezel J，2004. Cytogenetic and cytometric analysis of nuclear genome in *Musa* [M]// Jain S M，Swennen R. Banana improvement：cellular，molecular biology，and induced mutations. NH，USA：Science Publ Enfield：245 - 249.

Dolezel J，Dolezelova M，Novak F J，1994. Flow cytometric estimation of nuclear DNA amount in diploid bananas (*Musa acuminata* and *M. balbisiana*) [J]. Biol Plant，36：351 - 357.

Dolezelova M，Valarik M，Swennen R，et al，1998. Physical mapping of the 18S - 25S and 5S ribosomal RNA genes in diploid bananas [J]. Biol Plant，41：479 - 505.

Eden-Green S，2004. How can the advance of banana *Xanthomonas* wilt be halted [J]? InfoMusa，13 (2)：38 - 41.

Eden-Green S, 2007. Banana Xanthomonas wilt: a review of the disease, management strategies and future research directions [J]. African Journal of Biotechnology, 6: 953 - 962.

Elsen A, Stoffelen R, Tuyet R, et al, 2002. In vitro screening for resistance to *Radopholus similis* in *Musa* spp. [J]. Plant Science, 163: 407 - 416.

Engelberger L, 2003. Carotenoid-rich bananas in Micronesia [J]. InfoMusa, 12: 2 - 5.

FAO, 1999. The impact of banana supply and demand changes on income, employment and food security [C]. Gold Coast, Australia, 4 - 8 May 1999. (www. fao. org).

Fogain R, 1996. Screenhouse evaluation of *Musa* for susceptibility to *Radopholus similis*: evaluation of plantains AAB and diploids AA, AB and BB [M]//Frison E A, Horry J P, De Waele D. Proceedings of the workshop, new frontiers in resistance breeding for nematodes, *Fusarium* and sigatoka, Kuala Lumpur, Malaysia. Montpellier, France: INIBAP: 79 - 88.

Fogain R, Price N S, 1994. Varietal screening of some *Musa* genotypes for susceptibility to the banana weevil, *Cosmopolites sordidus* (Coleoptera: Curculionidae) [J]. Fruits, 49: 247 - 251.

Frison E, Sharrock S, 1999. The economic, social and nutritional importance of banana in the world [M]//Picq C, Foure E, Frison E A. Proceedings of the international symposium, bananas and food security, Douala, Cameroon. Montpellier, France: INIBAP: 21 - 35.

Frison E A, Escalant J V, Sharrock S, 2004. The Global *Musa* genomic consortium: a boost for banana improvement [M]//Mohan Jain S, Swennen R. banana improvement: cellular, molecular biology, and induced mutations. NH, USA: Science Publishers, Inc.: 341 - 349.

Fuller D Q, Madella M, 2001. Issues in Harappan archaeobotany: retrospect and prospect [M]//Settar S, Korisettar R. Indian archaeology in retrospect, vol Ⅱ: protohistory, archaeology of the Harappan civilization. Manohar, New Delhi, India: 317 - 390.

Fungo R, 2007. An assessment of β-carotene, iron and zinc in a range of banana cultivars [D]. MSc Thesis, Makerere Univ, Uganda.

Ganapathi T R, Higgs N S, Balint-Kurti P J, et al, 2001. Agrobacterium -mediated transformation of the embryogenic cell suspensions of the banana cultivars Rasthali (AAB)[J]. Plant Cell Rep, 20: 157 - 162.

Gawel N J, Jarret R L, Whittemore A P, 1992. Restriction fragment length polymorphism (RFLP) based phylogenetic analysis of *Musa* [J]. Theor Appl Genet, 84: 286 - 290.

Ge X J, Liu M H, Wang W K, et al, 2005. Population structure of wild bananas, *Musa balbisiana*, in China determined by SSR fingerprinting and cpDNA PCR-RFLP [J]. Mol Ecol, 14: 933 - 944.

Gepts P, Beavis W D, Brummer E C, et al, 2005. Legumes as a model plant family. Genomics for food and feed report of the cross-legume advances through genomics conference [J]. Plant Physiol, 137: 1228 - 1235.

Hakkinen M, 2004. *Musa voonii*. A new *Musa* species from Northern Borneo and discussion of the section Callimusa in Borneo [J]. Acta Phytotaxon Geobot, 55: 79-88.

Häkkinen M, 2009. *Musa chunii* Häkkinen, a new species (Musaceae) from Yunnan, China and taxonomic identity of *Musa rubra* [J]. J Syst Evol, 47: 87-91.

Hakkinen M, De Langhe E, 2001. *Musa acuminata* in Northern Borneo [M]. Montpellier, France: INIBAP Publ.

Hakkinen M, Meekiong K, 2004. A new species of the wild banana genus, *Musa* (Musaceae), from Borneo [J]. Syst Biodiver, 2: 169-173.

Hakkinen M, Wong W, 2007. New species and variety of *Musa* (Musaceae) from Yunnan, China [J]. Novon, 17: 440-446.

Häkkinen M, Yeh C L, Ge X J, 2010. A New combinaison and a new variety of *Musa itinerans* (Musaceae) [J]. Acta Phytotaxon Geobot (Jpn), 61: 41-48.

Hamill S D, Smith M K, Dodd W A, 1992. In vitro induction of banana autotetraploids by colchicine treatment of micropropagated diploids [J]. Aust J Bot, 40: 887-896.

Hardisson A, Rubio C, Baez A, et al, 2001. Mineral composition of the banana (*Musa acuminata*) from the island of Tenerife [J]. Food Chem, 73: 153-161.

Harris D R, 2006. The interplay of ethnographic and archaeological knowledge in the study of past human subsistence in the tropics [J]. J Roy Anthropol Inst (NS), 12: S63-S78.

Hautea D M, Molina G C, Balatero C H, et al, 2004. Analysis of induced mutants of Philippine bananas with molecular markers [M]//Mohan S Jain, Swennen R. Banana improvement: cellular, molecular biology, and induced mutations. NH, USA: Science Publishers, Inc.: 45-57.

Honfo F G, Kayodé P A P, Coulibaly O, et al, 2007. Relative contribution of banana and plantain products to the nutritional requirements for iron, zinc and vitamin A of infants and mothers in Cameroon [J]. Fruits, 62: 1-11.

Howell E C, Newbury H J, Swennen R, et al, 1994. The use of RAPDS for identifying and classifying *Musa* germplasm [J]. Genome, 37: 328-332.

Hughes Jd'A, Speijer P R, Olatunde O, 1998. Banana die-back virus—a new virus infecting banana in Nigeria [J]. Plant Dis, 82: 129.

Ittyeipe K, 1986. Studies on host preference of banana weevil borer, *Cosmopolites sordidus* Germ. (Curculionidae: Coleoptera) [J]. Fruits, 46: 375-379.

Jarret R L, Gawel N, Whittemore A, et al, 1992. RFLP-based phylogeny of *Musa* species in Papua New Guinea [J]. Theor Appl Genet, 84: 579-584.

Jones D R, 2000. Introduction to banana, abaca and enset [M]//Jones D R. Diseases of banana, abaca and enset. CAB International, Wallingford, UK: CABI Publishing.

Kajale M D, 1989. Mesolithic exploitation of wild plants in Sri Lanka: archaeobotanical study at the cave site of Beli-Lena [M]//Harris D R, Hillman G C. Foraging and farming: the evolution of plant exploitation. London, UK: Routledge: 269-281.

Kamate K, Brown S, Durand P, et al, 2001. Nuclear DNA content and base composition in 28 taxa of *Musa* [J]. Genome, 44: 622 - 627.

Kassim S, Nyine M, Pillay M, 2010. A screening method for banana weevil (*Cosmopolites sordidus* Germar) resistance using reference genotypes [J]. African Journal of Biotechnology, 9: 4725 - 4730.

Khanna H, Becker D, Kleidon J, et al, 2004. Centrifugation assisted *Agrobacterium tumefaciens*-mediated transformation (CAAT) of embryogenic cell suspensions of banana (*Musa* spp. Cavendish AAA and Lady finger AAB) [J]. Mol Breed, 14: 239 - 252.

Kiggundu A, Gold C S, Labuschagne M T, et al, 2003. Levels of host plant resistance to banana weevil, *Cosmopolites sordidus* (Germar) (Coleoptera: Curculionidae), in Uganda *Musa* germplasm [J]. Euphytica, 133: 267 - 277.

Kiggundu A, Gold C S, Labuschagne M T, et al, 2006. Components of resistance to banana weevil (*Cosmopolites sordidus*) in *Musa* in Uganda [J]. Entomol Exp Appl, 122: 27 - 35.

Kwon S J, Choi E Y, Choi Y J, et al, 2006. Proteomics studies of post-translational modifications in plants [J]. J Exp Bot, 57: 1547 - 1551.

Lawton-Rauh A, Robichaux R H, Purugganan M D, 2003. Patterns of nucleotide variation in homoeologous regulatory genes in the allotetraploid Hawaiian silversword alliance (Asteraceae) [J]. Mol Ecol, 12: 1301 - 1313.

Lebot V, 1999. Biomolecular evidence for plant domestication in Sahul [J]. Genet Resour Crop Evol, 46: 619 - 628.

Lebot V, Aradhya K M, Manshardt R, et al, 1993. Genetic relationships among cultivated bananas and plantains from Asia and the Pacific [J]. Euphytica, 67: 163 - 175.

Lee S K, Kader A, 2000. Preharvest and postharvest factors influencing vitamin C content of horticultural crops [J]. Postharvest Biol Technol, 20: 207 - 220.

Lejju J B, Robertshaw P, Taylor D, 2006. Africa's earliest bananas [J]? J Archaeol Sci, 33: 102 - 113.

Li W L, Berlyn G P, Ashton P M S, 1996. Polyploids and their structural and physiological characteristics relative to water deficit in *Betula papyrifera* (Betulaceae) [J]. Am J Bot, 83: 15 - 20.

Lynch M, 2002. Genomics. Genes duplication and evolution [J]. Science, 297: 945 - 947.

Lysak M A, Dolezelova M, Horry J P, et al, 1999. Flow cytometric analysis of nuclear DNA content in *Musa* [J]. Theor Appl Genet, 98: 1344 - 1350.

Mable B K, 2003. Breaking down taxonomic barriers in polyploidy research [J]. Trends Plant Sci, 8: 582 - 590.

Marriott H, Lancaster P A, 1983. Bananas and plantains [M] // Chan H T. Handbook of tropical foods. New York, USA: Marcel Dekker: 85 - 143.

Mason J B, Garcia M, 1993. Micronutrient deficiency—the global situation [J]. SCN News, 9: 11 - 16.

May G D, Rownak A, Mason H, et al, 1995. Generation of transgenic banana (*Musa acuminata*) plants via *Agrobacterium*-mediated transformation [J]. Bio/Technology, 13: 486 - 492.

Mbaka J, Ndumgu V, Mwangi M, 2007. Outbreak of Xanthomonas wilt (*Xanthomonas campestris* pv. *musacearum*) on banana in Kenya [J]. ISHS/ProMusa symposium. Recent advances in banana crop protection for sustainable production and improved livelihoods. Greenway Woods Resort, White River, South Africa.

Mbaka J N, Nakata V G, Auma J, et al, 2009. Status of banana Xanthomonas wilt in western Kenya and factors enhancing its spread [J]. Afr Crop Sci Conf Proc, 9: 673 - 676.

Ming R, Del Monte T A, Hernandez E, et al, 2002. Comparative analysis of QTLs affecting plant height and flowering among closely-related diploid and polyploid genomes [J]. Genome, 45: 794 - 803.

Nwakanma D C, Pillay M, Okoli B E, et al, 2003a. Sectional relationships in the genus *Musa* L. inferred from the PCR-RFLP of organelle DNA sequences [J]. Theor Appl Genet, 107: 850 - 865.

Nwakanma D C, Pillay M, Okoli B E, et al, 2003b. PCR-RFLP of the ribosomal DNA internal transcribed spacers (ITS) provides markers for the A and B genomes in *Musa* L. [J]. Theor Appl Genet, 108: 154 - 159.

Ortiz R, 1997. Occurrence and inheritance of 2*n* pollen in *Musa* [J]. Ann Bot, 79: 449 - 453.

Ortiz R, Vuylsteke D, Dumpe B, et al, 1995. Banana weevil resistance and corm hardness in *Musa* germplasm [J]. Euphytica, 86: 95 - 102.

Osborn T C, Pires J C, Birchler J A, et al, 2003. Understanding mechanisms of novel gene expression in polyploids [J]. Trends Genet, 19: 141 - 147.

Oselebe H O, Tenkouano A, Pillay M, et al, 2006. Ploidy and genome segregation in *Musa* breeding populations assessed by flow cytometry and randomly amplified polymorphic markers [J]. J Hort Sci, 131: 780 - 786.

Osuji J O, Vuylsteke D, Ortiz R, 1997a. Ploidy variation in hybrids from interploid $3x \times 2x$ crosses in *Musa* [J]. Tropicultura, 15: 37 - 39.

Osuji J O, Crouch J, Harrison G, et al, 1997b. Identification of the genomic constitution of *Musa* L. genotypes (bananas, plantains and hybrids) using molecular cytogenetics [J]. Ann Bot, 80: 787 - 793.

Osuji J O, Crouch J, Harrison G, et al, 1998. Molecular cytogenetics of *Musa* species, cultivars and hybrids: location of 18S - 5.8S - 25S and 5S rDNA and telomere-like sequences [J]. Ann Bot, 82: 243 - 248.

Pavis C, 1991. Etude des relatiuons plante-insecte chez le charancon du bananier *Cosmopolites sordidus* Germar (Coleoptera: Curculionidae) [M]//Gold C S, Gemmill B. Proc of a research coordination meeting on biological and integrated control of highland banana and plantain pests and diseases. Cotonou, Benin, IITA, Ibadan, Nigeria: 171 - 181.

Pietersen G, Thomas J E, 2001. Overview of *Musa* virus diseases [M/OL]//Plant virology in sub Saharan Africa. http://www.iita.org/info/virology/pdf_files/50-60.pdf.

Pillay M, Adeleke M T V, 2001. Silver Staining of *Musa* L. Chromosomes [J]. Cytologia, 66: 33-37.

Pillay M, Tripathi L, 2007. Banana breeding [M]//Kang M S, Priyadarshan P M. Breeding major food staples. Boston, MA, USA: Blackwell Science: 393-428.

Pillay M, Nwakanma D C, Tenkouano A, 2000. Identification of RAPD markers linked to A and B genome sequences in *Musa* [J]. Genome, 43: 763-767.

Pillay M, Tenkouano A, Hartman J, 2002. Future challenges in *Musa* breeding [M]//Kang M S. Crop improvement: challenges in the twenty-first century. New York, USA: Food Products Press: 223-252.

Pillay M, Tenkouano A, Ude G, et al, 2004. Molecular characterization of genomes in *Musa* and its applications [M]//Mohan Jain S, Swennen R. Banana improvement: cellular, molecular biology, and induced mutations. NH, USA: Science Publishers, Inc.: 271-286.

Pinochet J, Jaizme M C, Fernandez C, et al, 1998. Screening banana for root-knot (*Meliodogyne* spp.) and lesion nematode (*Pratylenchus goodeyi*) resistance for the Canary Islands [J]. Fund Appl Nematol, 21: 17-23.

Ploetz R C, 1990. Population biology of *Fusarium oxysporum* f. sp. *cubense* [M]//Ploetz R C. Fusarium wilt of banana. St Paul, Minnesota, USA: APS Press: 63-76.

Ploetz R C, 1994. Panama disease: return of the first banana menace [J]. Int J Pest Manag, 40: 326-336.

Ploetz R C, Vazquez A, Nagel J, et al, 1997. Current status of banana disease in Thailand [J]. Fruits, 51: 387-395.

Price N S, McLaren C G, 1996. Techniques for field screening of *Musa* germplasm [M]//Frison E A, Horry J P, De Waele D. Proceedings of the workshop, new frontiers in resistance breeding for nematodes, fusarium and sigatoka. Kuala Lumpur, Malaysia, INIBAP, Montpellier, France: 87-107.

Purseglove J W, 1972. Tropical crops: monocotyledons [M]. New York, USA: John Wiley.

Reynolds P K, 1927. The banana, its history, cultivation and place among staple foods [M]. Boston, MA, USA: Houghton Miffiin.

Rishi N, 2009. Significant plant virus diseases in India and a glimpse of modern disease management technology [J]. J Gen Plant Pathol, 75: 1-18.

Robinson J C, 1996. Bananas and plantains: crop production science in horticulture 5 [M]. Wallingford, Oxon, UK: CAB Intl.

Rossel G, 1991. The diffusion of plantain (*Musa* sp. AAB) and banana (*Musa* sp. AAA) in Africa: a case for linguists, taxonomists and historians, focused on Nigerian crop names [M]//Leakey R E, Slikkerveer L J. Origins and development of agriculture in East Africa: the ethnosystems approach to the study of early food production in Kenya. Ames, Iowa, USA: Iowa State University Research Foundation: 129-160.

Rossignol M, Peltier J B, Mock H P, et al, 2006. Plant proteome analysis: a 2004 - 2006 update [J]. Proteomics, 6: 5529 - 5548.

Roux N S, 2004. Mutation induction in *Musa*—a review [M] // Mohan Jain S, Swennen R. Banana improvement: cellular, molecular biology, and induced mutations. NH, USA: Science Publishers, Inc.: 23 - 32.

Roux N, Toloza A, Radecki Z, et al, 2003. Rapid detection of aneuploidy in *Musa* using flow cytometry [J]. Plant Cell Rep, 21: 483 - 490.

Roux N S, Toloza A, Dolezel J, et al, 2004. Usefulness of embrygenic cell suspension cultures for the induction and selection of mutants in Musa spp. [M]//Mohan Jain S, Swennen R. Banana improvement: cellular, molecular biology, and induced mutations. NH, USA: Science Publishers, Inc.: 33 - 43.

Rowe P, Rosales F, 1990. Breeding bananas and plantains with resistance to black Sigatoka [M]//Fullerton R A, Stover R H. Proceeding of an international workshop, sigatoka leaf spot disease of bananas. San Jose, Costa Rica: 243 - 251.

Rowe P, Rosales F, 1996. Bananas and plantains [M]//Janick J, Moore J N. Fruit breeding, vol Ⅰ: tree and tropical fruits. New York, USA: John Wiley and Sons, Inc.: 167 - 211.

Sagi L, Remy S, Panis B, et al, 1994. Transient gene expression in electroporated banana (*Musa* spp. cv. Bluggoe, ABB group) protoplasts isolated from regenerable embryogenetic cell suspensions [J]. Plant Cell Rep, 13: 262 - 266.

Sagi L, Panis B, Remy S, et al, 1995. Genetic transformation of banana (*Musa* spp.) via particle bombardment [J]. Bio/Technology, 13: 481 - 485.

Samyn B, Sergeant K, Carpentier S, et al, 2007. Functional proteome analysis of the banana plant (*Musa* spp.) using de novo sequence analysis of derivatized peptides [J]. J Proteome Res, 6: 70 - 80.

Sarah J L, Pinochet J, Stanton J, 1996. The burrowing nematode of bananas, Radopholus similis Cobb, 1913. *Musa* pest fact sheet No. 1 [M]. INIBAP, Montpellier, France.

Sasser J N, Freckman D W, 1987. A world perspective on nematology [M]//Dickson D W, Veech J A. Vistas in nematology. Hyattsville, Maryland, USA: Society of Nematologists: 7 - 14.

Sathiamoorthy S, Balamohan T N, 1993. Improvement of banana [J]. Adv Hort (India), 1: 303 - 335.

Sharrock S, 1989. Collecting *Musa* in Papua New Guinea [C] // Jarret R L. Identification of genetic diversity in the genus *Musa*, Los Baños, the Philippines. Montpellier, France: INIBAP: 140 - 157.

Shepherd K, 1996. Cytogenetics of the genus *Musa* [R]. International Network for the Improvement of Banana and Plantain, Montpellier, France.

Shepherd K, Ferreira F R, 1984. The PNG biological foundation banana collection at Laloki, Port Moresby, Papua New Guinea [J]. IBPGR Regional Committee for South East Asia

Newsl, 8: 28-34.

Shepherd K, Da Silva K M, 1996. Mitotic instability in banana varieties: aberrations inconventional triploid plants [J]. Fruits, 51 (2): 99-103.

Shepherd K, Dantas J L L, De Oliveira e Silva S, 1994. Breeding Prata and Maca cultivars in Brazil [C]//Jones D R. Proceedings of the first global conference of the international testing program, the improvement and testing of *Musa*: a global partnership, Honduras. Montpellier, France: INIBAP: 57-168.

Simmonds N W, 1948. Genetical and cytological studies of *Musa*, Ⅹ: stomata size and plant vigour in relation to polyploidy [J]. J Genet, 49: 57-68.

Simmonds N W, 1954. Isolation in *Musa* section Eumusa and Rhodochlamys [J]. Evolution, 8: 65-74.

Simmonds N W, 1956. Botanical results of the banana collecting expedition 1954-1955 [J]. Kew Bull, 3: 463-489.

Simmonds N W, 1962. Evolution of the bananas [M]. London, UK: Longmans.

Simmonds N W, 1995. Bananas *Musa* (Musaceae) [M]//Smartt J, Simmonds N W. Evolution of crop plants. 2nd ed. Essex, UK: Longmans Scientific and Technical: 370-375.

Simmonds N W, 1966. Bananas [M]. 2nd ed. Longman, London, UK: Tropical Agri Series.

Simmonds N W, Shepherd K, 1955. The taxonomy and origins of the cultivated bananas [J]. J Linn Soc Lond Bot, 55: 302-312.

Someya S, Yoshiki Y, Okubo K, 2002. Antioxidant compounds from bananas (*Musa Cavendish*) [J]. Food Chem, 79: 351-354.

Stoffelen R, Verlinden R, Xuyen N T, et al, 1999. Screening of Papua New Guinea bananas to root-lesion and root-knot nematodes [J]. InfoMusa, 8: 12-15.

Stover R H, Simmonds N W, 1987. Bananas. 3rd ed. London, UK: Longman.

Strosse H, Schoofs H, Panis B, et al, 2006. Development of embryogenic cell suspensions from shoot meristematic tissue in bananas and plantains (*Musa* spp.) [J]. Plant Sci, 170: 104-112.

Swennen R, Rosales F, 1994. Bananas [M]//Arntzen C J, Ritter E M. Encyclopedia of Agricultural Science A-D vol. New York, USA: Academic Press: 215-232.

Swennen R, Wilson G F, De Langhe E, 1984. Preliminary investigation of the effects of gibberellic acid (GA_3) on sucker development in plantain (*Musa* cv. AAB) under field conditions [J]. Trop Agri, 61: 253-256.

Tate J, Soltis D E, Soltis P S, 2005. Polyploidy in plants [M]//Gregory T R. The evolution of the genome. San Diego, USA: Elsevier: 371-426.

Taylor G, 2002. Populus: arabidopsis for forestry: do we need a model tree [J]. Ann Bot, 90: 681-689.

Tenkouano A, 2005. Breeding banana and plantain: integrating molecular techniques and conventional approaches to explore genetic polymorphisms and predict progeny performance [M]// Thangadurai D, Pullaiah T, Tripathi L. Genetic resources and biotechnology, vol 3. New Delhi, India: Regency Publications: 153 - 168.

Tenkouano A, Crouch J H, Crouch H K, et al, 1998. Ploidy determination in *Musa* germplasm using pollen and chloroplast characteristics [J]. HortScience, 33: 889 - 890.

Tezenas du Montcel H, 1988. *Musa acuminata* subspecies *banksii*: status and diversity [R]// Jarret R L. Identification of genetic diversity in the genus *Musa*. Proc Int Workshop held at Los Banos, the Philippines. Montpellier, France: INIBAP: 211 - 218.

Tezenas du Montcel H, Carreel F, Bakry F, 1995. Improve the diploids: the key for banana breeding [R]// New frontiers in resistance breeding for nematode, fusarium and sigatoka. Proc of the Workshop on New frontiers in resistance breeding for nematode, fusarium and sigatoka held at Kuala Lumpur, Malaysia.

Tripathi L, Tripathi J N, Hughes J, 2005. *Agrobacterium*-mediated transformation of plantain cultivar Agbagba (*Musa* spp.) [J]. Afr J Biotechnol, 4: 1378 - 1383.

Ude G, Pillay M, Nwakanma D, et al, 2002a. Analysis of genetic diversity and sectional relationships in *Musa* using AFLP markers [J]. Theor Appl Genet, 104: 1239 - 1245.

Ude G, Pillay M, Nwakanma D, et al, 2002b. Genetic diversity in *Musa acuminata* Colla and *Musa balbisiana* Colla and some of their natural hybrids using AFLP Markers [J]. Theor Appl Genet, 104: 1246.

US Department of Agriculture, Agricultural Research Service (USDA-ARS), 2004. USDA national nutrient database for standard reference, release 17 [M]. Nutrient Data Laboratory. http://www.nal.usda.gov/fnic/foodcomp.

Valmayor R, Danh L D, Hakkinen M, 2004. Rediscovery of *Musa splendida* A. Chevalier and description of two new species (*Musa viridis* and *Musa lutea*) [J]. Philipp Agri Sci, 87: 110 - 118.

Vandenhout H, Ortiz R, Vuylsteke D, et al, 1995. Effects of ploidy on stomatal and other quantitative traits in plantain and banana hybrids [J]. Euphytica, 83: 117 - 122.

Van Duren M, Morpurgo R, Dolezel J, et al, 1996. Induction and verification of tetraploids in diploid banana (*Musa acuminata*) by in vitro techniques [J]. Euphytica, 88: 25 - 34.

Vavilov N I, 1935. Theoretical basis for plant breeding [M]// Love D Moscow. Origin and geography of cultivated plants, vol 1: the phytogeographical basis for plant breeding. Cambridge, UK: Cambridge Univ. Press: 316 - 366.

Viaene N, Duran L F, Rivera J M, et al, 2003. Responses of banana and plantain cultivars, lines and hybrids to the burrowing nematode *Radopholus similis* [J]. Nematology, 5: 85 - 98.

Vuylsteke D, Ortiz R, Ferris R S B, et al, 1997. Plantain improvement [J]. Plant Breed Rev, 14: 267 - 320.

Wall M M, 2006. Ascorbic acid, vitamin A, and mineral composition of banana (*Musa* sp.) and papaya (*Carica papaya*) cultivars grown in Hawaii [J]. J Food Comp Anal, 19: 434 - 445.

Wehunt E J, Hutchinson D J, Edwards D I, 1978. Reaction of banana cultivars to the burrowing nematode (*Radopholus similis*) [J]. J Nematol, 10: 368 - 370.

Wendel J F, 2000. Genome evolution in polyploids [J]. Plant Mol Biol, 42: 225 - 249.

Wenkam N S, 1990. Food of Hawaii and the Pacific Basin, fruits and fruit products: raw, processed, and prepared, vol. 4: composition [Z]. Hawaii Agricultural Experiment Station Research and Extension Series 110. Hawaii.

WHO, 1992. National strategies for overcoming micronutrient malnutrition [C]. 45th World Health Assembly Provisional Agenda Item 21, doc A45/17. WHO, Geneva, Switzerland.

Wilson G B, 1946. Cytological studies in the *Musae*. Ⅱ. Meiosis in some diploid clones [J]. Genetics, 31: 475 - 482.

Wong C, Kiew R, Argent G, et al, 2002. Assessment of the Validity of the sections in *Musa* (Musaceae) using AFLP [J]. Ann Bot, 90: 231 - 238.

World Bank, 1994. Enriching lives: overcoming vitamin and mineral malnutrition in developing countries [M]. Washington DC, USA. World Bank.

Wullschleger S D, Sanderson M A, McLaughlin S B, et al, 1996. Photosynthetic rates and ploidy levels among different populations of switchgrass [J]. Crop Sci, 36: 306 - 312.

第 2 章　芭蕉属经典遗传学与传统育种

Michael Pillay[①]

摘　要：传统芭蕉属育种面临的一些障碍，主要是三倍体品种的不育特性。尽管已从芭蕉属植物的形态学、细胞学及同工酶标记性状中观察到性状的分离，但是利用它们对芭蕉属植物进行的传统和连锁图谱的工作尚未完成。缺乏这些图谱的主要原因在于难以培育出大量分离群体，特别是在三倍体栽培品种中。芭蕉属植物中分子作图技术正快速发展。基因组共线性可以用来连接芭蕉属和其他单子叶植物之间的遗传信息。与芭蕉属植物密切相关的种群的分子标记物可用于加速芭蕉属植物连锁图谱的绘制。芭蕉属植物的育种目标包括提高产量、提高生物和非生物胁迫的抗性、改变植株的形态和缩短作物生长周期。迄今为止，香蕉育种计划主要集中于开发抗病虫害的品种和保持当前的产量。传统育种成果和先进的分子育种原理将在本章进行介绍。

关键词：作图，传统育种成果，分子育种

2.1　引言

芭蕉属植物是大型的复合种质，包括二倍体、三倍体和某些四倍体合成品种。这些高度不育、营养繁殖的品种来源于 *M. acuminata* 提供的 A 基因组，仅仅是甜蕉品种，比如 Cavendish，或关联 B 基因组（*M. balbisiana*）的煮食蕉，如大蕉品种。传统芭蕉属植物育种面临的一些障碍，主要困难是三倍体品种不育（Pillay et al.，2002）。目前培育芭蕉属植物新品种有几种不同的方法（Pillay and Tenkouano，2011）。在 *M. acuminata* 品种中，结构杂合性被认为是造成不育的重要原因之一，虽然结构杂合性对于获得无籽果实必不可少，但也阻碍育种（Hippolyte et al.，2010）。还需了解更多关于芭蕉属植物基因组、

① Vaal University of Technology，Private Bag X021，Vanderbijlpark 1900，Gauteng，South Africa；e-mail：mpillay@vut. ac. za.

细胞学、作图和遗传学的知识，以便之后制定育种策略。本章将审视这方面所做的努力。

2.2 芭蕉属植物传统作图尝试

遗传作图的目的是确定遗传材料的基因线性序列。已知的两种主要图谱类型是遗传或连锁图谱和物理图谱。遗传连锁图谱反映出染色体中的基因排列和这些基因之间的距离，因此，该图谱的作用如同高速公路上的标牌或地标（Collard et al.，2005）。表示基因之间距离的单位是厘摩（cM），利用 Haldane 或 Kosambi 作图函数测出基因位点的重组率，从而得出基因之间的距离。物理图谱描绘了基因之间或标记之间的物理的、DNA 碱基对的距离。物理图谱由全基因组序列和细菌人工染色体（BAC）需求构成或在原位杂交的基础上绘制出。碱基对显示重组距离与物理图距没有特定关系，因为重组率随着染色体长度的变化而变化（Kumar，1999）。此外，在着丝粒等异染色质区域，重组频率大大降低。在异染色质和着丝粒中，交换被抑制，而即使在交换频发的常染色质中，交换仍不稳定（Anderson et al.，2004）。因此，不能简单地用连锁图谱覆盖在染色体上来确定基因的物理位置。这种情况下，一种被称为细胞遗传图或细胞学图的物理图谱显得十分有用，因为它是基于染色体的微细结构，并且可以提供补充信息。细胞遗传图让连锁群与染色体联系起来，能测定该连锁群在染色体形态上的具体位置（Schneider，2005）。最详细的物理图谱是序列图，它根据碱基对的数目来显示遗传标记和标记的序列（www. ncbinlm. nih. gov/about/primer/mapping. html）。缺少遗传标记将无法进行作图。遗传标记代表着生物个体或物种间的遗传差异。遗传标记主要有 3 类：①形态学，“传统”或“可见”标记；②生化标记，如称为同工酶的等位基因变异酶；③DNA 或分子标记（Kumar，1999；Collard et al.，2005）。

与在玉米和水稻等传统作物上取得的进展不同，利用形态学标记、细胞遗传标记和同工酶标记对芭蕉属植物进行连锁作图的研究工作还未完成。这并不意味着芭蕉属植物的形态学、细胞遗传学及同工酶性状不明显。虽然没有关于细胞学性状分离的信息发表，但对芭蕉属植物 5S rDNA 的位点数目的研究表明可能存在细胞学性状的分离（Dolezelova et al.，1998；Bartos et al.，2005）。已有关于芭蕉属植物形态学性状和同工酶标记分离的报道（Faure et al.，1993；Lebot et al.，1993；Shepherd，1999）。形态特征的差异构成了芭蕉属植物分类研究和分类系统的基础（Simmonds and Shepherd，1955）。例如，15 个形态学性状被用于 *M. acuminata* 和 *M. balbisiana* 的分离，这两个种是香蕉栽培种的祖先，它们形态不同但可以杂交。形态学特征的差异性已被广泛应用于芭蕉属的种质分类，亚种、组、基因组、克隆、克隆系和亚组的区分（Swennen and Vuylsteke，1987；Swennen et al.，1995；Ortiz，1997；Osuji et al.，1997；Ortiz et al.，1998a；Karamura，1999）。以大蕉为例，在其分类中，形态学分析清晰地将大蕉分为 4 个亚组：French、French Horn、False Horn 和 Horn（Tezenas du Montcel et al.，1983；Swennen and Vuylsteke，1987）。根据假茎高度，每一组可再划分为大、中、小 3 类（De Langhe，1964）。芭蕉属的形态学差异之大甚至导致产生了一系列形态描述符来对香蕉进行分类（Anon，1984）。需要指出的是，芭蕉属植物有丰富的形态学标记。无需专门的或分子技术就能监控形态学标记的遗传（Kumar，1999）。因此，需要被确定的是，到底是什么原因导致了芭蕉属植物没有形态学连锁图谱。主要原因可能在于难以培育出大量的分离群体，特别是难以在高度不育的三倍体品种中培育出。第二个原因是 20 世纪的香蕉研究人员将主要研究目标放在香蕉的遗传改良上。威胁香蕉出口的头号病害是镰孢枯萎病（Jones，2000）。该疾病在拉丁美洲-加勒比地区肆虐，严重影响该地区出口香蕉的产量，因而推动了 1922 年第一个香蕉育种计划在特立尼达岛的发展。然而，当第一种抗镰孢枯萎病的出口香蕉杂交种被培育出来时，出现了另一种破坏力更大的叶斑病。因而，香蕉研究人员将全部精力集中在研究病虫害上，只有少量的香蕉育种计划专注于培育出抗主要病虫害的香蕉新品种。现在的情况是，只有少部分研究人员开展香蕉作物的基础研究和传统图谱的绘制。

形态标记的一个问题是这些标记会受到环境和遗传因素，如基因上位性的影响。或许芭蕉属植物没有形态学图谱的另一原因在于形态学性状的遗传性质。形态标记的等位基因在显性、隐性的相互作用下使之无法将杂合子从纯合子个体中区分出来（Kumar，1999）。

目前关于香蕉的基础研究尤其是关于可育的并且能产生大量分离种群的野生二倍体的研究显著增加。现已培育出一些拥有某些重要性状的芭蕉属植物分离群体（http://www.musagenomics.org）。

2.3　细胞学图谱

细胞学图谱遗传图数据和染色体的实际细胞学特征结合在一起，这些细胞学特征包括着丝粒、染色体结、染色体带和原位与荧光原位杂交（FISH）信号（Harper and Cande，2000）。通过显微镜观察固定和着色的染色体的可见结构，细胞学图被绘制出来。细胞学图对以下方面具有巨大的帮助，绘制物理图谱，基于图谱的无性繁殖项目，在大量的异染色质区域定位基因，其中在异染色质区域由于低水平的减数分裂重组导致连锁距离失准（Lambie and

Roeder，1986；Zhong et al.，1999)。分子细胞遗传图对沿着染色体放置细菌人工染色体（BAC）和其他 DNA 序列也十分有帮助。细胞遗传图对基因组测序项目至关重要，因为这些图能提供整个基因组的异染色质、常染色质、着丝粒和基因的分布信息（Cheng et al.，2001a；Kulikova et al.，2001)。尝试绘制芭蕉属植物基因组细胞学图的研究相对较少（Dolezelova et al.，1998；Osuji et al.，1998；D'Hont et al.，2000)。绘制芭蕉属植物细胞学图的主要难题在于缺少合适的分析工具和芭蕉属植物基因组的均质核型。FISH 被用于研究许多作物的核型，因为该技术能够提供关于染色体结构、基因定位、物理和遗传图谱联系的更为详细的标记（Dolezel et al.，2004)。FISH 曾用来绘制芭蕉属植物染色体的重复 DNA 序列、端粒和移动遗传因子的图谱（Osuji et al.，1998；Dolezelova et al.，1998；D'Hont et al.，2000)。迄今为止，在所有被检测的香蕉基因型中发现了 18S-5.8S-26S rDNA 的两个杂交位点。这两个位点对应的是染色体的核仁组织区。在香蕉染色体银染后，这个区域还观察到了微卫星（Pillay and Adeleke，2001)。不同香蕉基因型的 5S rDNA 的位点数目不同，4～6 个不等，反映出某些 AA 基因型的结构性染色体杂合性的存在（Dolezelova et al.，1998)。在许多其他植物中，5S rDNA 的位点数目也有变化（Cerbah et al.，1999)。有趣的是，通过对 5S rDNA 位点数目不同的基因型互相杂交产生的后代的审查可以估算 5S rDNA 位点的遗传。FISH 研究证实，芭蕉属植物具有拟南芥状端粒（Dolezel et al.，2004)。通过对芭蕉属植物进行 FISH 分析，被称为“猴子”的可动遗传因子的 3 个片段显示不同的位点和染色体上存在差异杂交（Balint-Kurtiet al.，2000)。直到最近，芭蕉属植物的 FISH 分析全部应用到体细胞中期染色体上。中期染色体通常高度浓缩，因而将相邻 FISH 目标的分辨率限制在 DNA 的百万碱基而不是千碱基（Kulikova et al.，2001)。细胞遗传学研究依赖于准确一致的染色体识别，而这对具有小染色体的植物物种来说也一直是个挑战（Cheng et al.，2001b)。要克服染色体大小的限制，可以采取其他方法，如使用粗线期染色体。粗线期染色体长度是有丝分裂中期染色体的 10～40 倍，它们同时显示出常染色质和异染色质区域的不同模型（Albini and Schwarzacher，1992)。粗线期染色体的异色带、染色体长度和着丝粒的位置是为数不多的标识，使人们更容易地识别单个染色体。现在，新的技术使对芭蕉属植物粗线期染色体进行染色成为可能（Adeleke et al.，2002)。然而还需要进一步的研究来确定粗线期染色体和培育出芭蕉属植物的粗线核型。这将为详细的芭蕉属植物基因组的细胞学特性描述、基于 DNA 序列的图谱和细胞学图谱的融合铺平道路。18S-5.8S-26S rDNA 位点是芭蕉属植物中唯一的染色体特异分子标记（Dolezel et al.，2004)。Valarik 等（2002）首次尝试通过利用高度重复 DNA 的 FISH 分析来识别香蕉的每条染色体。自此，建成了几个关于香蕉的 BAC 文库，它们包括 *M. acuminata* cv. Calcutta 4 的 BAC 文库（MA4 BAC library；Vilarinhos et al.，2003）、*M. acuminata* cv. Tuu Gia 的 BIBAC 文库（TGBIBAC library；Ortiz-Vazquez et al.，2005）和 *M. balbisiana* cv. Pisang Klutuk Wulung 的 BAC 文库（MBP BAC library；Safar et al.，2004）。这些克隆能用来培育新的芭蕉属植物细胞遗传标记（Hribova et al.，2008）。一项研究利用 MA4 BAC 文库来识别低拷贝和单拷贝 BAC 克隆，从而识别出 *M. acuminata* cv. Calcutta 4 的一个染色体特异 BAC 克隆（Hribova et al.，2008）。BAC 克隆的 FISH 锚入遗传图谱能够描述 *M. acuminata* 中易位的特征（Vilarinhos et al.，2006）。识别易位将有助于作图，因为易位的存在一般会使遗传图绘制、研究农艺遗传和育种困难重重。最近的一项研究使高分辨率 BAC-FISH 有可能实现，这是香蕉育种的福音（De Capdeville et al.，2008）。这些作者讨论了该技术的用处，包括在芭蕉属染色体上定位 BAC 克隆、生成细胞基因图、确认标记的位置和检测出染色体的倒位与缺失。

2.4 同工酶

同工酶是带不同电荷的蛋白质分子。可通过电泳将其分离，并通过特异性染色将其检测出来（Marker and Moller，1959）。蛋白质标记揭示出基因序列的差异，功能上可作为共显性标记（Kumar，1999）。一些研究已经利用同工酶作为香蕉和大蕉的遗传标记（Bonner et al.，1974；Jarret and Litz，1986a，b；Lebot et al.，1993）。这些研究表明同工酶谱的变化同样发生在芭蕉属植物中。该研究也表明 *M. acuminata* 和 *M. balbisiana* 的幼苗后代在不同酶系统中等位酶的分离（Lebot et al.，1993）。Faure 等（1993）也观察到了同工酶谱的分离，并在芭蕉属植物第一张低密度连锁图中定位苹果酸脱氢酶（Mdh）。这些研究表明绘制芭蕉属植物的同工酶连锁图谱是可能的。唯一的问题就是，虽然同工酶系统检测出高多态性，但通常香蕉和植物中的酶系统很少被优化。迄今为止，只开发出 40～50 种试剂系统允许特定蛋白质染色。由于只有这么少的标记，因此只可能绘制出基因组小段的图谱。缺少同工酶位点和这些位点发生翻译后修饰，这些都限制了同工酶在作图中发挥作用（Staub et al.，1982）。同工酶也显示出组织变异性。一些同工酶在根部表达良好，而另一些最好从叶片采样。这意味着，一个分离群体不得不从植株的各个部分采样，从而记录所有可用的同工酶。

为了方便作图，理论上可以利用突变将形态学、细胞学和同工酶标记的数量大大增加（Kumar，1999）。例如，通过扩增片段长度多态性（AFLP）技术可实现对大麦的突变体作图（Castiglioni et al.，1998）。

2.5 传统作图的限制和分子作图的效用

芭蕉属植物和其他作物传统作图的限制已得到充分描述（Escalant and Panis，2002）。DNA 标记的产生为遗传作图打开了新思路。最能表现终极遗传图谱的是生物体基因组完整的核苷酸序列。现在已知几种植物的全基因组序列，这些植物包括拟南芥、水稻、大豆、木瓜、葡萄、桃、高粱和杨树，还有其他 20 多种植物的基因组正在测序。

2.5.1 分子作图

生物技术现在对农业具有重要贡献，因为被用于改良微生物、植物、动物和农业生产（Kumar，1999）。生物技术已广泛应用于农业，如培育细胞和组织来促进植物的快速繁殖；诊断学中，利用单克隆抗体和核酸探针来检测植物是否患病虫害；植物的基因工程中，引入新性状，利用分子标记来辅助传统植物育种项目（Persley，1992）。

随着 DNA 标记的出现，许多植物的遗传图谱被绘制出。与形态或生化标记不同，DNA 标记几乎没有限制，不会受到环境因素或植物生长阶段的影响（Winter and Kahl，1995）。基于检测方法的不同，DNA 标记可分为 3 类：①基于杂交的，②基于聚合酶链式反应（PCR）的，③基于 DNA 序列的（Gupta et al.，1994；Joshi et al.，1999）。基于杂交的标记包括限制性片段长度多态性（RFLP）和可变数目串联重复（VNTR）位点，其中随机基因组克隆或探针被杂交到包含已用限制性内切酶消化的 DNA 的滤膜上。PCR 是一项技术，它被用于扩增 DNA 特定区域的大量拷贝，从而产生足够多的 DNA 使之在凝胶中充分可见。PCR 技术产生的大量变体现在能在植物中产生多态性。决定使用哪种技术的唯一的局限是可用的设备和资金的多少。当前的一些 PCR 技术与其他相比似乎是最能揭示特定物种 DNA 多态性的技术。一些评论已充分描述了不同类型的分子标记和它们如何产生多态性（Kumar，1999；Gostimsky et al.，2005；Doveri et al.，2008）。同样的，大量论文也评述了这种最常用标记的优缺点（Barone，2004；Collard et al.，2005；Doveri et al.，2008）。

DNA 标记一直被认为在植物育种上有巨大潜力。DNA 标记对植物育种最有价值的应用包括构建遗传连锁图和标记辅助选择（MAS）。

2.5.2 构建遗传连锁图

连锁图一直被用于识别包含基因控制简单性状（由单基因控制）和数量性

状位点（QTL）分析的数量性状的染色体区域。遗传连锁图用图形方式来表示许多位点的分布，如形态、同工酶和DNA标记沿染色体的分布。

包括芭蕉属植物在内的大量作物的分子连锁图已被绘制出。芭蕉属植物的首张连锁图由Faure等（1993）绘制。这张图谱基于58个RFLP标记、4个同工酶标记和28个分离在F_2群体的92个个体的随机扩增多态性DNA（RAPD）标记。

最近，两张亲本图被绘制出来，该图包含167个简单重复序列（SSR）和322个多样性阵列技术标记（DArT），大小为1 197 cM，该图谱源自*M. acuminata*的两种基因差距很大的种质Borneo和Pisang Lilin（P. Lilin）杂交而得的180个F_1后代（Hippolyte et al.，2010）。在每个亲本配子重组的基础上，很大比例的标记（21.7%）偏离预期的孟德尔比率（$p \leqslant 0.05$）。这些偏离的标记分布在每个亲本不同的连锁组中。第一张饱和遗传图被视为“芭蕉属植物参考图谱”，可用于今后进一步的分析（Hippolyte et al.，2010）。

由于芭蕉属植物表现为异花授粉，因此杂合亲本可用来衍生出作图种群，如F_1和回交系。现在仍不知道芭蕉属种或品种的杂合度。但是，马铃薯这种营养繁殖作物的亲本系杂合度，经测算为57%～59%（Gebhardt et al.，1989）。

芭蕉属植物的作图研究并没有跟上其他作物在作图研究上所取得的进展。遗传连锁图最大的用途之一是比对远缘和杂交不亲和性物种的基因组（Kumar，1999）。根据预测，芭蕉属研究人员将利用基因组同线性，因为同线性是目前所研究的几种植物的主要特点。基因组同线性意味着植物基因组的大片区域具有优势，数百万年来未产生显著变化。禾本科植物（单子叶植物）中，已观察到同线性的大片区域。由于芭蕉属植物也是单子叶植物，因而基因组同线性关系可以用来连接芭蕉属和其他单子叶植物之间的遗传信息。芭蕉属植物近缘类群的分子标记可加快芭蕉属植物连锁图的绘制。相关类群的保守区域的DNA标记可作为异源探针，标记芭蕉属植物的目的基因。同线性关系有助于预测相关物种农艺学上直系同源的基因位置（Jena et al.，1994；Harrington et al.，1997）。例如，植物中大部分克隆的抗病基因（*R*基因）编码一大类核苷酸结合位点/富亮氨酸重复（NBS-LRR）蛋白。NBS-LRR基因之间在核苷酸序列、氨基酸序列和结构基序上有显著的同源性。因此，针对NBS的短保守区域采用PCR简并扩增引物，可从植物中识别出*R*基因类似物（RGAs）。通过这种方法，Pei等（2007）从芭蕉属植物的5个种质中分离出RGAs。该研究的有趣之处在于12个不同的芭蕉属植物RGAs中，发现芭蕉属植物RGA-I仅限于ABB品种（Zhongshandajiao）和AA品种（Gongjiao），而芭蕉属RGA-K仅限于Zhongshandajiao品种。

通过分子标记，遗传连锁图能有效地标记有重要经济作用的基因。识别联系某个基因的标记的可能性与该标记和该基因间的距离成反比（Kumar，1999）。使用多种统计方法可建立标记和基因的连锁关系。传统遗传改良计划中的选择是以表型为基础。除了在主基因影响的情况下，育种者并不知道实际被选择的是哪个基因。分子标记的发展是划时代的突破，它将大大变革育种和选择方式。

然而，大多数植物遗传改良中要考虑的性状，如产量、品质、成熟度和抗数个生物和非生物胁迫的性状由某些基因控制，这些性状被称为数量、“多基因的”、“多因子的”或“复合的”性状（Collard et al.，2005）。这些基因或位点的表型效果受到环境因素的影响。这种称为数量性状位点（QTLs）的多基因集群，对某个性状的表型终值产生微小的正效应或负效应。绘制连锁图和进行 QTL 分析来识别 QTLs 相关的基因组区域这一过程被称为 QTL 作图（Paterson，1996）。QTL 作图的最初目的是提供一个均匀涵盖所有染色体的“框架”连锁图，以识别控制目标性状的 QTLs 旁侧的标记（Collard et al.，2005）。通过对附加标记作图，框架图可饱和。高分辨率的 QTL 图谱对于培育出标记辅助选择的可靠标记非常必要。

一些分子标记并不是基因，也不具有任何生物效应，而是基因组中的标志。它们是在基因组特定位置发现的可识别的 DNA 序列，并遵循孟德尔遗传定律传播。现在存在大量的分子标记，包括 RFLPs、RAPDs、AFLPs、微随体和单核苷酸多态性（SNPs）。

在作物中，要对目的性状作图，图谱必不可少。然后，标记可与重要的农艺性状关联通过 MAS 用于遗传改良。

分子遗传图谱对于确定 QTLs 和潜在基因之间的联系来说必不可少。分子作图对基因的图位克隆也十分必要。图位克隆或定位克隆要利用基因任一侧紧密相连的标记。图位克隆需要的是物理图谱而非连锁图谱。

2.6 育种目标

和多数作物育种一样，香蕉育种的主要目标包括：

（1）提高作物产量和品质。

（2）抗/耐病虫害。

（3）改变植株形态，如矮化性状。

（4）缩短作物生长周期。

（5）提高作物环境适应性，如适应干旱、盐分、极端温度和金属离子。

迄今为止，香蕉育种计划主要集中于培育抗病虫害品种，保持种质当前产量。这是由于严重的真菌疾病肆虐几乎摧毁了香蕉的整个商业化种植（Tenkouano et al.，2011）。尽管香蕉育种计划的发起主要是为了改良商业品种，但是人们也意识到小农户种植的品种也同样受到病虫害攻击。在热带地区，香蕉和大蕉是数以百万人口的主要粮食来源，而香蕉的减产将威胁他们的生计。大量的香蕉育种计划已培育出一些杂交品种。然而，摆在眼前的主要问题是消费者不接受这些新杂交品种，这导致它们难以得到推广。但这一状况正在改变，新杂交品种现在正不断被许多国家的农民接受。随着全球气候变化的加剧和重视营养对人类健康所起的作用，香蕉育种的新目标将不断涌现以满足不断出现的新需求。

2.6.1 芭蕉属植物的抗逆性育种

关于生物胁迫的定义，最好理解的就是抑制包括植物在内的生物系统的正常运作和良好生长的不利条件或力量（Jones et al.，1998）。除了生物胁迫会影响作物，非生物胁迫，如干旱、低温、高温、盐碱化、洪水、重金属毒性和氧化胁迫也能在全球引起严重的作物损失（Mahajan and Tuteja，2005）。干旱和盐碱化已蔓延到许多地区，预计截至2050年，50%以上的耕地将会严重盐碱化（Ashraf，1994）。温室气体排放导致的全球气候变化预计将导致温度上升，改变降雨模式（Challinor et al.，2007）。

鉴于世界人口增长和影响植物生长的各种胁迫，需要技术创新来提高现有耕地的产量以养活世界人口。小农户最易受到非生物胁迫的影响，此外，一旦胁迫降临，他们通常束手无策（Kijne，2006）。相比其他干预方式，抗非生物胁迫的植物育种最有可能成功。

植物对生物胁迫的抗性主要由单个基因控制，而非生物胁迫与之不同，其遗传上的复杂应答是多基因的，因此更难控制，也难以用传统育种的方法来解决（Vinocur and Altman，2005）。转基因手段为包括香蕉在内的传统作物育种方式提供了一种最好的替代方法，因为转基因手段可以将一个或多个基因引入到基因组中而不影响它们的遗传背景（Jain and Jain，2000）。当对细胞/组织/整个植株施加胁迫时，大量信息表明出现生态/生理/细胞的改变（Grover et al.，1999）。当植物组织暴露在胁迫下时，碳水化合物、脂肪、蛋白质和核酸的代谢受到不利影响。所有主要的生理过程，如光合作用、氮代谢和呼吸，以及植物的发育阶段都受到胁迫的不利影响。非生物胁迫导致植物细胞内的多种应答。对不同非生物胁迫的抗性得益于一系列的生化/生理机制（Grover et al.，1999）。了解这些不同的机制，特别是主要机制，对培育出耐、抗胁迫的植物至关重要。非生物胁迫的相关研究有限，而且这些研究似乎只集中在低温对果实（因为它的经济价值）产生的影响（Caamal-Velaquez et al.，2007）。

用于鉴定香蕉果实对低温基因表达的差异显示，对病原体攻击、受伤和成熟应答的基因也受低温影响（Caamal-Velaquez et al.，2007），这表明耐非生物胁迫在整个植株和细胞水平上非常复杂（Ashraf and Foolad，2007）。Yano-Melo 等（2003）报道，在有丛枝菌根真菌的温室环境下，Pacovan 品种植株的耐盐性增强。这项研究能否扩展到野外中仍是个未知数。

利用遗传转化来提高植物抗逆性的工作取得了长足进步，然而，非生物胁迫抗性基因的复杂机制使得这一工作变得极其困难（Vinocur and Altman，2005）。限制基因工程广泛应用的主要因素在于靶基因的可用性。但与 10 年前相比，育种人员现在对于改良植物、培育抗逆性上有了新思路。不久的将来，利用 QTLs 标记和单基因改良植物将成为可能。

干旱被认为是威胁农业的最大环境胁迫。培育作物的耐旱性可使作物在干旱环境下增产。关于芭蕉属植物耐旱性培育的相关研究现在未有报道。作为栽培香蕉的祖先种之一，*M. balbisiana*（BB）被认为源自干燥季风区（Robinson，1995）。似乎含有两个 B 基因组的基因型，如 ABB 煮食蕉，能在较为干燥的地区生长，抵御长时间的旱期（Pillay 以前的观察）。另一方面，栽培香蕉需要每周 25～50 mm 的水量（Stover and Simmonds，1987）。

本节将简要概述除香蕉外其他作物的耐旱性育种进展，以探究这些方法是否适用于香蕉。

近年来作物生理学和基因组学在耐旱性上有新进展，这为育种人员进行植物改良提供了新知识与新工具（Tuberosa and Salvi，2006）。生理学研究已经确认植物的某些特点与其对干旱易发环境的适应性有关。这些特点包括植株体积更小、叶面积减小、成熟更早、气孔关闭时间更长和总蒸发蒸腾量减少。这些特征可能不易于应用在芭蕉属植物上，因为芭蕉属植物植株形态大，并且生殖期长。尽管育种者已将矮化列入香蕉育种目标，但减少叶片数可能对产量产生负面影响。在小麦和水稻中，未受胁迫下，为了高产潜力和高产稳定性而培育的基因型，显示出在轻度或中度干旱下仍然保持高产（Slafer et al.，2005）。大麦试验（Rizza et al.，2004）和甜菜试验（Ober et al.，2004，2005）也显示出类似结果。笔者未发表的在乌干达相对干燥地区进行的香蕉田间试验的结果显示，部分基因型相比其在不缺水地区的生长，没有遭受任何减产。这可能表明，当植物生长在受胁迫环境时，良好环境中的选择具有积极效果。对不同基因型耐旱程度的评估似乎难以一蹴而就。许多研究中，干旱耐受性/敏感性是基于与干旱应答相关的生理措施。一些指标已被用来描述基因型在胁迫和非胁迫条件下的表现（Cattivelli et al.，2008）。与芭蕉属植物耐旱性相关的育种工作应考虑到这一点。作物发育的不同阶段对干旱胁迫表现出不同的敏感性。

分子标记可用于鉴定耐旱/抗旱性状。耐旱是一个典型的数量性状。过去10年中，已有很多研究用QTL分析来确定生长在水分限制条件下出现的生理、形态和发育变化是由哪些染色体区域控制。这一信息可使耐旱性选择更为有效。

分子标记的缺点在于基因重组可打破标记与靶基因位点之间的连锁。耐旱的多基因遗传特性和G×E的影响使耐旱性QTLs的MAS非常困难（Francia et al.，2005）。当众多基因涉及一种多基因性状时，这意味着每个基因对表型有微小影响。而这也表明，必须对染色体（QTLs）的若干区域同时进行操作，以获得显著的影响，并且各个区域的效果不容易被识别（Cattivelli et al.，2008）。研究人员必须考虑分子标记技术的高成本和应用到QTLs的MAS的低效率，以比较QTL-MAS和抗旱/耐旱标准育种策略的有效性。

将分子标记应用到耐旱性培育中的一个重要步骤是克隆潜在QTLs的DNA序列（基因）。迄今为止，还没有耐旱性的QTL被克隆到作物中（Cattivelli et al.，2008）。然而，许多胁迫相关基因已被分离，并表现在一些作物的特征中（Cattivelli et al.，2008；Ramanjulu and Bartels，2002）。例如，Fu等（2007）鉴定了水稻耐旱性的功能候选基因。

据设想，模式植物的信息会对搜索植物的干旱应答调控元件有用（Shen et al.，2003；Li et al.，2005）。转录因子也被认为是构成复杂性状如胁迫耐受性的最佳靶标。然而，似乎参与干旱信号传导的转录因子并不全是生物技术的理想目标（Cattivelli et al.，2008）。转录因子的过度表达已经显示可激活非胁迫相关基因，并对作物的农艺性状产生不利影响（Wang et al.，2003）。当将*Arabidopsis*基因*GF14 lambda*引入棉花植株后，产生“保持绿色状”的表型。转基因植株表现出更耐水分胁迫和更高的光合速率，因为在缺水条件下气孔导度发生了改变（Yan et al.，2004）。因此培育有干旱回避机制的转基因植株是作物改良的良好策略。更健壮的根系可使植物在缺水状况下吸收更多水分。不同香蕉基因型的根系有很大不同（Swennen et al.，1987）。其直根带很浅，顶部100mm约有40％的根量，顶部300mm约有85％的根量（Robinson，1996）。此外，香蕉需要大量的水分以保证最大产量。

因此，在香蕉中培育耐旱的特性可能不太容易，应当利用野生种质资源来搜寻对耐旱性有用的新等位基因。数百年来对作物的选择育种已导致许多有用等位基因的丧失及失效。当传统育种和生理学、基因学真正融合，成功繁育在干旱条件下高产的作物将成为可能（Cattivelli et al.，2008）。

2.7　香蕉育种成果

尽管有一系列可用的新技术，但根据新品种年度注册记录，目前育种进展主要还是来自传统育种（Wenzel，2006）。尽管许多因素导致香蕉和大蕉的遗传改良步伐减缓，但两者的遗传改良已取得显著的成果和进步。

为了克服香蕉育种中存在的不育问题，提出了一系列育种策略，以充分利用部分栽培种的低的生育能力（Tomekpe et al.，2004）。所有育种计划的第一个主要任务都是确定用于育种的种质尤其是三倍体品种的生育力。这主要是通过将三倍体品种与可育的二倍体植株杂交来实现。然而，许多国家香蕉种质的特征都被描述为不育。许多野生种质被鉴定能抗病虫害。研究人员培育出了大量抗病和带理想农艺品质的二倍体（Rowe，1984；Rowe and Rosales，1993；Vuylsteke and Ortiz，1995；Tenkouano et al.，2003）。Tenkouano 等（1999a，b）报道，许多有经济价值的特性更可能遗传自二倍体亲本而不是高倍性水平的亲本。数量和质量育种工作产生了一些香蕉和大蕉杂交种，其中一些能高抗黑叶斑病并具有好的果穗特性（Rowe and Rosales，1993；Vuylsteke et al.，1993a，b；Vuylsteke et al.，1995；Ortiz and Vuylsteke，1998b，c），这些杂交种已在全球范围内进行测试，可能被采用作为新品种。大部分香蕉杂交种产量翻番，而 FHIA 的某些案例，杂交种的产量甚至是其亲本地方品种的 3 倍。这主要得益于与产量相关的性状，如矮化而健壮的植株体型、更强的吸收能力和更早成熟。相较于亲本，大蕉四倍体的高产量尤其得益于改良的截根苗解决了其亲本低吸收能力的问题。果实质量有长足提高，在三倍体后代中恢复了无籽性状（Tenkouano et al.，1998）。许多芭蕉属种质的结实率显著提高，并且通过组织培养技术发芽率也大大提高。因此，一些改良的基因型已得到广泛评估，同时了解了基因型与环境的相互作用和重要性状的稳定性。此外，还了解了结合能力、杂种优势群和数量及质量性状遗传学，将之应用于育种中提高了育种效率。在育种计划上已进行了广泛探索，结合传统和创新手段，从初级四倍体、次级三倍体和其他群体中培育出了潜在的品种（Tenkouano et al.，2003）。香蕉胚胎培养技术的发展将发芽率提高至少 10 倍以上（Vuylsteke et al.，1990）。包括微繁殖在内的组织培养技术对香蕉改良项目起了重要作用，因为该技术可将一粒种子培养出多个植株。体细胞无性系变异能够将有益变异引入香蕉种质（Vuylsteke et al.，1991）。

2.8 传统育种的局限与分子育种的基本原理

过去对传统芭蕉属植物育种的局限已有充分描述（Pillay et al.，2002；Pillay and Tripathi，2007）。主要局限总结如下：

（1）大多数可食用香蕉品种是三倍体（$2n=3x=33$），为雄性和/或雌性不育，或雄性、雌性具有低生育力。

(2) 大多数可食用品种缺少抗病虫害基因。

(3) 期望的性状如耐病虫害必须从野生和栽培二倍体（$2n=2x=22$）种质渗入。

(4) 因为野生种质具有许多不理想的特征，所以耗时长的回交程序对选择尽可能接近栽培品种的后代必不可少。

(5) 许多香蕉克隆系的天然不育使能用于育种项目的品种数量有限。

(6) 有雌性生育力的品种的果串结实率非常低，这些种子在土壤中的发芽率不到1%（Ortiz and Vuylsteke，1995）。

(7) 多数情况下，香蕉是18个月生作物，相比1年生作物，耗费更长的育种时间。

(8) 多倍体和同源多倍体的染色体行为除了能导致出现二倍、三倍和四倍整倍体，还可能导致出现非整倍体和并非期望的超多倍体（$>5x$），且出现频率难以预测（Simmonds，1966）。考虑到上述情况似乎有充分的理由在芭蕉属植物中采用分子育种。

2.9 结论

植物育种在增加粮食产量和提供可持续的营养方面的作用是公认的。随着人口的增加和土地资源的减少，它在培育不同区域多样的农业气候条件下保持产量的植物品种上发挥着越来越重要的作用（Ahloowalia and Maluszynski，2001）。

植物育种是一门多学科的应用科学，涉及多学科的知识，包括农学、植物学、遗传学、细胞遗传学、分子遗传学、生理学、病理学、昆虫学、生物化学和统计学（Schlegel，2003）。植物育种的最终目的主要是改良品种。世界各地作物大约50%的产量增加均来自遗传改良（Fehr，1984）。现今，植物育种是基于创造变异、评估和选择期望的基因型及这些基因型的增殖。芭蕉属的栽培品种易受多种病虫害攻击。与之相反，芭蕉属的野生种质对大多数的病原菌有抗性。野生种质的有用基因渗入栽培品种中受到先前描述的若干因素的阻碍。常规育种涉及大量的后代筛选，这将耗费多年时间才能找到具有期望性状的少数植物。因此，现代植物生物技术被看作是加速植物育种技术发展的最新工具。植物生物技术几乎为作物生产的所有领域提供重大改进的可能。虽然芭蕉属植物的分子应用领域已取得一定进展，但该领域的潜力还未全部开发。一些国家的多个实验室联手对其他作物品种（如水稻、小麦、高粱、玉米）开展密集的基因组学研究。大量资源不断投入这些作物的基因组学研究上。相反，进行芭蕉属植物的基因组学研究的实验室不多，且资源匮乏。

尽管过去100年来，植物育种在人类赖以生存的食品生产上取得了巨大的进步，但世界的目光越来越关注植物育种计划的损耗（Gepts and Hancock，2006）。传统植物育种计划的损耗可能来自现代植物技术的显著发展。

或许，这仅仅是科学界多年来的波澜之一。我们认为，虽然生物技术目前非常流行，但它无法完全取代传统育种。

缩略语

AFLP：扩增片段长度多态性

BAC：细菌人工染色体

FISH：荧光原位杂交

MAS：标记辅助选择

PCR：聚合酶链式反应

QTL：数量性状位点

RAPD：随机扩增多态性DNA

RFLP：限制性片段长度多态性

VNTR：可变数目串联重复

参考文献

Adeleke M T V，Pillay M，Okoli B E，2002. An improved procedure for examining meiotic chromosomes in *Musa* [J]. HortScience，37：959-961.

Ahloowalia B S，Maluszynski M，2001. Induced mutations a new paradigm in plant breeding [J]. Euphytica，118：167-173.

Albini S M，Schwarzacher T，1992. In situ localization of two repetitive DNA sequences to surface-spread pachytene chromosomes of rye [J]. Genome，35：551-559.

Anderson L K，Salameh N，Bass H W，et al，2004. Integrating genetic linkage maps with pachytene chromosome structure in maize [J]. Genetics，166：1923-1933.

Anonymous，1984. Revised banana descriptors IBGR [M]. Rome，Italy：AGPG：IBPGR.

Ashraf M，1994. Breeding for salinity tolerance in plants [J]. Crit Rev Plant Sci，13：17-42.

Ashraf M，Foolad M R，2007. Roles of glycine betaine and proline in improving plant abiotic stress resistance [J]. Environ Exp Bot，59：206-216.

Balint-Kurti P J，Clendennen S K，Dolezelova M，et al，2000. Identification and chromosomal localization of the monkey retrotransposon in *Musa* sp. [J]. Mol Gen Genet，263：908-915.

Barone A，2004. Molecular marker-assisted selection for potato breeding [J]. Amer J Potato Res，81：111-117.

Bartos J, Alkhimova O, Dolezelová M, et al, 2005. Nuclear genome size and genomic distribution of ribosomal DNA in *Musa* and *Ensete* (Musaceae): taxonomic implications [J]. Cytogenet and Genome Res, 109: 50 - 57.

Bonner J W, Warner R M, Brewbaker J L, 1974. A chemosystematic study of *Musa* cultivars [J]. HortScience, 9: 325 - 328.

Caamal-Velazquez J H, Chi-Manzanero B H, Canche-Yam J J, et al, 2007. Low temperature induces differential expression genes in banana fruits [J]. Sci Hort, 114: 83 - 89.

Castiglioni P, Pozzi C, Heun M, et al, 1998. An AFLP-based procedure for the efficient mapping of mutations and DNA probes in barley [J]. Genetics, 149: 2039 - 2056.

Cattivelli L, Rizza F, Badeck F-W, et al, 2008. Drought tolerance improvement in crop plants: an integrated view from breeding to genomics [J]. Field Crop Res, 105: 1 - 14.

Cerbah M, Kevei Z, Siljak-Yakovlev S, et al, 1999. FISH chromosome mapping allowing karyotype analysis in *Medicago truncatula* lines Jemalong J5 and R-108-1 [J]. Mol Plant-Micr Interact, 12: 947 - 950.

Challinor A, Wheeler T, Garforth C, et al, 2007. Assessing the vulnerability of food crop systems in Africa to climate change [J]. Climatic Change, 83: 381 - 399.

Cheng Z, Buell C R, Wing R A, et al, 2001a. Toward a cytological characterization of the rice genome [J]. genome res, 11: 2133 - 2141.

Cheng Z, Presting G G, Buell C R, et al, 2001b. High-resolution pachytene chromosome mapping of bacterial artificial chromosomes anchored by genetic markers reveals the centromere location and the distribution of genetic recombination along chromosome 10 of rice [J]. Genetics, 157: 1749 - 1757.

Collard B C Y, Jahufer M Z Z, Brouwer J B, et al, 2005. An introduction to markers, quantitative trait loci (QTL) mapping and marker-assisted selection for crop improvement: the basic concepts [J]. Euphytica, 142: 169 - 196.

D' Hont A, Paget-Goy A, Escoute J, et al, 2000. The interspecific genome structure of cultivated banana, *Musa* spp. revealed by genomic DNA in situ hybridization [J]. Theor Appl Genet, 100: 177 - 183.

De Capdeville G, Souza Junior M T, Szinay D, et al, 2009. The potential of high-resolution BAC-FISH in banana breeding [J]. Euphytica, 166: 431 - 443.

De Langhe E, 1964. The origin of variation in the plantain banana [J]. State Agricultural University of Ghent, Ghent, Belgium, 39: 45 - 80.

Dolezel J, Valarik M, Vrana J, et al, 2004. Molecular cytogenetics and cytometry of banana (*Musa* spp.) [M]//Mohan J S, Swennen R. Banana improvement: cellular, molecular biology, and induced mutations. Enfield, NH, USA: Science Publishers, Inc.: 229 - 244.

Dolezelova M, Valarik M, Swennen R, et al, 1998. Physical mapping of the 18S-25S and 5S ribosomal RNA genes in diploid bananas [J]. Biol Plant, 41: 497 - 505.

Escalant J V, Panis B, 2002. Biotechnologies toward the genetic improvement in *Musa* [C].

Colombia: Memorias XV Reunión Internacional ACORBAT Cartagena de Indias.

Faure S, Noyer J L, Horry J P, et al, 1993. A molecular marker-based linkage map of diploid bananas (*Musa acuminata*) [J]. Theor Appl Genet, 87: 517-526.

Fehr W R, 1984. Genetic contributions to yield gains of five major crop plants [M]. Madison, WI, USA: CSSA Spl Publ 7. ASA and CSSA.

Francia E, Tacconi G, Crosatti C, et al, 2005. Marker assisted selection in crop plants [J]. Plant Cell, Tissue and Organ Culture, 82: 317-342.

Fu B Y, Xiong J H, Zhu L H, et al, 2007. Identification of functional candidate genes for drought tolerance in rice [J]. Mol Genet Genom, 278: 599-609.

Gebhardt C, Ritter E, Debener T, et al, 1989. RFLP analysis and linkage mapping in *Solanum tuberosum* [J]. Theor Appl Genet, 78: 65-75.

Gepts P, Hancock J, 2006. The future of plant breeding [J]. Crop Sci, 46: 1630-1634.

Gostimsky S A, Kokaeva Z G, Konovalov F A, 2005. Studying plant genome variation using molecular markers [J]. Russ J Genet, 41: 378-388.

Grover A, Sahi C, Sanan N, et al, 1999. Taming abiotic stresses in plants through genetic engineering: current strategies and perspective [J]. Plant Sci, 143: 101-111.

Gupta M, Chyi Y S, Romero-Severson J, et al, 1994. Amplification of DNA markers from evolutionarily diverse genomes using single primers of simple-sequence repeats [J]. Theor Appl Genet, 89: 998-1006.

Harper L C, Zacheus Cande W, 2000. Mapping a new frontier; development of integrated cytogenetic maps in plants [J]. Funct Integr Genom, 1: 89-98.

Harrington S E, Bligh H F J, Park W D, et al, 1997. Linkage mapping of starch branching enzyme Ⅲ in rice (*Oryza sativa* L.) and prediction of location of orthologous genes in other grasses [J]. Theor Appl Genet, 94: 564-568.

Hippolyte I, Bakry F, Seguin M, et al, 2010. A saturated SSR/DArT linkage map of *Musa acuminata* addressing genome rearrangements among bananas [J]. BMC Plant Biol, 10: 65.

Hribova E, Dolezelova M, Dolezel J, 2008. Localization of BAC clones on mitotic chromosomes of *Musa acuminata* using fluorescence in situ hybridization [J]. Biol Plant, 52: 445-452.

Jain R K, Jain S, 2000. Transgenic strategies for genetic improvement of Basmati rice [J]. Indian J Exp Biol, 38: 6-17.

Jarret R L, Litz R E, 1986a. Enzyme polymorphism in *Musa acuminata* Colla [J]. J Hered, 77: 183-188.

Jarret R L, Litz R E, 1986b. Isozymes as genetic markers in bananas and plantains [J]. Euphytica, 35: 539-549.

Jena K K, Khush G S, Kochert G, 1994. Comparative RFLP mapping of a wild rice, *Oryza officinalis*, and cultivated rice, *O. sativa* [J]. Genome, 37: 382-389.

Jones D R, 2000. Introduction to banana, abaca and enset [M]//Jones D R. Diseases of banana，abaca and enset. Wallingford，UK：CAB International：1－36.

Jones H G，Flowers T J，Jones M B，1998. Introduction：some terminology and common mechanisms [M]//Jones H G，Flowers T J，Jones M B. Plants under stress. Cambridge，UK：Cambridge University Press：1－10.

Joshi S，Ranjekar P，Gupta V K，1999. Molecular markers in plant genome analysis. Curr Sci，77：230－240.

Karamura D A，1999. Numerical taxonomic studies of the East African Highland bananas (*Musa* AAA-East Africa) in Uganda [M]. France：INIBAP.

Kijne J W，2006. Abiotic stress and water scarcity：identifying and resolving confl icts from plant level to global level [J]. Field Crops Res，97：3－18.

Kulikova O，Gualtieri G，Geurts R，et al，2001. Integration of the FISH pachytene and genetic maps of *Medicago truncatula* [J]. Plant J，27：49－58.

Kumar L S，1999. DNA markers in plant improvement：an overview [J]. Biotechnol Adv，17：143－182.

Lambie E J，Roeder G S，1986. Repression of meiotic crossing over by a centromere (CEN3) in *Saccharomyces cerevisiae* [J]. Genetics，114：769－789.

Lebot V，Aradhya K M，Manshardt R，et al，1993. Isozymes and classification of Pacific bananas and plantains [M]//Ganry J. Breeding banana and plantain for resistance to disease and pests. Montpellier：CIRAD in collaboration with INIBAP：43－53.

Li X P，Tian A G，Luo G Z，et al，2005. Soybean DRE-binding transcription factors that are responsive to abiotic stresses [J]. Theor Appl Genet，110：1355－1362.

Mahajan S，Tuteja N，2005. Cold，salinity and drought stresses：an overview [J]. Arch Biochem Biophys，444：139－158.

Markert C L，Moller F，1959. Multiple forms of enzymes：tissue，ontogenetic，and species specific patterns [J]. Proc Natl Acad Sci USA，45：753－763.

Ober E S，Clark C J A，Bloa M L，et al，2004. Assessing the genetic resources to improve drought tolerance in sugar beet：agronomic traits of diverse genotypes under droughted and irrigated conditions [J]. Field Crop Res，90：213－234.

Ober E S，Bloa M L，Clark C J A，et al，2005. Evaluation of physiological traits as indirect selection criteria for drought tolerance in sugar beet [J]. Field Crop Res，91：231－249.

Ortiz R，1997. Morphological variation in *Musa* germplasm [J]. Genet Resour Crop Evol，44：393－404.

Ortiz R，Vuylsteke D，1995. Factors influencing seed set in triploid *Musa* spp. L. and production of euploid hybrids [J]. Ann Bot，75：151－155.

Ortiz R，Vuylsteke D，1998a. 'BITA-3'：a starchy banana with partial resistance to black and tolerance to streak virus [J]. Hort Sci，33：358－359.

Ortiz R，Vuylsteke D，1998b. 'PITA-14'：a black sigatoka-resistance tetraploid plantin hybrid with virus tolerance [J]. HortScience，33：360－361.

Ortiz R, Madsen S, Vuylsteke D, 1998. Classifi cation of African plantain landraces and banana cultivars using a phenotypic distance index of quantitative descriptors [J]. Theor Appl Genet, 96: 904 - 911.

Ortiz-Vazquez E, Kaemmer D, Zhang H B, et al, 2005. Construction and characterization of a plant transformation-competent BIBAC library of the black Sigatoka-resistant banana *Musa acuminata* cv. Tuu Gia (AA) [J]. Theor Appl Genet, 110: 706 - 713.

Osuji J O, Okoli B E, Vuylsteke D, et al, 1997. Multivariate pattern of quantitative trait variation in triploid banana and plantain cultivars [J]. Sci Hort, 71: 197 - 202.

Osuji J O, Crouch J H, Harrison G, et al, 1998. Molecular cytogenetics of *Musa* species, cultivars and hybrids: location of 18S-5. 8S-25S and 5S rDNA and telomere-like sequences [J]. Ann Bot, 82: 243 - 248.

Paterson A H, 1996. Genome mapping in plants [M]. Austin, USA: Academic Press, California and R. G. Landes Company.

Pei X, Li S, Jiang Y, et al, 2007. Isolation, characterization and phylogenetic analysis of the resistance gene analogues (RGAs) in banana (*Musa* spp.) [J]. Plant Sci, 172: 1166 - 1174.

Persley G J, 1992. Beyond mendel's garden: biotechnology in agriculture [M]//Biotechnology enhancing research on tropical crops in Africa. CTA/IITA co-publication: 11 - 19.

Pillay M, Adeleke M T V, 2002. Silver staining of *Musa* chromosomes [J]. Cytologia, 66: 33 - 37.

Pillay M, Tripathi L, 2007. Banana breeding [M]//Kang M S, Priyadarshan P M. Breeding major food staples. Boston MA, USA: Blackwell Publishing: 393 - 428.

Pillay M, Tenkouano A, 2011. Genomes, cytogenetics and flow cytometry of *Musa* [M]// Pillay M, Tenkouano A. Banana breeding: progress and challenges. Boca Raton, FL, USA: CRC Press: 53 - 70.

Pillay M, Hartman J, Tenkouano A, 2002. Future challenges in *Musa* breeding [M]//Kang M. Crop improvement for the 21st century. New York, USA: Food Products Press: 223 - 252.

Ramanjulu S, Bartels D, 2002. Drought-and desiccation-induced modulation of gene expression in plants [J]. Plant Cell Environ, 25: 141 - 151.

Rizza F, Badeck F-W, Cattivelli L, et al, 2004. Use of water stress index to identify barley genotypes adapted to rainfed and irrigated conditions [J]. crop Sci, 44: 2127 - 2137.

Robinson J C, 1996. Bananas and plantains [M]. Wallingford, UK: CABI.

Rowe P, 1984. Breeding banana and plantains [J]. Plant Breed Rev, 2: 135 - 155.

Rowe P, Rosales F, 1993. Diploid breeding at FHIA and the development of Goldfinger [J]. InfoMusa, 2: 9 - 11.

Safar J, Noa-Carrazana J C, Vrana J, et al, 2004. Creation of a BAC resource to study the structure and evolution of the banana (*Musa balbisiana*) genome [J]. Genome, 47: 1182 - 1191.

Schlegel R H J, 2003. Dictionary of plant breeding [M]. New York, USA: Food Products

Press, The Haworth Reference Press.

Schneider K, 2005. Mapping populations and principles of genetic mapping [M] // Meksem K, Kahl G. The handbook of plant genome mapping genetic and physical mapping. Weinheim, Germany: Wiley-VCH Verlag GmbH & Co.

Shen Y G, Zhang W K, He S J, et al, 2003. An EREBP/AP2-type protein in *Triticum aestivum* was a DRE-binding transcription factor induced by cold, dehydration and ABA stress [J]. Theor Appl Genet, 106: 923 - 930.

Shepherd K, 1999. Cytogenetics of the genus *Musa* [M]. Montpellier, France: INIBAP.

Simmonds N W, 1966. Bananas [M]. London, UK: Longmans.

Simmonds N W, Shepherd K, 1955. The taxonomy and origins of the cultivated bananas [J]. J Linn Soc (London) Bot, 55: 302 - 312.

Slafer G A, Araus J L, Royo C, et al, 2005. Promising eco-physiological traits for genetic improvement of cereal yields in Mediterranean environments [J]. Ann Appl Biol, 146: 61 - 70.

Staub J E, Kuhns L J, May B, et al, 1982. Stability of potato tuber isozymes under different storage regimes [J]. HortScience, 107: 405 - 408.

Stover R H, Simmonds N W, 1987. Bananas [M]. 3rd ed. London, UK: Longmans.

Swennen R, Vuylsteke D, 1987. Morphological taxonomy of plantains (*Musa* cultivars AAB) in West Africa [C] // Persley G J, De Langhe E A. Banana and plantain breeding strategies proceedings of a workshop held at Cairns, Australia. Cairns, Australia: ACIAR Proceedings: 165 - 171.

Swennen R, Vuylsteke D, Ortiz R, 1995. Phenotypic diversity and patterns of variation in west and central African plantains (*Musa* spp., AAB group Musaceae) [J]. Econ Bot, 49: 320 - 327.

Swennen R, De Langhe E, Janssen J, et al, 1987. Study of the root development of some *Musa* cultivars in hydroponics [J]. Fruits, 41: 515 - 524.

Tenkouano A, Ortiz R, Vuylsteke D, 1998. Combining ability for yield and plant phenology in plantain-derived populations [J]. Euphytica, 104: 151 - 158.

Tenkouano A, Crouch J H, Crouch H K, et al, 1999a. Comparison of DNA marker and pedigree-based methods of genetic analysis in plantain and banana (*Musa* spp.) clones. Ⅱ. Predicting hybrid performance [J]. Theor Appl Genet, 98: 69 - 75.

Tenkouano A, Crouch J H, Crouch H K, et al, 1999b. Comparison of DNA marker and pedigree-based methods of genetic analysis of plantain and banana (*Musa* spp.) clones. Ⅰ. estimation of genetic relationships [J]. Theor Appl Genet, 98: 62 - 68.

Tenkouano A, Pillay M, Ortiz R, 2011. Breeding techniques [M] // Pillay M, Tenkouano A. Banana breeding: progress and challenges. Boca Raton, FL, USA: CRC Press: 181 - 202.

Tezenas de Montcel H, De Langhe E, Swennen R, 1983. Essaide classification des bananiers plantains (AAB) [J]. Fruits, 38: 461 - 474.

Tomekpe K, Jenny C, Escalant J V, 2004. A review of conventional improvement strategies for *Musa* [J]. InfoMusa, 13: 2-6.

Tuberosa R, Salvi S, 2006. Genomics-based approaches to improve drought tolerance of crops [J]. Trends Plant Sci, 11: 405-412.

Valarik M, Simkova H, Hribova E, et al, 2002. Isolation, characterization and chromosome localization of repetitive DNA sequences in bananas (*Musa* spp.) [J]. Chromosome Research: An International Journal on the Molecular, Supramolecular and Evolutionary Aspects of Chromosome Biology, 10: 89-100.

Vilarinhos A D, Piffanelli P, Lapoda P, et al, 2003. Construction and characterization of a bacterial artificial chromosome library of banana (*Musa acuminata* Colla) [J]. Theor Appl Genet, 106: 1102-1106.

Vilarinhos A D, Carreel F, Rodier M, et al, 2006. Characterization of translocations in Banana by FISH of BAC clones anchored to a genetic map [C]. Plant Animal Genome XIV Conference. San Diego, CA, USA.

Vinocur B, Altman A, 2005. Recent advances in engineering plant tolerance to abiotic stress: achievements and limitations [J]. Curr Opin Biotechnol, 16: 123-132.

Vuylsteke D, Ortiz R, 1995. Plantain-derived diploid hybrids (TMP2*x*) with black sigatoka resistance [J]. HortScience, 30: 147-149.

Vuylsteke D, Swennen R, De Langhe E, 1990. Tissue culture technology for the improvement of African plantains [M]//Fullerton R A, Stover R H. Sigatoka leaf spot disease of bananas. Montpellier, France: INIBAP: 316-337.

Vuylsteke D, Swennen R, De Langhe E, 1991. Somaclonal variation in plantains (*Musa* spp., AAB group) derived from shoot-tip culture [J]. Fruits, 46: 429-439.

Vuylsteke D, Swennen R, Ortiz R, 1993a. Registration of 14 improved tropical *Musa* plantain hybrids with black sigatoka resistance [J]. HortScience, 28: 957-959.

Vuylsteke D R, Swennen R L, Ortiz R, 1993b. Development and performance of balck sigatokaresistant tetraploid hybrids of plantain (*Musa* spp., AAB group) [J]. Euphytica, 65: 33-42.

Vuylsteke D, Ortiz R, Ferris S, et al, 1995. Pita-9—a black-sigatoka-resistant hybrid from the false horn plantain gene pool [J]. HortScience, 30: 395-397.

Wang W, Vinocur B, Altman A, 2003. Plant responses to drought, salinity and extreme temperatures: towards genetic engineering for stress tolerance [J]. Planta, 218: 1-14.

Wang Y, Chen J Y, Jiang Y M, et al, 2007. Cloning and expression analysis of phenylalanine ammonia-lyase in relation to chilling tolerance in harvested banana fruit [J]. Postharvest Biology and Technology, 44: 34-41.

Wenzel G, 2006. Molecular plant breeding: achievements in green biotechnology and future perspectives [J]. Applied Microbiology and Biotechnology, 70: 642-650.

Winter P, Kahl G, 1995. Molecular marker technologies for plant improvement [J]. World J Microbiol Biotechnol, 11: 438-448.

Yan J，He C，Wang J，et al，2004. Overexpression of the *Arabidopsis* 14-3-3 protein GF14 lambda in cotton leads to a "stay-green" phenotype and improves stress tolerance under moderate drought conditions [J]. Plant & Cell Physiology，45：1007－1014.

Yano-Melo A M，Saggin O J，Costa Maia L，2003. Tolerance of mycorrhized banana (*Musa* sp. cv. Pacovan) plantlets to saline stress [J]. Agriculture，Ecosystems & Environment，95：343－348.

Zhong X B，Bodeau J，Fransz P F，et al，1999. FISH to meiotic pachytene chromosomes of tomato locates the root-knot nematode resistance gene *Mi-1* and the acid phosphatase gene *Aps-1* near the junction of euchromatin and pericentromeric heterochromatin of chromosome arms 6S and 6L，respectively [J]. Theor Appl Genet，98：365－37.

第3章 破译香蕉品种起源的多样性分析

KornelBurg[①,*] **Ratri Boonruangrod**[②]

摘 要： 由当地农民历经多年选出的数千香蕉品种已成为世界各地人们饮食的重要部分。其中，大多数可食用香蕉基因型都没有种子，并通过根状茎无性繁殖。芭蕉属最早的描述和分类系统是基于形态学观察的。20 世纪 50 年代，香蕉的黄酮类化合物被用于香蕉分类。之后，同工酶被用于对香蕉的无性系进行分类。直接研究基因组技术的发展促进了遍布全基因组的标记的发展。本章总结了近年来利用细胞核和细胞器基因组标记来描述芭蕉属植物遗传多样性的报道。

关键词： 标记，细胞核基因组，细胞器基因组，多样性，芭蕉属

3.1 引言

由当地农民历经多年选出的数千香蕉品种已成为世界各地人们饮食的重要部分。其中，大多数可食用的香蕉基因型都没有种子，并通过根状茎无性繁殖。如果没有人工干预，它们是无法在自然中存活并遍布各地的。1747 年，Georgius Everhardus Rumphius 对可食用香蕉进行描述，这是最早关于香蕉植株的描述。1820 年，Luigi Aloysius Colla 首次记录了有种子的野生香蕉植株，他意识到作为野生种的 *M. balbisiana* 和 *M. acuminata* 与当时已知的其余香蕉种质明显不同（Cheesman，1948）。数百年来，这些野生基因型和栽培品种之间的关系一直让育种人员和科学家困惑不解。Kurz（1865）提出，大多数 *Eumusa* 组的可食用香蕉源自 *M. acuminata* 和 *M. balbisiana* 这两个野生种的直接遗传或者多种杂交组合。这些栽培品种普遍倍性较高，多为三倍体。而他的这

① Health & Environment Department Bioresources，AIT Austrian Institute of Technology GmbH，2444 Seibersdorf，Austria；e-mail：kornel. burg@ait. ac. at.

② Department of Horticulture，Faculty of Agriculture at Kamphaengsaen，Kasetsart University，Kamphaengsaen，Nakhon Pathom，73140 Thailand；e-mail：ratri. b@ku. ac. th.

* 通信作者。

一想法之后被人们遗忘了近一个世纪，直到 1948 年由 Cheesman 证实，才得以振兴。紧接着 Simmonds 和 Shepherd（1955）创立了以 *M. acuminata* 与 *M. balbisiana* 的形态学性状为基础的评分系统。这一系统在描述不同倍性水平的各种杂交基因型的特征上非常成功（Stover and Simmonds，1987），它将栽培香蕉分为 5 种主要基因组（AA、AB、AAA、AAB、ABB）（Simmonds and Shepherd，1955；Simmonds，1962）。新证据表明其他基因组，如 S（*M. schizocarpa*）和 T（*M. textilis*）基因组，也促成了现今栽培香蕉的基因组。A、B 和 S 基因组代表单倍体组染色体数为 11 的 *Eumusa*（新命名为芭蕉属）种，而 T 基因组则代表单倍体组染色体数为 10 的 *Australimusa*。由于大多数栽培香蕉源自 A 和 B 基因组，因此本章的重点将主要集中在这两种基因组。

3.2　标记历史

芭蕉属最早的描述和分类系统是基于形态学观察。然而，与所有表型特征一样，形态学特征也受到环境因素对其形态学/表型描述的影响（例如，当地田间状况大大影响假茎颜色）。虽然对形态标记进行仔细、有意识的选择可能有助于精确分析上述问题。例如，Simmonds 和 Shepherd（1955）提出了形态学特征系统，该系统基于 15 个能清楚地区别 *M. acuminata* 和 *M. balbisiana* 野生种的性状，描述了 *M. acuminata*/*M. balbisiana* 杂交种基因组构成的特征，并显示所有这些表型标记在杂交种中按一定比例显现。在该系统中，13 个检测性状局限于花序表型。7 个标记与雄蕾的苞片相关（苞肩、苞片卷曲情况、苞片形状、苞尖、苞片颜色、苞片是否褪色和苞片缢痕），2 个标记与雄花相关（游离花瓣和花色），2 个标记与雌花相关（柱头颜色和胚珠排列），2 个标记与花梗和花序梗相关，而另外 2 个标记与假茎（颜色）和叶子的叶柄沟槽相关。根据这一评分系统，不同倍性水平的各种杂交基因型可以被很好地描述（Stover and Simmonds，1960），其将栽培香蕉分为 5 种主要基因组（AA、AB、AAA、AAB、ABB）（Simmonds and Shepherd，1955；Simmonds，1962）。之后对野生种详细的形态学分析揭示出 *M. balbisiana* 变化不大，而 *M. acuminata* 中 9 种具有明显地理分布区域的亚种（ssp. *banksii*、ssp. *burmannica*、ssp. *burmannicoides*、ssp. *errans*、ssp. *malaccensis*、ssp. *microcarpa*、ssp. *siamea*、ssp. *truncata* 和 ssp. *zebrina*）被识别出（图 3-1）。形态特征的缺点在于对描述性状的数量或精确测量十分困难。此外，多数形态特征可能存在多基因遗传，因而基因型的微小变化不一定体现在植株的表型上，这使识别突变体和体细胞无性系变异体十分困难。因此，建立定量测量性状和发展芭蕉属种质描述的标准方法十分必要。

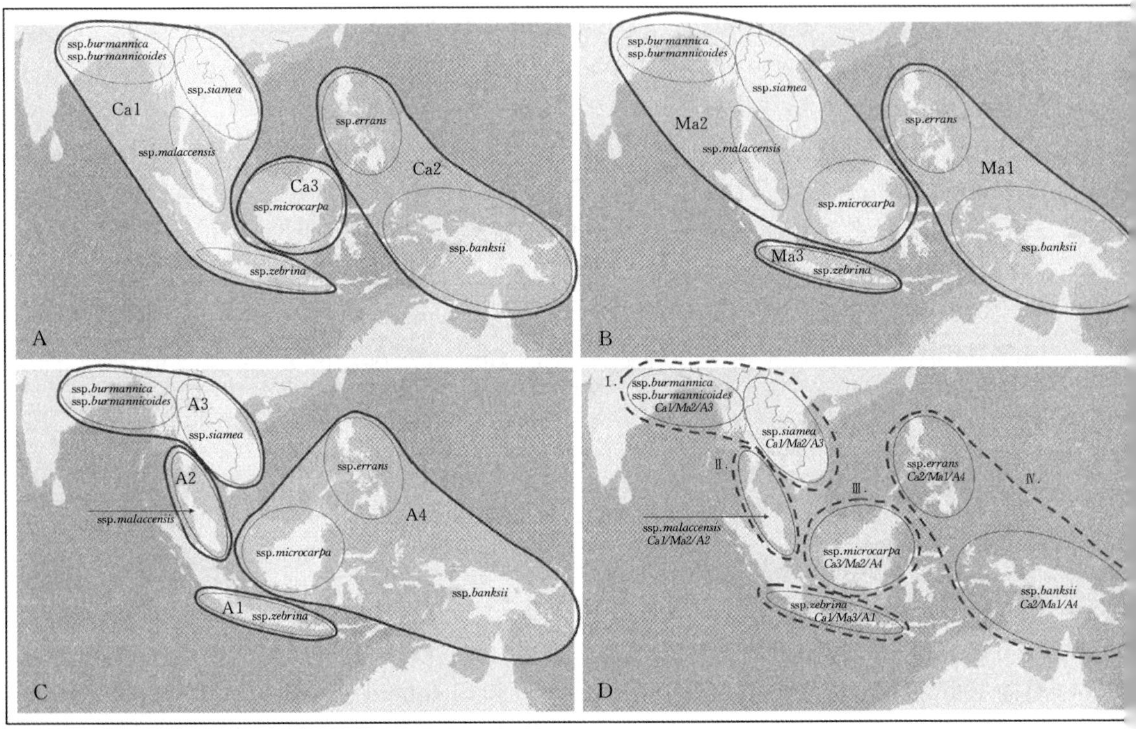

图 3－1　芭蕉属植物叶绿体单倍型与线粒体单倍型的分布

A. *M. acuminata* 亚种的叶绿体单倍型分布　B. *M. acuminata* 亚种的线粒体单倍型分布

C. *M. acuminata* 的核糖体 45S 基因型分布　D. 形态学亚种的合并基因型，列为：叶绿体/线粒体/核糖体 45S

Ⅰ～Ⅴ. 建议根据形态亚种的合并基因型进行分组

M. acuminata ssp. *truncata* 的分布在 ssp. *malaccensis* 区域内（未显示）

引进细胞学方法，即中期染色体计数和减数分裂染色体配对分析，为了解多倍体品种的基因组结构提供更多的细节。染色体计数证实了上述 5 个主要遗传群体的预期倍性，而减数分裂染色体配对和染色体结构变化的分析证实 7 种 *M. acuminata* 亚种的存在（Shepherd，1990，1999）。

由于形态特征是数量性状，并且易受环境条件的影响，定量测量性状为香蕉多样性分析提供了下一代标记。生物化学的出现使分析某些代谢产物/代谢途径和蛋白质成为可能，而分子遗传学的出现使直接分析基因组成为可能。对代谢产物和蛋白质多样性的分析不仅能反映遗传多样性，还能反映生物体对当地环境的适应性，这种适应性通过改变基因表达，可能会影响观察到的变化。不过，直接分析 DNA 的多样性能对生物体进行最精确的区分，揭示出 DNA 碱基对在水平上的差异。此外，因为代谢和蛋白质标记仅反映一部分表达的遗传信息，所以，它们的可用标记数有限，而 DNA 标记则仅由基因组大小决定。

3.2.1　代谢组——类黄酮标记

类黄酮是次生代谢产物，通过依靠遗传背景的黄酮醇骨架的羟基化、甲基化和糖基化显示出高度的结构多样性，如修饰酶的基因可用性。这组基因大多为种/亚种特异，通过分析类黄酮模型可评估完整的进化谱系。该系统的缺点在于参与类黄酮生物合成的基因仅代表芭蕉属植物基因组的小部分基因。香蕉类黄酮化合物的分析始于 20 世纪 50 年代，Simmonds（1954）注意到类黄酮标记可能在香蕉分类中有用，因为其显示出对无色花青素的分类学价值。基于这一标记系统，Horry 和 Jay（1988）通过对花青素和黄酮醇的分析，将野生种 *M. acuminata* 和 *M. balbisiana* 明显地区分开，这一分析也用于对 *M. acuminata* 的亚种进行分类，分析建议将其分为 3 组：①ssp. *siamea* 与 ssp. *burmannica*；②ssp. *malaccensis*、ssp. *microcarpa* 与 ssp. *burmannicoides*；③ssp. *zebrina*（Horry and Jay，1990）。该研究基于 *M. acuminata* 亚种中修饰酶的有或无揭示了花青素生物合成的历程，并提出 ssp. *zebrina* 可能是最接近假定祖先的物种。根据复杂的花青素模型，*M. balbisiana* 被认定为是最不发达的基因组。对花青素代谢物或代谢路径的详细分析揭示了可食用二倍体中花青素色素的变异，并提出 ssp. *malaccensis*、ssp. *banksii* 和 ssp. *zebrina* 为卡文迪什香蕉品种的假定祖先种（Horry and Jay，1990）。

3.2.2　蛋白质组——酶多态性

异型酶/同工酶分析是基于蛋白质/酶的多态性，这种多态性存在于它们的氨基酸成分中，最终反映为基因组的多态性。异型酶代表同一基因的等位基因变异，而同工酶指具有相似模板特异性但代表不同基因的酶（认为是基因家族）。因为这两种酶很难区分，所以大多数学者称它们为同工酶。该技术是基于蛋白质的分离和天然电泳，其中天然电泳是根据蛋白质的大小、电荷和结构，然后对特定酶的活动进行凝胶可视化，再根据酶的活性进行适当的染色。产生的带型代表着这些经过分析的酶/蛋白质的遗传多样性。然而，要指出的一点是产生的谱带可能因不同的组织而不同，或产生季节性波动。最早关于多种香蕉无性系和种质的同工酶研究（Bonner et al.，1974）未能反映出过氧化物酶同工酶的任何差异。之后，Rivera（1983）的研究利用过氧化物酶和多酚氧化酶的多态性区分了 BBB 和 ABB 基因组。Jarret 和 Litz（1986）利用 8 个同工酶位点对 *M. acuminata* 和 *M. balbisiana* 进行比较分析，揭示了 *M. acuminata* 和 *M. balbisiana* 的物种差异，也揭示了 *M. acuminata* ssp. *microcarpa*、ssp. *burmannica*、ssp. *errans* 和 ssp. *zebrina* 亚种等位基因特异性，表明了这一标记系统在阐明杂交品种的基因结构上卓有成效。之后 Lebot

等（1993）提出假设，*M. acuminata* ssp. *banksii* 是太平洋大蕉的 A 基因组供体。

3.3 基因组——基于 DNA 的香蕉遗传多样性分析

直接研究基因组技术的出现促进了遍布全基因组标记的发展，使得研究人员可以详细描述基因组中人们感兴趣的特定区域的表征。由于 DNA 序列变异数目非常多，因而标记发展所需的可用位点仅由基因组大小决定。聚合酶链式反应（PCR）的发现使标记产生革命性变化，促进了大量标记系统发展，引发的标记密度前所未有。这些技术包括随机扩增多态性 DNA（RAPD）、扩增片段长度多态性（AFLP）及其相关技术，以及涵盖等位基因变异程度高的区域，如简单重复序列（SSR）区域。其他技术如 PCR 限制性片段长度多态性（PCR-RFLP）和单核苷酸多态性（SNP）允许分析基因组的特定部位或被选定的基因。基于 PCR 技术的方法成本相对较低，易于在任何实验室采用，因而可对有大量标记位点的大量样本进行高通量分析。基于 DNA 序列的标记也独立用于分析植物细胞的细胞器和细胞核基因组。

植物细胞内有 3 种基因组：叶绿体基因组、线粒体基因组和核基因组。*M. acuminata* 的核基因组大小为 613 Mb，而 *M. balbisiana* 的核基因组较小，为 564 Mb（Dolezel et al.，1994）。细胞器基因组要小得多，叶绿体基因组大小为 0.13～0.14 Mb（Kornel Burg，未发表），线粒体基因组的确切大小现在还不得而知，估计为 0.4～0.5 Mb。芭蕉属植物中，叶绿体基因组强烈偏向母系遗传，而线粒体基因组则是父系遗传（Fauré et al.，1994）。单亲遗传排除了重组的可能性，其对应的基因组仅依靠突变来演化。因此，细胞器基因组为分别利用叶绿体和线粒体标记来遵循母系和父系遗传提供了绝佳的可能。

鉴于 Kurz 的假设，认为现今的栽培品种是古老野生类型的后代，所以应特别关注现存野生类型的遗传多样性。现存野生类型代表着原始祖先种的后代，它们通过杂交生产现代不育品种，最终，这些古老的野生基因型以或多或少未改变的形式存在于栽培品种中。将栽培品种基因型和现存野生类型进行分类可能识别出为栽培品种提供基因组元件的亚种/地理区域。

3.3.1 单亲遗传细胞器基因组的遗传分析

监测香蕉细胞器基因组的多态性，基本采用 3 种方法。首先，采用 Southern 杂交，利用不同的 DNA 探针识别香蕉细胞器的多态性。然后，基于植物中叶绿体基因组（Dumolin-Lapègue et al.，1997）和线粒体基因组（Duminil et al.，2002）的高水平序列相似性，开发出了通用的细胞器特异性 PCR

引物系统。根据这一组引物，采用 PCR-RFLP 分析或生成片段的直接测序来鉴定多态性区域/序列。

Gawel 和 Jarret（1991a）首次利用不同的叶绿体探头（*Lactuca sativa* 及 *Oncidium excavatum*）和 Southern 分析发表了第一个基于叶绿体 DNA 的标记和香蕉系统发育的分析。该分析首次基于细胞质分组，明确区分了 *M. acuminata* 和 *M. balbisiana* 这两个野生种。尽管它们中的一些有相似的 *acuminata* 核基因组，另一些却含有 *balbisiana* 型叶绿体基因组，但几个假定的 *M. acuminata*×*M. balbisiana* 三倍体杂交种仍无法区分。此外，卡文迪什的一种类型（Grande Naine）的叶绿体基因组显示出类似 *M. acuminata* spp. *malaccensis* 的叶绿体基因组。这些结果表明，利用细胞质标记可能对重建生成当今品种祖先的杂交或突变机制有帮助。之后这些学者（Gawel and Jarret，1991b）利用同样的方法比较芭蕉属种质，揭示出 *M. textilis*（*Australimusa*）的叶绿体基因组与 *M. balbisiana*（*Eumusa*）具有较近的亲缘关系，提出它们有同一个祖先。之后 Carreel 等（2002）发表了一项综合研究，是根据大量香蕉种质的叶绿体和线粒体基因组的 Southern 分析，明确了 *acuminata*、*balbisiana* 和 *schizocarpa* 基因组的 9 种叶绿体和 9 种线粒体模型。Umali（2002）利用 PCR 扩增，通过对叶绿体的 *rpl16* 和 *rpl14* 基因的基因间隔区测序识别 SNPs，区分了多种卡文迪什亚组品种。随后，Umali 和 Nakamura（2003）报道了一个 SNP 标记，该标记可以区分出 *acuminata* 和 *balbisiana* 的叶绿体基因组上的 *trn*L-F 基因间隔序列，并描述了 *M. acuminata* spp. *banksii* 的特异性缺失。Nwakanma 等（2003）根据叶绿体和线粒体 PCR-RFLP 标记系统，建立了多种芭蕉属种质的组间关系，结果表明，与 *M. acuminata* 的亚种相比，*M. balbisiana* 细胞器基因组在进化上更能代表祖先种。进化路线有两种，一种是 *Australimusa* 和 *Callimusa*，另一种是 *Eumusa* 和 *Rhodoclamys*，*M. balbisiana* 属于 *Australimusa*/*Callimusa* 系，这和它们的细胞器基因组有共同的假定起源。部分根据叶绿体（PCR-RFLP）标记，Ge 等（2005）发表了 *M. balbisiana* 的详细种群分析，揭示了分布在中国南方两个地理区域的截然不同的叶绿体基因组。Swangpol 等（2007）通过分析叶绿体基因组中的 SNP 变异，对栽培品种进行谱系分析，发现杂交品种的形成至少涉及 2 个 *M. balbisiana* 供体，也证实了之前 Gawel 和 Jarret（1991a）的观察，即卡文迪什品系的 Grande Naine 可能拥有一个源自 *M. acuminata* ssp. *malaccensis* 的叶绿体基因组。

上述研究表明，*M. acuminata* 和 *M. balbisiana* 的细胞器基因组差别很大，而 *M. acuminata* 的多样性对于区分部分形态学亚种十分有用。*M. balbisiana* 还未发现有亚种。然而，Ge 等（2005）已经识别出中国 *M. balbisiana* 种群内的分支。

单亲遗传基因组在进化中易发生突变，且大多数香蕉品种因单性结实而不育。因此，一个可能的假设是，最初的栽培品种形成时，古细胞器基因组困在现存品种中，未发生巨大改变。鉴于现存的野生类型是古老种质的后代，将这些野生类型与现今栽培品种比较可能识别共同起源的基因库。Boonruangrod 等（2008）利用这一假设，识别出 *M. acuminata* 和 *M. balbisiana* 的 3 个叶绿体基因库（图 3－1B）。同样，也观察到线粒体基因组的 3 个基因库（图 3－1C）。然而，叶绿体/线粒体组合显示不同的地理分布（图 3－1D）。对某些杂交种的分析，显示它们中无一有类似野生类型的叶绿体/线粒体组合。杂交品种中最常见的细胞型结合了西部最丰富的叶绿体类型和东部最丰富的线粒体类型，表明“西部”（*malaccensis*，*siamea*，*zebrina*，*burmannica*，*burmannicoides*）和“东部”（*banksii*，*errans*）野生种质的杂交（Boonruangrod et al.，出版中）。

3.3.2 基于 DNA 序列的核基因组多样性分析

大量基于 DNA 的方法已被用于揭示香蕉个体/栽培品种或野生种/亚种之间核 DNA 序列的差异。产生 PCR 多带最简单的方法是使用随机 PCR 引物（RAPD），或驱动基因组扩增的限制性内切酶切割位点（AFLP）。这些最常用的显性标记系统不需要事先知道基因组序列，但在这些系统中，不可能排除图谱中细胞器标记带的存在，因为大多数情况下，从香蕉细胞中分离的用于分析的是总 DNA（包括核 DNA 及细胞器 DNA）。然而，大多数标记带代表的是核基因组区域。另一方面，利用高度重复核序列元件（Jarret et al.，1993）和转座子序列基序（Nair et al.，2005），核基因组特异性带谱被开发，用来提供一种显性标记系统。不过，后述这些标记系统需要事前有基因组分析区域的序列信息。同样，一些标记系统也需要事先掌握序列信息，如 SSR、RFLP 和 SNP，这些为共显性标记系统。最近出现一些高通量手段，如多样性阵列技术（DArT；Risterucci et al.，2009）和定向诱导基因组局部突变技术（TILLING；Till et al.，2010），用于大规模鉴定有序列多样性的基因组区域，这些区域最终可用作遗传标记。

许多研究人员利用显性标记系统如 RAPD 和 AFLP，尝试阐明核基因组的遗传变异性。这些易于使用的技术频繁用于评估香蕉种群的局部变异性（可参见 Crouch et al.，1955；Ude et al.，2002，2003；Wang et al.，2007；Brown et al.，2009），也用于区分种质（Pillay et al.，2000）。利用 AFLP 来分析 *M. acuminata* 的野生种和栽培品种（Ude et al.，2002），能将形态描述的亚种明显区分为 3 组/个基因库：ssp. *microcarpa*、ssp. *malaccensis* 和 ssp.

burmannica。之前 Carreel 等（1994）曾提出分为 4 组形态亚种（库 1 *banksii*，*errans*；库 2 *zebrina*，*microcarpa*；库 3 *burmannica*，*burmannicoides*，*siamea*；库 4 *malaccensis*），该结论采用 Southern-RFLP 技术，基于能代表 *M. acuminata* 基因组全部 11 对染色体的 30 个核标记。基于 SSRs 的标记也常用于芭蕉属植物遗传多样性的研究，然而，尚未检测出种质的特异性等位基因。Grapin 等（1998）将形态亚种划分为 4 组/个基因库（ssp. *malaccensis*、ssp. *zebrina*、ssp. *banksii* 和 ssp. *burmannica*），而 Perrier 等（2009）利用核 SSR 标记将形态亚种划分为 5 个基因库（ssp. *banksii*、ssp. *siamea*/*burmannica*、ssp. *malaccensis*、ssp. *zebrina* 和 ssp. *microcarpa*）。利用 45S 核糖体区域，*M. acuminata* 和 *M. balbisiana* 的基因组特异序列变异被明显区分（Nwakanma et al.，2003）。Boonruagrod 等（2009）开发了种质特异或 *M. acuminata* 亚种特异的标记。之后的研究证实并结合 Carreel（1994）、Grapin 等（1998）、Ude 等（2002）和 Perrier 等（2009）的发现，将形态亚种分成 4 个基因库，而这些基因库呈现明显的地域分布（图 3－1C）。然而，种内分析显示 *M. balbisiana* 变异不大，尽管 Ge 等（2005）基于 SSR 标记描述 *M. balbisiana* 中存在分支。综上所述，采用几种不同的方法对 *M. acuminata* 核基因组进行的分析，将地缘接近的形态亚种分为 4 个基因库，如①*banksii*，*errans*，*microcarpa*；②*burmannica*，*burmannicoides*，*siamea*；③*zebrina*；④*malaccensis*。

3.3.3 在描述 *M. acuminata* 和 *M. balbisiana* 遗传多样性中细胞核和细胞器标记的合并利用

早前，形态数据和分子数据的结合大多是利用细胞器或细胞核标记。最近，Boonruangrod 等（2011）建议基因型的鉴定应结合叶绿体、线粒体和细胞核的基因型。依据这一方法，利用分子标记区分了 *M. acuminata* 的所有主要形态亚种（图 3－1D），而 *M. balbisiana* 的多样性基本上可通过细胞器基因组识别出。根据这一研究并且考虑到它们的地域分布，9 种形态亚种可能减少至 5 种，保留 ssp. *malaccensis* 和 ssp. *zebrina*，而 ssp. *banksii* 和 ssp. *errans*，ssp. *microcarpa* 和 ssp. *truncata*，ssp. *burmannica*、ssp. *burmannicoides* 和 ssp. *siamea* 将被分别合并为 3 组（图 3－1D）。

3.4 多倍体品种基因组结构的分析

自从 Kurz 认为 *Eumusa* 组大多数可食用香蕉品种源自 *M. acuminata*（AA）和 *M. balbisiana*（BB）这两个野生种的直接遗传或各种杂交组合后，Carreel 等（1994）揭示，S 和 T 基因组也存在于可食用品种中，并被 D'Hont

等（2000）利用基因组原位杂交（GISH）将其可视化。生化标记系统的发展支持香蕉基于形态标记的主要分类，还确定了 A、B 基因组在栽培品种中所占的比例（Simmonds，1962）。随后，对表型化学特征的分析为栽培种的基因组结构提供了更详细的信息，揭示了卡文迪什品种形成过程中的假定供体亚种（Horry and Jay，1990），而之后 Lebot 等（1993）利用同工酶标记确定 *M. acuminata* ssp. *banksii* 是太平洋大蕉的 A 基因组的供体。DNA 分子标记的出现使得栽培种能产生与野生种一样精确的基因组“指纹”。Raboin 等（2005）发表了利用 30 个单拷贝核标记来鉴定野生类型基因型假定供体的首次详细的分析，并揭示了三倍体出口香蕉品种卡文迪什和 Gros Michel 的假定祖先。

形态学和分子工具显示，大多数可食用栽培品种来自 *M. acuminata* 和 *M. balbisiana* 的种内或种间杂交。由于 *M. acuminata* 及其亚种都显示出高度的多样性，且这些亚种显示出特异的地理分布，因此监测这些亚种在现今品种中的参与情况，将为香蕉育种提供有价值的信息。最近，Boonruangrod 等（2009）开发了基于 45S rDNA 位点的亚种特异性 PCR 方法用于杂交种的分析，这使得研究人员可以确定 *M. acuminata* 亚种在品种形成中所起的作用。细胞器与核遗传数据的结合可以确定在现今品种进化过程中发生的推定杂交事件（Boonruangrod et al.，2011）。常用的形态标记表明在 *acuminata*×*balbisiana* 杂交种中出现了意外的不平衡的表观。对此的解释可能是在栽培品种发展的早期阶段，基因组之间的信息发生改变。根据细胞器和核多样性数据，De Langhe 等（2010）认为在香蕉杂交种早期进化中，可能出现的回交有重要意义，因为这是基因组失衡发展的推定来源。

3.5 香蕉分子分类学

芭蕉科种质的分布范围涵盖亚洲、澳大利亚和非洲的热带地区。根据形态标记及染色体数目，芭蕉科有公认的 3 个属：芭蕉属、地涌金莲属和象腿蕉属。芭蕉属植物的单倍体组染色体数目为 $n=10\sim11$，而地涌金莲属和象腿蕉属植物的为 9。芭蕉属作为最大的属，包括超过 60 种的野生种（OECD，2009），可分为 4 组：*Australimusa*、*Callimusa*、*Musa*（之前称为 *Eumusa*）和 *Rhodochlamys*。*Musa* 和 *Rhodochlamys* 的单倍体组染色体数目为 $n=11$，而 *Australimusa* 和 *Callimusa* 的则为 10。最近基于叶绿体和核基因组区域比较序列分析的研究（Li et al.，2010；Liu et al.，2010）揭示了两个主要分支，一个包括地涌金莲属和象腿蕉属，而另一个包括芭蕉属的所有物种。取代之前提到的 4 组分组法，芭蕉属可以分成两个分支：一个包括 *Musa* 和

Rhodochlamys 组，而另一个包括 *Australimusa* 和 *Callimusa*。染色体数的分布证实了这种分类法的可行性。

3.6　结论

芭蕉属种质和 *M. acuminata* 亚种的形态学分类已完全在分子水平上进行分析，特别是基于 DNA 标记系统的证实。该标记系统还证实了形态学家所做的出色工作。根据基于 DNA 分子的研究结果，认为芭蕉科下各属和芭蕉属下各组结构应简化。此外，基于分子数据，应重新思考 *M. acuminata* 的亚种分类。

分子标记，特别是基于 DNA 的标记，证实了 *M. acuminata* 在亚种/种水平上对杂交基因型所做的贡献，这也为香蕉育种提供了有价值的信息。

缩略语

AFLP：扩增片段长度多态性
GISH：基因组原位杂交
PCR：聚合酶链式反应
RAPD：随机扩增多态性 DNA
RFLP：限制性片段长度多态性
SNP：单核苷酸多态性
SSR：简单重复序列

参考文献

Bonner J W, Warner R M, Brewbaker J L, 1974. A chemosystematic study of *Musa* cultivars [J]. HortScience, 9: 325 - 328.

Boonruangrod R, Desai D, Fluch S, et al, 2008. Identification of cytoplasmic ancestor genepools of *Musa acuminata* Colla and *Musa balbisiana* Colla and their hybridsby chloroplast and mitochondrial haplotyping [J]. Theor Appl Genet, 118: 43 - 55.

Boonruangrod R, Desai D, Berenyi M, et al, 2011. Application of cytoplasmic and nuclear DNA-based marker systems for elucidation of the phylogenetic relationships of *Musa acuminata* and *Musa balbisiana* and their hybrids Proc [J]. Acta Hort, 897: 133 - 137.

Boonruangrod R, Fluch S, Burg K, 2009. Elucidation of origin of the present day hybrid banana cultivars using the 5'ETS rDNA sequence information [J]. Mol Breed, 24: 77 - 91.

Brown N, Venkatasamy S, Khittoo G, et al, 2009. Evaluation of geneticdiversity between 27 banana cultivars (*Musa* spp.) in Mauritius using RAPD markers [J]. Afr J Biotechnol,

8：1834－1840.

Carreel F，Faura S，de Lean D，et al，1994. Evaluation de la diversite genetique chez les bananiers diploids（*Musa* sp.）[J]. Genet Sel Evol，26：1－12.

Carreel F，Gonzalez de Leon D，Lagoda P，et al，2002. Ascertaining maternal and paternal lineage within *Musa* by chloroplast and mitochondrial DNA RFLP analyses [J]. Genome，45：679－692.

Cheesman E E，1948. Classification of the bananas Ⅲc *Musa paradisiaca* L. sp.，*Musa sapientum* L. [J]. syst Kew Bull，3：145－153.

Crouch J H，Crouch H K，Constandt H，et al，1999. Comparison of PCR-based molecular marker analyses of *Musa* breeding populations [J]. Mol Breed，5：233－244.

D'Hont A，Paget-Goy A，Escoute J，et al，2000. The interspecific genome structure of cultivated banana，*Musa* spp. revealed by genomic DNA in situ hybridization [J]. Theor Appl Genet，100：177－183.

De Langhe E，Hribova E，Carpentier S，et al，2010. Did backcrossing contributeto the origin of hybrid edible bananas [J]? Ann Bot，106：849－857.

Dolezel J，Dolezelova M，Novak F J，1994. Flow cytometric estimation of nuclear DNA amountin diploid bananas（*Musa acuminata* and *M. balbisiana*）[J]. Biol Planta，36：351－357.

Duminil J，Pemonge Ma H，Petit R J，2002. A set of 35 consensus primer pairs amplifying genes and introns of plant mitochondrial DNA [J]. Mol Ecol Notes，2：428－430.

Dumolin-Lapegue S，Demesure B，Fineschi S，et al，1997. Phylogeographic structure of white oaks throughout the European continent [J]. Genetics，146：1475－1487.

Faure S，Noyer J L，Carreel F，et al，1994. Maternal inheritance of chloroplast genome and paternal inheritance of mitochondrial genome in bananas（*Musaacuminata*）[J]. Curr Genet，25：265－269.

Gawel N，Jarret R L，1991a. Cytoplasmic genetic diversity in bananas and plantains [J]. Euphytica，52：19－23.

Gawel N J，Jarret R L，1991b. Chloroplast DNA restriction fragment length polymorphisms（RFLPs）in *Musa* species [J]. Theor Appl Genet，81：783－786.

Ge X J，Liu M H，Wang W K，et al，2005. Population structure of wild bananas，*Musa balbisiana*，in China determined by SSR fingerprinting and cpDNA PCR-RFLP [J]. Mol Ecol，14：933－944.

Grapin A，Noyer J L，Carreel F，et al，1998. Diploid *Musa acuminata* genetic diversity assayed with sequence-tagged microsatellite sites [J]. Electrophoresis，19：1374－1380.

Horry J，Jay M，1988. Distribution of anthocyanins in wild and cultivated banana varieties [J]. Phytochemistry，27：2667－2672.

Horry J，Jay M，1990. An evolutionary background of bananas as deduced from flavonoid diversification [C]//Jarret R L. Identification of genetic diversity in the genus *Musa* proceedings of an international workshop, Los Banos (1989), the Philippines. Montpellier, France: INIBAP: 41-55.

Jarret R L, Litz R E, 1986. Enzyme polymorphism in *Musa acuminata* Colla [J]. J Hered, 77: 183-188.

Jarret R L, Vuylsteke D R, Gawel N J, et al, 1993. Detecting genetic diversity in diploid bananas using PCR and primers from a highly repetitive DNA sequence [J]. Euphytica, 68: 69-76.

Kurz S, 1865. Note on plantains of the Indian Archipelago [J]. J Agric Hort Soc India, 14: 295-301.

Lebot V, Aradhya K M, Manshardt R, et al, 1993. Genetic relationships among cultivated bananas and plantains from Asia and the Pacific [J]. Euphytica, 67: 163-175.

Li L F, Hakkinen M, Yuan Y M, et al, 2010. Molecular phylogeny and systematics of the banana family (Musaceae) inferred from multiple nuclear and chloroplast DNA fragments, with a special reference to the genus *Musa* [J]. Mol Phylogenet Evol, 57: 1-10.

Liu A Z, Kress W J, Li D Z, 2010. Phylogenetic analyses of the banana family (Musaceae) based on nuclear ribosomal (ITS) and chloroplast (trnL-F) evidence [J]. Taxon, 59: 20-28.

Nair A S, Teo C H, Schwarzacher T, et al, 2005. Genome classification of banana cultivars from South India using IRAP markers [J]. Euphytica, 144: 285-290.

Nwakanma D C, Pillay M, Okoli B E, et al, 2003. Sectional relationships in the genus *Musa* L. inferred from the PCR-RFLP of organelle DNA sequences [J]. Theor Appl Genet, 107: 850-856.

OECD, 2009. Series on harmonisation of regulatory oversight in biotechnology No. 48 consensus document on the biology of bananas and plantains (*Musa* spp.) [Z].

Perrier X, Bakry F, Carreel F, et al, 2009. Combining biological approaches to shed light on the evolution of *Musa* complex [J]. Ethnobot Res Appl, 7: 199-216.

Pillay M, Nwakanma D C, Tenkouano A, 2000. Identification of RAPD markers linked to A and B genome sequences in *Musa* L. [J]. Genome, 43: 763-767.

Raboin L M, Carreel F O, Noyer J L, et al, 2005. Diploid ancestors of triploid export banana cultivars: molecular identification of 2*n* restitution gamete donors and *n* gamete donors [J]. Mol Breed, 16: 333-341.

Risterucci A M, Hippolyte I, Perrier X, et al, 2009. Development and assessment of diversity arrays technology for high-throughput DNA analyses in *Musa* [J]. Theor Appl Genet, 119: 1093-1103.

Rivera F N, 1983. Protein and isoenzyme banding patterns among Philippine cooking bananas and their wild parents (*Musa* species) [J]. Paradisiaca, 6: 7-12.

Rumphius G E, 1749. Herbarium amboinense [M].

Shepherd K, 1990. Observations on *Musa* taxonomy: a note on *Musa* germplasm in the Philippines [C]//Jarret R L. Identification of genetic diversity in the genus *Musa* proceedings of an international workshop held at Los Banos (1989), the Philippines. Montpellier, France: INIBAP: 158-165.

Shepherd K, 1999. Cytogenetics of the genus *Musa* [M]. Montpellier, France: INIBAP.

Simmonds N W, 1954. Anthocyanins in Bananas [J]. Nature, 173: 402-403.

Simmonds N W, 1962. The Evolution of the bananas [M]. London, UK: Longmans.

Simmonds N W, Shepherd K, 1955. The taxonomy and origins of the cultivated bananas [J]. J Linn Soc (London) Bot, 55: 302-312.

Stover R H, Simmonds N W, 1987. Bananas, tropical agriculture series [M]. 3rd ed. London, UK: Longman.

Swangpol S, Volkaert H, Sotto R C, et al, 2007. Utility of selected non-coding chloroplast DNA sequences for lineage assessment of *Musa* interspecific hybrids [J]. Journal of Biochemistry and Molecular Biology, 40: 577-587.

Till B J, Jankowicz-Cieslak J, Sagi L, et al, 2010. Discovery of nucleotide polymorphisms in the *Musa* gene pool by EcoTILLING [J]. Theor Appl Genet, 121: 1381-1389.

Ude G, Pillay M, Nwakanma D, et al, 2002. Genetic diversity in *Musa acuminate* Colla and *Musa balbisiana* Colla and some of their natural hybrids using AFLP markers [J]. Theor Appl Genet, 104: 1246-1252.

Ude G, Pillay M, Ogundiwin E, et al, 2003. Genetic diversity in an African plantain core collection using AFLP and RAPD markers [J]. Theor Appl Genet, 107: 248-255.

Umali R P, 2002. Development of PCR-based fingerprinting tool in banana (*Musa* sp., AAA) and conversion of negative to positive DNA marker [J]. HortScience, 37: 1108-1111.

Umali R, Nakamura I, 2003. Identification of dCAPS markers that discriminate A and B cytoplasms in banana (*Musa* spp.) [J]. Plant Biotechnol, 20: 159-164.

Wang X L, Chiang T Y, Roux N, et al, 2007. Genetic diversity of wild banana (*Musa balbisiana* Colla) in China as revealed by AFLP markers [J]. Genet Resour Crop Evol, 54: 1125-1132.

第4章 芭蕉属植物基因组研究的分子标记技术

Michael Pillay[①*] **Kaliyaperumal Ashokkumar**[②] **Andrew James**[③]
Silvas Jabekumar Prince Kirubakaran[④] **Robert Miller**[⑤]
Rodomiro Ortiz[⑥] **Elayabalan Sivalingam**[⑦]

摘 要：香蕉是最重要的亚热带作物之一，其遗传系统因为种间和种内杂交、杂合性和多倍体而较为复杂，不为人所知。因此分子标记被广泛应用于芭蕉属种质来检测遗传变异及遗传关系、鉴定重复种质、监测遗传稳定性、分析辐照基因型和鉴定用于育种的特性标记。现已存在基于形态学、蛋白质和DNA的各种标记。其中，适用于标记辅助选择（MAS）的最常用标记是RFLP、随机扩增多态性DNA（RAPD）、扩增片段长度多态性（AFLP）和简单重复序列（SSR）。本章比较可用标记的效率和其在芭蕉属植物基因组分析上的应用和发展。基于单核苷酸多态性（SNPs）或小规模插入和缺失的高通量技术是有效的替代技术，因为其理想的剂量敏感性并且能将单倍型从多倍体作物中区分出来。然而，选择不同的标记，取决于检测

① Vaal University of Technology, Private Bag X021, Vanderbijlpark 1900, Gauteng, South Africa; e-mail: mpillay@vut. ac. za.

② Department of Plant Sciences, University of Saskatchewan, 51 Campus Drive, Saskatoon, SK, Canada, S7N 5A8; e-mail: ashok. kumar@usask. ca.

③ Centro de Investigacion Cientifi ca de Yucatan, Calle 43 No. 130, Col. Churburna de Hidalgo, Merida, C. P. 97200, Yucatan, Mexico; e-mail: andyj007@cicy. mx.

④ Department of Plant Molecular Biology and Biotechnology, Center for Plant Molecular Biology, Tamil Nadu Agricultural University, Coimbatore-641 003-India; e-mail: symsil07@gmail. com.

⑤ Laboratório de Microbiologia: Interação Planta-Praga, Instituto de Ciências Biológicas, Departamento de Biologia Celular, Universidade de Brasília, Campus Universitário Darcy Ribeiro, Asa Norte, 70910-900, Brasília, D. F. , Brasil; and Universidade Católica de Brasília, SGAN Qd. 916 Módulo B, Av. W5 Norte CEP: 70. 790-160, Brasília/DF, Brazil; e-mail: rmiller @ pos. ucb. br or bruna @ tba. com. br.

⑥ Department of Plant Breeding and Biotechnology, Swedish University of Agricultural Sciences, Box 101, SE-230 53, Alnarp, Sweden; e-mail: rodomiroortiz@gmail. com.

⑦ Crop Tech Ltd. , P. O. Box 1367, Arusha, Tanzania; e-mail: balabiotech@gmail. com.

* 通信作者。

所需人力、揭示单个或多个位点的可能性、显性或共显性性质和成本。据报道，最近在多种作物研究上所取得的进展如高分辨率 DNA 熔解（HRM）分析，使得品种鉴定、候选基因鉴定和作图（基因组和关联作图）研究具有巨大潜力。因此，利用这些方法鉴定的香蕉基因组区域有可能显著提升芭蕉属植物遗传改良的速度和效率。

关键词：分子标记系统，芭蕉属，遗传改良

4.1 引言

分子生物学的发展产生了若干新技术，其给作物育种带来了巨大希望。这些创新技术之一是可标记目的基因的分子标记技术。遗传标记是染色体的位置标记，辅助基因组的分析（Lefebvre et al.，1995）。标记通常被描述为“遗传标签”，其能识别植株 DNA 序列上的特定位置并遵循遗传法则从一代传给下一代子孙（Drew，1997）。研究人员利用基于 DNA 的标记，可以标记与遗传性状，如与寄主植物对病原菌和害虫的抗性相关的基因和染色体区域（Drew，1997）。遗传标记一旦确定，它们可直接用于育种，因为研究人员可根据这些标记的有无来预测表型。除此之外，一旦利用遗传标记技术鉴定出特定性状的基因，这些基因可被克隆。这些技术的特别优势在于可对复杂的多基因性状进行分析。完全根据基因的遗传行为，作图技术也可用于分离基因。一旦这些基因被鉴定、测序并克隆，转基因技术可以将它们转移到其他物种中（Drew，1997）。

分子标记是基于 DNA 序列中天然存在的多态性（如碱基对缺失、取代或添加等）。分子标记有多种类型，一些是形态标记或基于蛋白质的标记，而最常见的是基于 DNA 的标记。揭示蛋白质水平上多态性的遗传标记被称为生化标记，而 DNA 标记揭示的是 DNA 水平上的多态性。生化标记是蛋白质，由基因表达产生，可通过电泳进行分离从而鉴定其等位基因。最常用的蛋白质标记是同工酶，它们是同一种酶的变体形式。蛋白质标记作为共显性标记能揭示基因序列和功能上的差异。然而，由于它们在任一作物种质中的数量有限，且易发生翻译后修饰，它们的应用受到限制（Staub et al.，1982）。根据多态性的表达方式，DNA 标记可分为两类：基于杂交的多态性和基于 PCR 的多态性。基于杂交的多态性包括限制性片段长度多态性（RFLP）（Sambrook et al.，1989）和易变的顺序重复数目（VNTR）位点（Rogstad，1993；Weising et al.，1998）。在此，随机基因组克隆等探针和微卫星探针（或简单重复

序列，SSR）与小卫星序列杂交到含有已被限制性内切酶消化的 DNA 的滤膜上。对于 VNTR 位点，产生多态性是由于重复次数的差异，而产生 RFLPs 则是由于发生诸如点突变、倒位、缺失或易位等事件。

分子标记的利用极大地改进了育种方法，因为通过 DNA 分析这些标记能直接揭示遗传变异性（Staub et al.，1996），并且这种检测不受环境因素影响。用于植物基因组分析的众多分子标记的发展使研究人员可以挑选基因型而不是表型，因而产生了标记辅助选择或 MAS 的概念（Paterson et al.，1991）。适用于标记辅助选择（MAS）的最常用标记是 RFLP、随机扩增多态性 DNA（RAPD）、扩增片段长度多态性（AFLP）和简单重复序列（SSR）（Barone，2004）。然而，MAS 最适合标记的选择取决于检测所需人力、揭示单个或多个位点的可能性、显性或共显性特性及成本。

分子标记已被广泛应用于芭蕉属植物研究，同时用于检测香蕉种质的遗传变异和遗传关系，鉴定田间和组织培养种质库中的重复种质，监测组织培养材料的遗传稳定性，分析辐照后的香蕉基因型和鉴定用于育种的性状标记。本章将探讨在芭蕉属植物中最常用的标记，并提及可能对芭蕉属植物研究有用的新标记系统。

4.2　同工酶标记

许多关于香蕉和大蕉的研究描述作为遗传标记的同工酶可以评估属内遗传多样性（Bhat et al.，1992a），测量其他的体细胞无性系变异，鉴定原生质体融合产物和鉴定无性系（Bhat et al.，1992b）。

然而，这种技术的灵敏度是可解析的且起作用的多态性位点数目。使用同工酶作为遗传标记来鉴定香蕉品种的历史可追溯到 20 世纪 80 年代，这使同工酶得到广泛关注（Tanksley and Orton，1983）。较少研究进行了各种香蕉无性系和种质过氧化物酶同工酶的检测，指出其缺乏多态性（Bonner et al.，1974）。Rivera（1983）检测了过氧化物酶和多酚氧化酶系统的多态性，并能明显地区别 Saba（ABB/BBB）和 Bluggoe（ABB）香蕉基因组。Jarret 和 Litz（1986a，b）研究了多种同工酶的效率和应用，这些同工酶包括莽草酸脱氢酶（SKDH）、苹果酸脱氢酶（MDH）、过氧化物酶（PRX）、葡萄糖磷酸变位酶（PGM）和谷氨酸草酰乙酸转氨酶（GOT）。一些同工酶在区分某些基因组的无性系上十分有效。Simmonds（1966）应用酯酶（EST）、四唑氧化酶（TO）、6-磷酸葡萄糖酸脱氢酶（6-PGDH）、苹果酸酶（ME）、磷酸葡萄糖异构酶（PGI）、醇脱氢酶（ADH）、磷酸丙糖异构酶（TPI）、半乳糖脱氢酶（GDH）、琥珀酸脱氢酶（SUDH）、山梨醇脱氢酶（SDH）和谷氨酸脱氢酶

(GUDH)对香蕉进行分类。Espino 和 Pimentel(1990)指出，利用 MDH 可有效地将 AAB 与 ABB 品种从 BB 或 BBB 基因组品种中区分开。同工酶 MDH、PRX 和 GOT 被用来分析属于 AA、AAA、AAB、ABB 和 BB 基因组的 100 种印度尼西亚芭蕉属品种的变异(Megia et al.，2001)。MDH 和 PRX 被公认为多态性好。GOT 是多态性最低的同工酶。展现出特异性同工酶谱的栽培品种 MDH 有 16 种，PRX 有 20 种，GOT 有 8 种，这些同工酶谱可区分 31 个品种。同工酶也能用于评估香蕉通过 γ 辐射诱导的遗传变异(Megia et al.，2001)，且被用来鉴定对香蕉苞片花叶病毒(BBrMV)的抗性(Dhanya et al.，2006)。被 BBrMV 感染的样品明确显示存在单过氧化物酶带，而非感染样本中则不存在。

在香蕉研究中使用的大多数同工酶标记有助于在系统发育研究中阐明无性系的鉴定。不过，使用同工酶作为遗传标记可估计芭蕉属栽培种与野生近缘种的遗传多样性，促进无性系鉴定，并为三倍体品种的起源提供有价值的证据。同工酶分析偶尔能检测来自亲本无性系的体细胞突变体的变异(Brettell et al.，1986)。

4.3 基于杂交的系统

4.3.1 限制性片段长度多态性(RFLP)

RFLPs 是经限制性酶切的 DNA 样品中 DNA 片段长度电泳条带的变化。这些变化常常是由于某一个体基因组在某一位点存在限制性内切酶酶切位点，而另一个体中在该位点没有限制性内切酶酶切位点。RFLP 是共显性标记，这意味着它能区分纯合子和杂合子。RFLP 强健，易于在实验室之间转移，并且事先不需要提供物种的序列信息。因此，它适用于构建遗传连锁图，特别是在芭蕉属植物中。

由于它的位点特异性，对保守共线性的估计是可能的。使用单位点探针可对种群水平进行辨别，但缺点是成本高昂，需要大量的 DNA，与 AFLP、多样性阵列技术(DArT)或 SSRs 多项技术相比无法自动化，需要合适的探针库，可能还需要放射性标记，并且费时费力。至今，RFLPs 已被应用于所有的芭蕉属植物图谱项目，且最近已应用于抗性基因类似物多态性的连锁(Hippolyte et al.，2010)。

自从开始使用已知序列的探针，RFLPs 被认为其重复性高、特异性好。不同种质的 RFLPs 已被用于研究芭蕉属种质的分类和系统发育(Gawel and Jarret，1991a，b；Gawel et al.，1992；Jarret et al.，1992)，以及叶绿体基因组的变异(Gawel and Jarret，1991a，b；Nwakanma et al.，2003)。然而，

关于 RFLPs 区分近缘材料有效性的报道却仅有一篇（Bhat et al.，1995）。该技术相对高昂的成本和高要求使之不适用于常规育种项目。

4.3.2　数目可变串联重复序列（VNTR）

VNTRs 是 DNA 短序列的串联重复序列，包括小卫星（10～45 bp）和微卫星（2～6 bp）。虽然最初检测 VNTRs 需要放射性探针，现在却只需利用 PCR 技术。真核基因组中遍布 VNTRs（Tourmente et al.，1994）。这些序列可与多个位点杂交，并在自然界中变异度高，因而能为遗传分析提供大量信息。串联重复序列常被归类为微卫星（重复单元长度等于或小于 5 bp）和小卫星（重复单元长度大于 5 bp，不超过 100 bp）（Buard and Vergnaud，1994）。Crouch 等（1998，1999a，b）利用 VNTR 分析微卫星位点，从而比较全同胞二倍体、四倍体大蕉-香蕉杂交种的遗传相似性和它们的亲本基因型。

4.4　基于 PCR 的标记系统

所有基于 PCR 的分子标记似乎都能在一系列的芭蕉属植物育种群体中检测出高水平的多态性。PCR 的测定方法适于有高通量筛选育种群体需要的项目。因此，研究人员专注于 PCR 在芭蕉属植物基因组分析中的应用。各种各样的分子标记，特别是基于 PCR 的不同的分子标记，包括 AFLP、RAPD 和微卫星，常被用于评估野生香蕉与栽培种质的遗传多样性和系统发育（Grapin et al.，1998；Loh et al.，2000；Wong et al.，2001；Carreel et al.，2002；Udeet al.，2002a，b，2003；Creste et al.，2003a，b；Nwakanma et al.，2003）。

自 PCR 扩增方法出现伊始，涌现了大量的 PCR 扩增方法。这些方法包括单引物扩增反应（SPAR）（Gupta et al.，1994；Weising et al.，1995）。用于小卫星分析的 SPAR 方法一直被视为直接扩增小卫星区域 DNA（DAMD），即利用小卫星核心序列作为单一引物来直接扩增富含小卫星重复的区域（Heath et al.，1993；Zhou et al.，1997）。单链构象多态性分析（SSCP）是一种快速检测 DNA 聚合酶链式反应扩增中小序列变化的方法（Orita et al.，1989）。RAMPO（随机扩增微卫星多态性）将随机或半特异 PCR 引物与微卫星杂交，从而在每块电泳凝胶上产生几条单一或多态的遗传指纹（Richardson et al.，1995）。最近，旨在扩增开放阅读框（ORFs）的相关序列扩增多态性（SRAP）技术被用来分析 40 种香蕉种质的遗传变异及关系（Muhammad et al.，2010）。

4.4.1 随机扩增多态性DNA（RAPD）

RAPD技术的优势在于易开展，费用较低，但有时重复性较差，可靠性取决于操作者的技术。即使使用相同的PCR参数和条件，不同实验室进行的RAPD分析也会产生不同的结果。由于RAPD技术依靠所使用的随机短引物，因而无需事先知道生物体的基因组。RAPD技术的优势包括速度快，DNA模板要求低和技术简单。它是检测种质遗传变异的便捷工具。RAPDs已被广泛用于检测香蕉种质的遗传关系（Uma et al.，2006；Jain et al.，2007；Nsabimana and van Staden，2007），从而鉴定田间和组织培养种质资源库中种质的重复和监测组织培养材料的遗传稳定性（如体细胞无性系变异）（Bairu et al.，2006，2008；Ray et al.，2006；Lakshmanan et al.，2007），分析和区分被辐照的香蕉基因型（Toruan-Mathius and Haris，1999；Finalet et al.，2000；Imelda et al.，2001；Hautea et al.，2004），并鉴定用于杂交和突变育种项目的性状标记（Damasco et al.，1996）。DNA分析对于监管植物专利和对像芭蕉属这种无性繁殖作物的新育成品种有着重要的法律保护作用（Kaemmer et al.，1997）。信息量大的RAPD标记可被转换为序列特征性扩增区域（SCAR）引物进行特异性扩增。RAPD分析已被证明在标记辅助渗入和回交育种方面是强大而有效的。芭蕉属植物A与B基因组的特异性RAPD标记已得到鉴定，并且过去常被使用（Pillay et al.，2000；Oselebe et al.，2006；Pillay et al.，2006）。

然而，RAPD分析有几个缺点，包括标记系统的显性特性和重复性问题，这些问题会限制它们在标记辅助选择上的应用。RAPD技术已被成功用于区分不同的芭蕉属种质（Howell et al.，1994；Bhat and Jarret，1995；Pillay et al.，2001；Uma et al.，2006；Jain et al.，2007；Nsabimana and van Staden，2007）。此外，利用包括RAPD在内的多种标记系统绘制出了分子连锁图谱（Faure et al.，1993）。RAPD分析也被用来区分芭蕉属植物基因组（Howell et al.，1994；Pillay et al.，2000）、近缘香蕉种质（Bhat and Jarret，1995）和大蕉育种群体中的全同胞杂交种（Crouch et al.，1998，2000）。

克隆类型的保真度是任何香蕉品种进行微繁的最重要的先决条件之一。亲本系的继代培养中出现体细胞突变是在体外培养中常遇到的主要问题。RAPD和简单重复序列区间（ISSR）技术被用来评估香蕉苗的无性系保真度（Bennici et al.，2004；Carvalho et al.，2004；Martins et al.，2004；Ray et al.，2006）。通过基于PCR技术验证的体细胞无性系相对比较稳定，没有恢复回正常表型（Ramage et al.，2004）。一项严谨的被称为高退火温度（HAT）-RAPD技术被用于泰国香蕉栽培品种（Sitthipron et al.，2005）。尽管该项技术不乏批评声，但是近年来的著作证明 RAPD 技术现在仍应用于香蕉基因组分析中（Das et al.，2009；Miri et al.，2009；Khatri et al.，2009）。

4.4.2　简单重复序列（SSR）

RAPD 标记的局限性促进了芭蕉属植物微卫星引物的发展和应用（Jarret et al.，1994；Kaemmer et al.，1997；Grapin et al.，1998；Creste et al.，2003a，b）。微卫星或 SSR 位点为高丰度、随机分布、位点特异、共显性和多等位基因的标记。它们为串联重复 DNA 序列（一般小于 4 bp 或等于 4～6 bp），总长度约为几十碱基对。据报道，微卫星或 SSRs 数量丰富，随机分布于许多植物的整个基因组中（Crouch et al.，1998）。基序的变异被认为是在 DNA 复制过程中发生了移位的错误。SSRs 因其高丰度、多态性和可靠性而成为其他标记系统中可重复的标记（Campbell et al.，2003）。

微卫星广泛用于检测植物种质的遗传多样性。这些标记已广泛用于植物的指纹图谱、作图和遗传分析。简单重复序列长度多态性（SSRLP）分析能检测出芭蕉属育种群体个体间高水平的多态性（Crouch et al.，1998，2000）。然而，分离微卫星不仅耗时长，而且费用高。尽管如此，随着自动化 DNA 测序设备的出现，构建富含 SSR 的基因文库技术的改进和筛选合适克隆技术的改进，SSR 的分离已经常态化。最近，这项技术已能将数百微卫星从芭蕉属植物 B 基因组中快速分离（Buhariwalla et al.，2005）。迄今为止，已通过 *M. acumiata* 和 *M. balbisiana* 材料鉴定出几百个 SSR 标记（Jarret et al.，1994；Kaemmer et al.，1997；Crouch et al.，1998；Creste et al.，2003a，b；Wang et al.，2009），随着基因组技术的进步，标记数目可能还会增加。

SSR 标记用途广泛，但是可能最重要的应用是遗传图谱的加密和 MAS，利用特异性状共定位。例如，最近在 *M. acuminata* Calcutta 4 的细菌人工染色体（BAC）末端序列发现有 352 个 SSR 位点（Cheung and Town，2007）。然而与其他作物相比，用于芭蕉属植物遗传分析的特异 SSR 位点还相对有限，因为在品种中应用时等位基因可能会缺失或为单型。最近，从 5 个全基因组测序的 *M. acuminata* Calcutta 4 BAC consensi 数据集中鉴定出 40 多个微卫星位点（基因库登录号 AC186748，AC186749，AC186954，AC186747 和 AC186750），并在对叶斑病有抗性的植物中进行了不同基因型的多态性验证（Miller et al.，2010）。

4.4.3　反转录转座子之间扩增多态性（IRAP）

IRAP 是替补的显性标记系统，用于检测逆转录转座子插入位点间多态性。逆转录转座子在大多数植物的基因组中广泛存在且丰度高，在许多物种核

DNA 中的含量超过 50%（Kumar，1999）。

它们作为诱变因子，为生物多样性提供假定来源（Heslop-Harrison，2000）。逆转录因子的主要类别包括长散在重复元件（LINES）、短散在重复元件（SINES）、copia 和 gypsy 类元件及逆转录病毒。Teo 等（2005）利用 IRAP 标记对香蕉品种进行识别和特性描述，并对芭蕉属植物基因组构成进行分类。Ty1-copia 类逆转录转座子广泛用于其他植物的分子标记（Kalender et al.，1999；Pearce et al.，2000；Yu and Wise，2000）。Balint-Kurti 等（2000）认为，Ty3-gypsy 类逆转录转座子引入芭蕉属植物的时间先于 *M. acuminata*、*M. balbisiana* 和 *Musavellutina* 的分化，此外，这类转座子可区别 A 与 B 基因组。Aert 等（2004）从二倍体 *M. acuminata* Pisang Mas 和 *M. acuminata* Calcutta 4 中发现与“monkey”序列高度同源的序列，并指出这 3 个栽培品种的 A 基因组的关系。Nair 等（2005）利用 IRAP 标记来识别印度香蕉基因组的构成，并鉴定出 B 基因组基于 PCR 技术的特异分子标记。这个 B 特异条带在 AA 和 AAA 品种（除了一个 AAA 和 AAB 品种）中均没有出现，但在其他所有 AB、AAB 和 ABB 品种中都存在。ABB 基因组的 4 个品种被鉴定为 AAB，而一个 AAA 品种的带型与 ABB 基因组品种的带型相似。笔者认为之前对这些植物的鉴定有误。

4.4.4 扩增片段长度多态性（AFLP）

AFLP 是一种 DNA 指纹技术，是基于严格条件下 DNA 限制性片段的选择性 PCR 扩增（Vos et al.，1995）。该技术可用于任何来源与复杂度的 DNA。据报道其重复性好，可靠性高（Vos et al.，1995）。AFLP 结合了 RFLP 的可靠性与 PCR 的高效性。AFLP 技术是基于来自基因组总 DNA 消化的限制片段的选择性扩增。鉴于其显性和双等位基因的性质，AFLP 标记已被越来越多地应用到各种植物中，主要原因是其单个测定可以检测大量多态性，有很大的可重复性，并能合理覆盖基因组（Vos et al.，1995；Cervera et al.，1998；Vuylsteke et al.，1999；Shim and Jørgensen，2000）。AFLP 已被用于检测芭蕉属栽培种质和野生祖先的遗传多样性（Wong et al.，2001；Ude et al.，2002a，b，2003；Opara et al.，2010）。不幸的是，由于它们最初必须被视为显性标记，所以这些带型的信息量有限。不过，当 AFLP 用于分析大的群体时，它们在单个群体中可被检测为共显性标记（Masiga and Turner，2004）。基于谱带强度区分个体特性（纯合子或杂合子）的软件已被开发。AFLP 同样也是技术要求严格、费用高昂的测验，其需要大量的 DNA 操作和复杂的可视化程序。此外，它们还需要相对大量的适度高质量的 DNA，因为低质量的 DNA 可能造成不完全消化，进而导致假多态性。利用 AFLP 标记，

Lheureux等（2003）发现芭蕉属杂交种不论是否感染香蕉条纹病，有10个标记会产生共分离。AFLPs和SSRs现今被用于鉴定香蕉和大蕉的果实单性结实、矮小症和顶端优势。微卫星标记和AFLP分析似乎是最适合芭蕉属植物标记辅助育种的技术（Crouch et al.，1999；Hautea et al.，2004）。

现已发展出一项基于AFLP的技术，用来测量CCGG位点的胞嘧啶甲基化（Reyna-Lopez et al.，1997；Fraga and Esteller，2002）。DNA甲基化通过影响基因转录，在调节植物发育中起重要作用，并且参与许多特定生物过程，如基因沉默或可移动元件控制（Finnegan et al.，1996）。这些基于甲基化的分子标记与其他标记在遗传学、转录特性和进化/突变过程上都不同（Xiong et al.，1999；Cervera et al.，2002）。这些标记可以定位不同基因组的区域，帮助了解表观遗传的过程，而这一过程可能涉及大蕉地方品种的表型多样性。然而，由于大蕉的主要表型与基因型的性状之间不存在强关联（Crouch et al.，2000），并且大蕉有甲基化多样性，因此利用以大蕉作为亲本的杂交种来预测性状的问题仍未得到解决。即使在甲基化-表型相关性的假说中，甲基化的进化、传递和控制过程仍不明确，不同于已被揭示的随机突变过程（Cervera et al.，2002）。AFLP谱带也可被测序，利用琼脂糖凝胶电泳转化为共显性序列特征性扩增区域（SCAR）标记以进行PCR分析（Bradeen and Simon，1998；Xu et al.，2001）。这些标记也可用于筛选合并的BAC文库。AFLP技术已被证明是构建芭蕉属植物连锁图谱的重要工具，也被用于构建MM53（自交）和AFCAM（AFCAM指源自Calcutta 4与Madang杂交种的一个F_2群体）的作图群体。

4.4.5　多样性微阵列技术（DArT）

DArT是基于DNA杂交的基因分型技术，它可以低成本地构建作物的全基因组图谱而无需预知序列信息。DArT技术能降低代表性样本的复杂度（比如合并地表现芭蕉属植物多样性的DNA），遵循的原则是基因组的“代表”，包含两类片段：恒定片段，存在于给定的栽培品种或种质个体的DNA样本的“代表”中；可变（多态性）片段，又称为分子标记，只存在于某些而非全部“代表”中（www.diversityarray.com）。DArT标记为双等位基因，可能是显性（存在或不存在）或共显性（2剂量对1剂量或不存在）。这种方法可以利用数百个多态性标记，快速分离种群，进行数量性状位点（QTL）分析。测序后的DArT标记也可锚入BAC序列，不但速度快，而且费用低，因而可推动香蕉基因组优质物理图谱的绘制，从而在测序项目上迈开关键一步。在世代挑战计划（Generation Challenge Program）框架下，利用芭蕉属种质的宽阵列（“宏基因组”）已发展出约1 500个DArT标记（Kilian，2007），目前这些标记正被法国国际农业中心（CIRAD）用于绘制芭蕉属植物框架图，而最近其中 380 个标记，被 CIRAD 用来绘制 BORLI 图（Hippolyte et al.，2008，2010）。DArT 技术的分群分析应用已在大麦属阵列上得以演示（Wenzl et al.，2007）。该技术的不足在于其依赖于阵列、微阵列打印机和扫描仪的可用性，还需要计算机设备来分析、存储和管理产生的数据。然而，由于标记的序列已预先准备好，因此，一旦测序，就可以使用标准的电泳设备将其开发来进行 PCR 分析。

4.4.6 单核苷酸多态性（SNPs）

一个生物体 DNA 发生 SNPs 的平均时间超过 1%（www.biotech.iastate.edu）。大部分 SNPs 出现在目标基因区域外，因为生物体中有 3%～5%的 DNA 编码蛋白质（Sherry et al.，2001）。目标基因中发现的 SNPs 引起了研究人员的特别兴趣，因为这些 SNPs 直接关系到研究人员期望的性状。因为近些年的技术革新，SNPs 在遗传性状选择和分析上发挥着越来越重要的作用。SNPs 现已被收录至有用标记名单，特别是当它们出现在编码区而含有信息时（所以称为表达 SNPs 或 eSNPs，或属于 cDNA 一部分的 SNPs 的 cSNPs），因为只有那时，它们才被认为对蛋白质的功能产生影响，从而影响生物体的表型（Brookes，1999）。迄今为止，尚未有人对香蕉基因组的 SNPs 进行检测。

4.4.7 EcoTILLING

Till 等（2010）利用 EcoTILLING 技术来发现芭蕉属植物基因组的多态性。他们这样解释 EcoTILLING，它是一项用于发现 SNPs 与小插入/缺失（插入缺失）和描述它们特征的高通量手段（Till et al.，2010）。该技术是由酶错配酶切和荧光检测方法演变而来，是一种最初用于定向诱导基因组局部突变（TILLING）的反向遗传学方法（Colbert et al.，2001；Comai et al.，2004）。该技术最先用于拟南芥生态型的描述（因此被命名为 EcoTILLING）。这项技术准确度高、成本低、通量高，用于发现和评估核苷酸多样性，已用于多种生物体中。对于 EcoTILLING，700～1 600 bp 的基因靶区域需要使用带有荧光标记的基因特异性引物进行扩增，然后用错配酶进行切割。PCR 扩增后，样本变性和复性，通过多态性扩增子的杂交而产生杂合双链。然后利用含有单链特异性核酸酶 CELI 的芹菜汁粗提取物切割其他双链核酸分子的错配区。通过变性聚丙烯酰胺凝胶电泳（PAGE）对切割产物进行分离，并通过荧光检测进行观察（Till et al.，2006）。

变性 PAGE 产生的碱基对的分离将以单倍型分组的共同带型为基础进行种质分组（Comai et al.，2004）。序列验证仅可在一个或少数样本中实现，为全组提供碱基多态性数据，并且测序方法成本低，信息加载量小。或者仅需利用带型来评估基因特异型种质之间的遗传多样性与相似性。单独筛选样品时，EcoTILLING 技术能整理出涵盖样本中杂合核苷酸多样性的目录。在进行揭示纯合子基因多态性的筛选之前，可将参考 DNA 加入每一样本。该检测的高灵敏度能测定多个样本，因而能发现罕见的多态性。

4.5　分子标记系统的效率

在关于芭蕉属植物分子育种的研究中，Crouch 等（1999）指出，来自不同类型标记的遗传相似性估算值间相关性较小。主要原因在于数据不一致，这是由于不同的分子技术存在差异，只选择性地筛选基因组互补的区域而没有筛选重叠的区域。因此，建议将不同分子技术的遗传估计值结合起来，这样能对芭蕉属植物的遗传关系有更清楚的了解，在种质分析中形成高度精确的遗传相似性估计值（Crouch et al.，1999；Wong et al.，2001）。因此，同时使用不同的标记系统能提供更多的信息，以助于更好地了解芭蕉属的种群结构。因而，在遗传多样性的研究中，为了得出高度精确的遗传相似性估计值，有必要采用一系列的标记系统。

各分子标记的相对优缺点总结于表 4－1。

表 4－1　分子标记、多态性水平和影响它们使用的因素

（来自 Pillay 和 Tripathi，2007）

标记	DNA 质量要求	多态性水平	位点特异性	重复性	劳动强度	操作成本	开发成本	自动化顺应性
同工酶	—	低	有	高	低	低	低	无
RFLP	高	中	有	高	高	高	中高	无
小卫星	高	高	无/有	高	高	高	中高	无
RAPD	低	中	无	低	低	低	低	有
SSR	低	高	有	高	低	中低	高	有
SNPs	低	低	有	中	中低	中低	低	无
AFLP	中	中	无	高	中	中	中	有

4.6　结论

DNA 新技术，特别是用于人类基因组研究和一些用于植物研究的新技术在不断发展（Pillay et al.，2011）。基于 SNPs 或小规模插入缺失标记的高通量技术因其高丰度、高多态性、易测量和具有其他技术不具备的揭示隐性多态性的能力，成为传统标记技术（RFLP、RAPD或AFLP技术）的良好替代选择（Dillon et al.，2007）。SNPs还能快捷、清楚地识别等位基因或单倍型。

用于多倍体作物研究的良好标记系统应对剂量敏感，能用多种单倍型区分出杂合基因型（de Koeyer et al.，2010）。高分辨率DNA熔解（HRM）分析已显现出优于其他基因分型方法的几个优点（Montgomery et al.，2007；Reed et al.，2007；Erali et al.，2008）。这些优点包括分析时间短，不需要PCR后的样本处理或分离（de Koeyer et al.，2010）。De Koeyer等（2010）讨论了3种使用HRM技术进行基因分型和（或）变体筛选的方法。HRM技术已用于多种作物，在品种鉴定上潜力巨大，特别是在多倍体、作图、多态性发现、绘制候选基因图、与QTL结合或鉴定包含重要性状的基因组区域的关联研究。这些技术尚未应用于芭蕉属植物研究，但科学家们热切期待。

缩略语

6-PGDH：6-磷酸葡萄糖酸脱氢酶
ADH：醇脱氢酶
AFLP：扩增片段长度多态性
BAC：细菌人工染色体
BBrMV：香蕉苞片花叶病毒
CIRAD：法国国际农业中心
EST：酯酶
GDH：半乳糖脱氢酶
GOT：谷氨酸草酰乙酸转氨酶
GUDH：谷氨酸脱氢酶
LINES：长散在重复元件
MAS：标记辅助选择
MDH：苹果酸脱氢酶
ME：苹果酸酶
PAGE：聚丙烯酰胺凝胶电泳
PCR：聚合酶链式反应
PGI：磷酸葡萄糖异构酶
PGM：磷酸葡萄糖变位酶
PRX：过氧化物酶
QTL：数量性状位点
RAPD：随机扩增多态性DNA

RFLP：限制性片段长度多态性
SCAR：序列特征性扩增重复
SDH：山梨醇脱氢酶
SINES：短散在重复元件
SKDH：莽草酸脱氢酶
SSR：简单重复序列
SUDH：琥珀酸脱氢酶
TILLING：定向诱导基因组局部突变
TO：四唑氧化酶
TPI：磷酸丙糖异构酶
VNTR：数目可变串联重复序列

参考文献

Aert R，Volckaert G，Sagi L，2004. Gene content and density in banana（*Musa acuminata*）as revealed by genomic sequencing of BAC clones [J]. Theor Appl Genet，109：129 - 139.

Bairu M W，Fennell C W，Van Staden J，2006. The effect of plant growth regulators on somaclonal variation in Cavendish banana（*Musa* AAA cv. 'Zelig'）[J]. Sci Hort，108：347 - 351.

Bairu M W，Stirk W A，Dolezal K，et al，2008. The role of topolins in micropropagation and somaclonal variation of banana cultivars 'Williams' and Grand Naine（*Musa* spp. AAA）[J]. Plant Cell Tiss Org Cult，95：373 - 379.

Balint-Kurti P J，Clendennen S K，Dolezelova M，et al，2000. Identifi cation and chromosomal localization of the monkey retrotransposon in *Musa* sp. [J]. Mol Gen Genet，263：908 - 915.

Barone A，2004. Molecular marker-assisted selection for potato breeding [J]. Amer J Potato Res，81：111 - 117.

Bennici A，Anzidei M，Vendramin G G，2004. Genetic stability and uniformity of *Foeniculumvulgare* Mill. regenerated plants through organogenesis and somatic embryogenesis [J]. Plant Sci，166：221 - 227.

Bhat K V，Jarret R L，1995. Random amplified polymorphic DNA and genetic diversity in Indian *Musa* germplasm [J]. Genet Resour Crop Evol，42：107 - 118.

Bhat K V，Bhat S R，Chandel K P S，1992a. Survey of isozyme polymorphism for clonal identification in *Musa*. Ⅱ. Peroxidase，superoxide dismutase，shikimate dehydrogenase and malate dehydrogenase [J]. J Hort Sci，67：737 - 743.

Bhat K V，Bhat S R，Chandel K P S，1992b. Survey of isozyme polymorphism for clonal identification in *Musa*. Ⅰ. Esterase，acid phosphatase and catalase [J]. J Hort Sci，61：507 - 507.

Bonner J W，Warner R M，Brewbaker J L，1974. A chemosystematic study of *Musa* cultivars [J]. HortScience, 9: 325 - 328.

Bradeen J M, Simon P W, 1998. Conversion of an AFLP fragment linked to the carrot Y 2 locusto a simple, codominant, PCR-based marker form [J]. Theor Appl Genet, 97: 960 - 967.

Brettell R I S, Dennis E S, Scowcroft W R, et al, 1986. Molecular analysis of a somaclonal-mutant of maize alcohol dehydrogenase [J]. Mol Gen Genet, 202: 235 - 239.

Brookes A J, 1999. The essence of SNPs [J]. Gene, 234: 177 - 186.

Buard J, Vergnaud G, 1994. Complex recombination events at the hypermutable minisatellite CEB1 (D2S90) [J]. The EMBO Journal, 13: 3203 - 3210.

Buhariwalla H K, Jarret R L, Jayashree B, et al, 2005. Isolation and characterization of microsatellite markers from *Musa balbisiana* [J]. Mol Ecol Notes, 5: 327 - 330.

Campbell D, Duchesne P, Bernatchez L, 2003. AFLP utility for population assignment studies: analytical investigation and empirical comparison with microsatellites [J]. Mol Ecol, 12: 1979 - 1991.

Carreel F, Gonzalez de Leon D, Lagoda P, et al, 2002. Ascertaining maternal and paternal lineage within *Musa* by chloroplast andmitochondrial DNA RFLP analyses [J]. Genome, 45: 679 - 692.

Carvalho L C, Goulao L, Oliveira C, et al, 2004. RAPD assessment for identifi cation of clonal identity and genetic stability of *in vitro* propagated chestnut hybrids [J]. Plant Cell, Tiss Org Cult, 77: 23 - 27.

Cervera M T, Cabezas J A, Sancha J C, et al, 1998. Applicationof AFLPs to the characterization of grapevine *Vitis vinifera* L. genetic resources: a casestudy with accessions from Rioja (Spain) [J]. Theor Appl Genet, 97: 51 - 59.

Cervera M T, Ruiz-Garcia L, Martanez-Zapater J M, 2002. Analysis of DNA methylation in Arabidopsis thaliana based on methylation-sensitive AFLP markers [J]. Mol Genet Genom, 268: 543 - 552.

Cheung F, Town C D, 2007. A BAC end view of the *Musa acuminata* genome [J]. BMC Plant Biol, 7: 29 - 29.

Colbert T, Till B J, Tompa R, et al, 2001. High-throughput screening for induced point mutations [J]. Plant Physiol, 126: 480 - 484.

Comai L, Young K, Till B J, et al, 2004. Efficient discovery of DNA polymorphisms in natural populations by EcoTILLING [J]. Plant J, 37: 778 - 786.

Creste S, Neto T A, Silva S O, et al, 2003. Genetic characterization of banana cultivars (*Musa* spp.) from Brazil using microsatellite markers [J]. Euphytica, 132: 259 - 268

Creste C, Neto T A, Sebastino O S, et al, 2003. Genetic characterization of banana cultivars (*Musa* spp.) from Brazil using microsatellite markers [J]. Euphytica, 132: 259 - 268.

Crouch H K, Crouch J H, Jarret R L, et al, 1998. Segregation at microsatellite loci in haploid and diploid gametes of *Musa* [J]. Crop Science, 38: 211 - 217.

Crouch J H, Crouch H K, Tenkouano A, et al, 1999a. VNTR-based analysis of $2x$ and $4x$

full-sib *Musa* hybrids [J]. Elect J Biotechnol, 2: 99 - 108.

Crouch J H, Crouch H K, Constand H, et al, 1999b. Comparison of PCR-based molecular marker analysis of *Musa* breeding populations [J]. Mol Breed, 5: 233 - 244.

Crouch J H, Ortiz R, Crouch H K, 2000. Utilization of molecular genetic techniques in supportof plantain and banana improvement [J]. Acta Hort, 540: 185 - 191.

Damasco O P, Graham G C, Henry R J, et al, 1996. Random amplifi ed polymorphic DNA (RAPD) detection of dwarf off-types in micropropagated Cavendish (*Musa* spp. AAA) bananas [J]. Plant Cell Reports, 16: 118 - 123.

Das B K, Jena R C, Samal K C, 2009. Optimization of DNA isolation and PCR protocol for RAPD analysis of banana/plantain (*Musa* spp.) [J]. Int J Agric Sci, 1: 21 - 25.

De Koeyer D, Douglass K, Murphy A, et al, 2010. Application of high-resolution DNA melting for genotyping and variant scanning of diploid and autotetraploid potato [J]. Mol Breed, 25: 67 - 90.

Dhanya M K, Rajagopalan B, Umamaheswaran K, et al, 2006. Isozyme variation in banana (*Musa* sp.) in response to bract mosaic virus infection [J]. Indian J Crop Sci, 1: 140 - 141.

Dillon S L, Shapter F M, Henry R J, et al, 2007. Domestication to crop improvement: genetic resources for *Sorghum* and *Saccharum* (Andropogoneae) [J]. Ann Bot, 100: 975 - 989.

Drew R A, 1997. The Application of biotechnology to the conservation and improvement of tropical and subtropical fruit species [C]. Food and Agriculture Organization of the United Nations. Rome, Italy.

Erali M, Voelkerding K V, Wittwer C T, 2008. High resolution melting applications for clinical laboratory medicine [J]. Exp Mol Pathol, 85: 50 - 58.

Espino R R C, Pimentel R B, 1990. Electrophoretic analysis of selected isozymes in BB cultivars of Philippine bananas [C]//Jarret R L. Identification of genetic diversity in the genus *Musa* INIBAP, Los Banos, the Philippines.

Faure S, Noyer J L, Horry J P, et al, 1993. A molecular marker-based linkage map of diploid bananas (*Musa acuminata*) [J]. Theor Appl Genet, 87: 517 - 526.

Finalet J, Peteira B, Leon O, et al, 2000. Application of RAPD markers to the differentiation of *Musa* spp. mutants [J]. Revista de Proteccion Vegetal, 15: 60 - 63.

Finnegan E J, Peacock W J, Dennis E S, 1996. Reduced DNA methylation in *Arabidopsis thaliana* results in abnormal plant development [J]. Proc Natl Acad Sci USA, 93: 8449 - 8554.

Fraga M F, Esteller M, 2002. DNA methylation: a profile of methods and applications [J]. BioTechniques, 33: 632 - 634.

Gawel N J, Jarret R L, 1991a. Cytoplasmic genetic diversity in bananas and plantains [J]. Euphytica, 52: 19 - 23.

Gawel N J, Jarret R L, 1991b. Chloroplast DNA restriction fragment length polymorphisms (RFLPs) in *Musa* species [J]. Theor Appl Genet, 81: 783 - 786.

Gawel N J, Jarret R L, Whittemore A P, 1992. Restriction fragment length polymorphism

(RFLP) -based phylogenetic analysis of *Musa* [J]. Theor Appl Genet, 84: 286 - 290.

Grapin A, Noyer J L, Carreel F, et al, 1998. Diploid *Musa* acuminata genetic diversity assayed with sequence-tagged microsatellite sites [J]. Electrophoresis, 19: 1374 - 1380.

Gupta M, Chyi Y S, Romero-Severson J, et al, 1994. Amplification of DNA markers from evolutionarily diverse genomes using single primers of simple-sequence repeats [J]. Theor Appl Genet, 89: 998 - 1006.

Hautea D M, Molina G C, Balatero C H, et al, 2004. Analysis of induced mutants of Philippine with molecular markers [M]//Jain S, Swennen R. Banana improvement, cellular, molecular biology and induced mutations. Dordrecht, The Netherlands: Kluwer Academic Publishers: 45 - 57.

Heath D D, Lwama G K, Devlin R H, 1993. PCR primed with VNTR core sequences yields species specific patterns and hypervariable probes [J]. Nucl Acid Res, 21: 5782 - 5785.

Heslop-Harrison J S, 2000. Comparative genome organization in plants: from sequence and markers to chromatin and chromosomes [J]. Plant Cell, 12: 617 - 636.

Hippolyte I, Seguin M, Bakry F, et al, 2008. *Musa* genetic mapping [C]. XVI International plant and animal genome conference (PAG), San Diego, CA, USA.

Hippolyte I, Bakry F, Seguin M, et al, 2010. A saturated SSR/DArT linkage map of *Musa acuminata* addressing genome rearrangements among bananas [J]. BMC Plant Biol, 10: 65 - 65.

Howell E C, Newbury H J, Swennen R L, et al, 1994. The use of RAPD for identifying and classifying *Musa* germplasm [J]. Genome, 37: 328 - 332.

Imelda M, Estiati A, Hartati N S, 2001. Induction of mutation through gamma radiation in three cultivars of banana [J]. Ann Bogorien, 7: 75 - 82.

Jain P K, Saini M K, Pathak H, et al, 2007. Analysis of genetic variation in different banana (*Musa* species) variety using random amplified polymorphic DNAs (RAPDs) [J]. Afri J Biotechnol, 6: 1987 - 1989.

Jarret R L, Litz R E, 1986a. Enzyme polymorphism in *Musa acuminata* Colla [J]. J Hered, 77: 183 - 188.

Jarret R L, Litz R E, 1986b. Isozymes as genetic markers in bananas and plantains [J]. Euphytica, 35: 539 - 549.

Jarret R, Gawel N, Whittemore A, et al, 1992. RFLP-based phylogeny of *Musa* species in Papua New Guinea [J]. Theor Appl Genet, 84: 579 - 584.

Jarret R L, Bhat K V, Cregan P B, et al, 1994. Isolation of microsatellite DNA markers in *Musa* [J]. Info Musa, 3: 3 - 4.

Kaemmer D, Fischer D, Jarret R L, et al, 1997. Molecular breeding in the genus *Musa*: a strong case for STMS marker technology [J]. Euphytica, 96: 49 - 63.

Kalendar R, Grob T, Regina M, et al, 1999. IRAP and REMAP: two new retrotransposon-based DNA fingerprinting techniques [J]. Theor Appl Genet, 98: 704 - 711.

Khatri A, Dahot M U, Khan I, et al, 2009. Use of RAPD for the assessment of genetic diversity among exotic and commercial banana clones [J]. Pak J Bot, 41: 2995 - 2999.

Kilian A, 2007. Towards effective deployment of diversity arrays technology (DArT) in banana genomics and sequencing [C]. Plant and Animal Genomes XV Conference, January 13 - 17, San Diego, USA.

Kumar L S, 1999. DNA markers in plant improvement: an overview [J]. Biotechnol Adv, 17: 143 - 182.

Lakshmanan V, Venkataramareddy S R, Neelwarne B, 2007. Molecular analysis of genetic stability in long-term micropropagated shoots of banana using RAPD and ISSR markers [J]. Elec J Biotechnol, 10: 106 - 113.

Lefebvre V, Palloix A, Caranta C, et al, 1995. Construction of an intraspecific integrated linkage map of pepper using molecular markers and doubled-haploid progenies [J]. Genome, 38: 112 - 121.

Lheureux F, Carreel F, Jenny C, et al, 2003. Identification of genetic markers linked to banana streak disease expression in inter-specific *Musa* hybrids [J]. Theor Appl Genet, 106: 594 - 598.

Loh J P, Kiew R, Set O, et al, 2000. Amplifi ed fragment length polymorphism fingerprinting of 16 banana cultivars (*Musa* cvs.) [J]. Mol Phylogenet Evol, 17: 360 - 366.

Martins M, Sarmento D, Oliveira M M, 2004. Genetic stability of micropropagated almond plantlets, as assessed by RAPD and ISSR markers [J]. Plant Cell Rep, 23: 492 - 496.

Masiga D K, Turner C M R, 2004. Amplified (restriction) fragment length polymorphism (AFLP) analysis [M]. Melville S E. Methods in Molecular Biology. USA: Humana Press: 173 - 186.

Megia R, Hadisunars, Sulistyaningsih Y C, et al, 2001. Isozyme polymorphisms for cultivar identification in Indonesian banana [J]. Hayati, 8: 81 - 85.

Miller R N G, Passos M A N, Menezes N N P, et al, 2010. Characterization of novel microsatellite markers in *Musa acuminata* subsp. *burmannicoides*, var. Calcutta 4 [J]. BMC Res. Notes, 3: 149.

Miri S M, Mousavi A, Naghavi M R, et al, 2009. Analysis of induced mutants of salinity resistant banana (*Musa acuminata* cv. Dwarf Cavendish) using morphological and molecular markers [J]. Iran J Biotechnol, 7: 86 - 92.

Montgomery J, Wittwer C T, Palais R, et al, 2007. Simultaneous mutation scanning and by high-resolution DNA meltinganalysis [J]. Nature Protocols, 2: 59 - 66.

Muhammad Y, James A C, Rivera-Madrid R, et al, 2010. *Musa* genetic diversity revealed by SRAP and AFLP [J]. Mol Biotechnol. (August) . doi: 10. 1007/s12033-010-9328-8.

Nair A S, Teo C H, Schwarzacher T, et al, 2005. Genome classifi cation of banana cultivars from South India using IRAP markers [J]. Euphytica, 144: 285 - 290.

Nsabimana A, Van Staden J, 2007. Assessment of genetic diversity of highland bananas from the National Banana Germplasm Collection at Rubona, Rwanda using RAPD markers [J]. Sci Hort, 113: 293-299.

Nwakanma D C, Pillay M, Okoli B E, et al, 2003. PCR-RFLP of the ribosomal DNA internal transcribed spacers (ITS) provides: markers for the A and B genomes in *Musa* L [J]. Theor Appl Genet, 108: 154-159.

Opara U L, Jacobson D, Al-Saady N A, 2010. Analysis of genetic diversity in banana cultivars (*Musa* cvs.) from the South of Oman using AFLP markers and classification by phylogenetic, hierarchical clustering and principal component analyses [J]. J Zhejiang Univ Sci, 11: 332-341.

Orita M, Suzuki Y, Sekiya T, et al, 1989. Rapid and sensitive detection of point mutations and DNA polymorphisms using the polymerase chain reaction [J]. Genomics, 5: 874-879.

Oselebe H O, Tenkouano A, Pillay M, et al, 2006. Ploidy and genome segregation in *Musa* breeding populations assessed by flow cytometry and randomly amplified polymorphic DNA markers [J]. HortScience, 131: 780-786.

Paterson A H, Damon S, Hewitt J D, et al, 1991. Mendelian factors underlying quantitative traits in tomato: comparison across species, generations, and environments [J]. Genetics, 127: 181-197.

Pearce S R, Knox M, Ellis T N M, et al, 2000. Pea *Tyl-copia* group retrotransposon: transitional activity and use as makers to study genetic diversity in *Pisum* [J]. Mol Genet-Genomics, 263: 898-907.

Pillay M, Tripathi L, 2007. Banana breeding [M]//Kang M S, Priyadarshan P M. Breeding major food staples. Boston, USA, Asia: Blackwell Science: 393-428.

Pillay M, Nwakanma D C, Tenkouano A, 2000. Identifi cation of RAPD markers linked to A and B genome sequences in *Musa* L [J]. Genome, 43: 763-767.

Pillay M, Ogundiwin E, Nwakanma D C, et al, 2001. Analysis of genetic diversity and relationships in East African banana germplasm [J]. Theor Appl Genet, 102: 965-970.

Pillay M, Ogundiwin E, Tenkouano A, et al, 2006. Ploidy and genome composition of *Musa* germplasm at the International Institute of Tropical Agriculture (IITA) [J]. Afr J Biotechnol, 5: 1224-1232.

Pillay M, Tenkouano A, Ortiz R, 2011. Molecular breeding of other vegetatively propagated crops: lessons for banana [M]//Pillay M, Tenkouano A. Banana breeding: progress and challenges. Boca Raton, FL, USA: CRC Press: 321-350.

Ramage C M, Borda A M, Hamill S D, et al, 2004. A simplified PCR test for early detection of dwarf off-types in micropropagated Cavendish bananas (*Musa* spp. AAA) [J]. Sci Hort, 103: 145-151.

Ray T, Dutta I, Saha P, et al, 2006. Genetic stability of three economically important micropropagated banana (*Musa* spp.) cultivars of lower Indo-Gangetic plains, as assessed by RAPD and ISSR markers [J]. Plant Cell Tiss Org Cult, 85: 11-21.

Reed G H, Kent J O, Wittwer C T, 2007. High-resolution DNA melting analysis for simple and efficient molecular diagnostics [J]. Pharmacogenomics, 8: 597 - 608.

Reyna-Lopez G E, Simpson J, Ruiz-Herrera J, 1997. Differences in DNA methylation patterns are detectable during the dimorphic transition of fungi by amplifi cation of restriction polymorphisms [J]. Mol Gen Genet, 253: 703 - 710.

Richardson T, Cato S, Ramser J, et al, 1995. Hybridization of microsatellites to RAPD: a new source of polymorphic markers [J]. Nucl Acid Res, 23: 3798 - 3799.

Rivera F N, 1983. Protein and isoenzyme banding patterns among Philippine cooking bananas and their wild parents (*Musa* species) [J]. Paradisiaca, 6: 7 - 12.

Rogstad S H, 1993. Surveying plant genomes for variable number of tandem repeat loci [J]. Meth Enzymol, 224: 278 - 294.

Sambrook J, Fritsch E F, Maniatis T, 1989. Molecular cloning—a laboratory course manual [M]. Cold Spring Harbor, New York. USA: Cold Spring Harbor Laboratory Press.

Sherry S T, Ward M H, Kholodov M, et al, 2001. dbSNP: the NCBI database of genetic variation [J]. Nucl Acid Res, 29: 308 - 311.

Shim S I, Jorgensen R B, 2000. Genetic structure in cultivated and wild carrots (*Daucus carota* L.) revealed by AFLP analysis [J]. Theor Appl Genet, 101: 227 - 233.

Simmonds N W, 1966. Bananas [M]. 2nd ed. London, UK: Longmans.

Sitthipron S, Anutalabhochai S, Dumampai N, et al, 2005. Investigation of genetic relationships and hybrid detection in Longan by high annealing temperature RAPD [J]. Acta Hort, 665: 161 - 169.

Staub J E, Kuhns L J, May B, et al, 1982. Stability of potato tuber isozymes under different storage regimes [J]. J Amer Society Hort Sci, 107: 405 - 408.

Staub J E, Sequen F C, Gupta M, 1996. Genetic markers, map construction, and their application in plant breeding [J]. HortScience, 31: 742 - 762.

Tansley S D, Orton T J, 1983. Isozymes in plant genetics and breeding, part A and B [M]. Amsterdam, The Netherlands: Elsevier.

Teo C H, Tan S H, Ho C L, et al, 2005. Genome constitution and classifi cation using retrotransposon-based markers in the orphan crop banana [J]. J Plant Biol, 48: 96 - 105.

Till B J, Zerr T, Comai L, et al, 2006. A protocol for TILLING and EcoTILLING in plants andanimals [J]. Nat Protocols, 1: 2465 - 2477.

Till B J, Jankowicz-Cieslak J, Sagi L, et al, 2010. Discovery of nucleotide polymorphisms in the *Musa* gene pool by EcoTILLING. Theor Appl Genet, 121: 1381 - 1389.

Toruan-Mathius N, Haris N, 1999. Induction of genetic variation of banana cv. Nangka by gamma Co-60 irradiation and fusaric acids [J]. Menara Perkebunan, 67: 13 - 22.

Tourmente S, Deragon J M, Lafl euriel J, et al, 1994. Characterization of minisatellites in *Arabidopsis thaliana* with sequence similarity to the human minisatellite core sequence [J]. Nucl Acid Res, 22: 3317 - 3321.

Ude G, Pillay M, Nwakanma D, et al, 2002a. Analysis of genetic diversity and sectional relationships in *Musa* using AFLP markers [J]. Theor Appl Genet, 104: 1239 - 1245.

Ude G, Pillay M, Nwakanma D, et al, 2002b. Genetic Diversity in *Musa acuminate* Colla and *Musa balbisiana* Colla and some of their natural hybrids using AFLP markers [J]. Theor Appl Genet, 104: 1246 - 1252.

Ude G, Pillay M, Ogundiwin E, et al, 2003. Genetic diversity in an African plantain core collection using AFLP and RAPD markers [J]. Theor Appl Genet, 107: 248 - 255.

Uma S, Siva S A, Saraswathi M S, et al, 2006. Variation and intraspecific relationships in Indian wild *Musa balbisiana* (BB) population as evidenced by random amplified polymorphic DNA [J]. Genet Resour Crop Evol, 53: 349 - 355.

Vos P, Hogers R, Bleeker M, et al, 1995. AFLP: a new technique for DNA fingerprinting [J]. Nucl AcidRes, 23: 4407 - 4414.

Vuylsteke M, Mank R, Antonise R, et al, 1999. Two high-density AFLP linkage maps of *Zea mays* L. : analysis of distribution of AFLP markers [J]. Theor Appl Genet, 99: 921 - 935.

Wang J Y, Zheng L S, Huang B Z, et al, 2009. Development, characterization, and variability analysis of microsatellites from a commercial cultivar of *Musa acuminata* [J]. Genet Resour Crop Evol, 57: 553 - 563.

Weising K, Weigand F, Driesel A J, et al, 1989. Polymorphic simple GATA/GACA repeats in plant genomes [J]. Nucl Acid Res, 17: 10128.

Weising K, Atkinson R G, Gardner R C, 1995. Genomic fingerprinting by microsatellite-primed PCR: a critical evaluation [J]. Genome Res, 4: 249 - 255.

Wenzl P, Raman H, Wang J, et al, 2007. A DArT platform for quantitative bulked segregant analysis [J]. BMC Genomics, 8: 196 - 196.

Wong C, Kiew R, Loh J P, Gan et al, 2001. Genetic diversity of the wild banana *Musa acuminata* Colla in Malaysia as evidenced by AFLP [J]. Ann Bot, 88: 1017 - 1025.

Xiong L Z, Xu C G, Saghai Maroof M A, et al, 1999. Patterns of cytosine methylation in an elite rice hybrid and its parental lines, detected by a methylation-sensitive amplification-polymorphism technique [J]. Mol Gen Genet, 261: 439 - 446.

Xu M, Huaracha E, Korban S S, 2001. Development of sequence-characterized amplified regions (SCARs) from amplified fragment length polymorphism (AFLP) markers tightly linkedto the *Vf* gene in apple [J]. Genome, 44: 63 - 70.

Youssef M, James A C, Rivera-Madrid R, et al, 2010. *Musa* genetic diversity revealed by SRAP and AFLP [J]. Mol Biotechnol, 47: 189 - 199.

Yu G X, Wise R P, 2000. An anchored AFLP-and retrotransposon-based map of diploid *Avena* [J]. Genome, 43: 736 - 749.

Zhou Z, Bebeli P J, Somers D J, et al, 1997. Direct amplification of minisatellite-region DNA with VNTR core sequences in the genus *Oryza* [J]. Theor Appl Genet, 95: 942 - 949.

第 5 章　分子连锁图谱：策略、资源和成果

Cory Johnson[①,a]　**Christopher Cullis**[①,b,*]

摘　要：遗传图谱最早源自孟德尔的发现，即表型性状遗传可被预测。在之后的 70 年，实现了利用表型性状作图。一对父母的多个性状的组合往往是不可能的，因为一个性状会屏蔽或干扰其他性状的表型评分。使用基于 DNA 的标记能消除这些限制，因为只要是个体的表型，大多数变异体都沉默，所以单交中可检测的标记数不受限制。DNA 标记对绘制遗传图谱整个过程产生了革命性的影响。本章重点介绍分子连锁图谱的策略、资源和成果。

关键词：标记，连锁图谱，作图

5.1　引言

遗传图谱最早源自孟德尔的发现，即表型性状遗传可被预测。孟德尔通过研究豌豆原始性状独立分离而发现的遗传规律，被后人大量沿用。第一张实际意义上的遗传图谱于 1913 年发表（Sturtevant，1913），该图谱为果蝇 6 个性连锁基因。遗传作图的重要方面是要在概念上明确基因的遗传距离和物理距离不同，并了解干扰的概念。干扰是指重组事件的出现影响附近另一交叉的可能性。在 70 年之后，实现了利用表型性状作图。对植物作图往往涉及颜色变化和容易计分的形态学标记。然而，过去无法对单对亲本中的多个性状进行组合，因为一个性状会屏蔽或干扰其他性状的表型评分。因此，遗传图谱由不同亲本的多种杂交数据汇聚而成。此外，许多性状的替代形式是突变的结果。使用基于 DNA 的标记能消除这些限制，因为只要是个体表型，大多数变异体沉

---

① Department of Biology，Case Western Reserve University，10900 Euclid Avenue，Cleveland，Ohio 44106-7080.

a. e-mail：clj18@case. edu.

b. e-mail：cac5@case. edu.

*　通信作者。

默，所以一次杂交中可检测的标记数不受限制。

自此以后，DNA 标记对绘制遗传图谱的整个过程产生了革命性影响，因为这是历史上首次能在单一分离群体中追踪大量位点。可用的遗传标记包括限制性片段长度多态性（RFLP）(Botstein et al.，1980)、随机扩增多态性 DNA（RAPD）(Williams et al.，1990)、扩增片段长度多态性（AFLP）(Vos et al.，1995)、单核苷酸多态性（SNPs）和简单重复序列（SSRs）或微卫星（Senior and Heun，1993）。任何情况下，确定多态性后，多态性的效用将取决于变异被观察到的频率。因此，如果在某一特定区域有 3 个多态性状态，其中一个出现的概率为 99%，那么它就不是有用的多态性，因为在这一区域两个个体不太可能出现差异。多态性位点的作用主要在于多态性信息含量（PIC），即一个位点上任两个个体相异的概率。例如，如果 PIC 为 0.5，则代表任意两个个体在该位点上相异的概率为 50%。识别多态性仅是构建分子图谱的第一步，同样重要的是识别那些被用于作图的，将在种质中分离的标记（表 5-1）。此外还要确定这一种质不仅包含差异很大的个体，否则图谱在实际应用上，标记的适用性可能受限。

表 5-1　不同作图种群在标记位点上的遗传分离比率

标记	类型	遗传分离比率				
		F_2	RIL	DH	NIL	BC_1
RFLP	共显性	1∶2∶1	1∶1	1∶1	1∶1	1∶1
RAPD	显性	3∶1	1∶1	1∶1	1∶1	1∶0
AFLP	显性	3∶1	1∶1	1∶1	1∶1	1∶0
SSR	共显性	1∶2∶1	1∶1	1∶1	1∶1	1∶1
SNP	共显性	1∶2∶1	1∶1	1∶1	1∶1	1∶1

作图的效力取决于对个体分类的准确性。精确描述分子和物理特性对于正确了解标记和基因之间的关系至关重要。若一个错误的个体进入作图软件将毁掉整个数据，而之后也很难找出错误在哪儿。因此计分和数据录入的准确性是重中之重。

5.2　标记类型的演变：从 RFLPs 到 SNPs

分子工具的发展使得利用这些工具来区分个体成为可能。这些工具包括同工酶的性状描述、区别性次生产物的识别和 DNA 片段的利用。本章重点讨论 DNA 片段的利用。

5.2.1　限制性片段长度多态性

第一代 DNA 标记以限制性片段长度多态性（RFLP）为基础。

顾名思义，该标记通过识别限制性内切酶的变异为基础。如图 5-1 所示，多态性可能由限制性位点上的单碱基变异导致，或由相邻限制性位点之间的插入或缺失导致。传统方法利用单（低）拷贝探针探测杂交来检测此多态性。如果多态性是由限制性位点的丢失所致，那可能出现两种结果。如果探针被限制在基因组 DNA 的一个区域，而该基因组 DNA 又完全在某一限制性片段中，之后会观察到更大的条带（探针 y，图 5-1）。然而，如果探针跨越改变的限制性位点（探针 x，图 5-1），将观察到新的条带，大小是其他两条之和。若是插入或缺失导致的 RFLP，无论使用哪个探针只能识别出一条条带，大小随着个体的不同而不同（假设它们是纯合的）。RFLP 由插入或缺失导致的另一迹象是 RFLP 可被具有多个限制性内切酶的单个探针识别出。

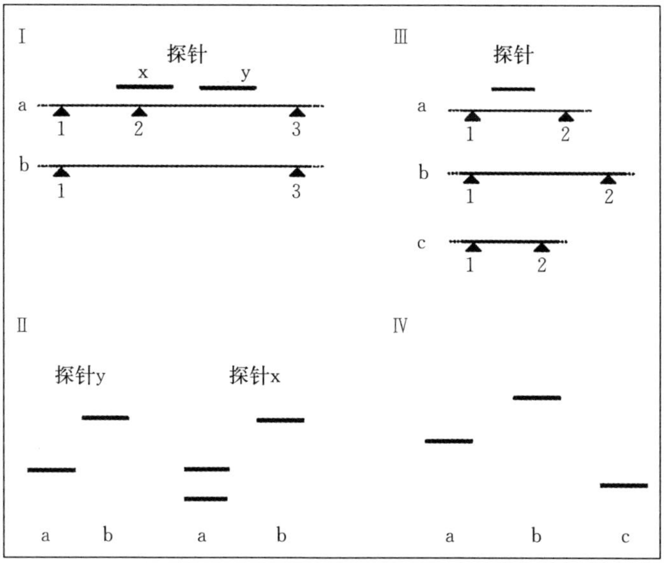

图 5-1　不同探针杂交的结果确定各种 RFLPs 产生的原因

Ⅰ. 限制性位点丢失导致的 RFLP

Ⅱ. 对 x 和 y 探针进行 Southern 印迹法与杂交后得出的带型

Ⅲ. 原等位基因 a 中插入 b 或缺失 c 所致的 RFLPs

Ⅳ. 对这 3 种等位基因的探针进行 Southern 印迹法与杂交后得出的带型

传统 RFLPs 技术利用 Southern 印迹法和杂交鉴定，不仅检测周期长，还需要大量的起始 DNA，因此现已很少用于标准鉴定中。取而代之的是聚合酶链式反应（PCR）扩增技术，如随机扩增多态性 DNA（RAPD）、扩增片段长度多态性（AFLP）和简单重复序列（SSR）标记，它们常常作为 RFLPs 技术的替代选择。

5.2.2 随机扩增多态性 DNA

聚合酶链式反应（PCR）的可用性为之后大量的应用铺平了道路。PCR 的缺点在于反应中所用引物需要序列信息。统计计算一个 10 bp（碱基对）序列的存在是每 10^6 个核苷酸中它应该出现一次。PCR 扩增要求扩增片段的每个末端都有引物。因此单一 10 个碱基引物可能出现在两个位点之间 50～2 000 bp的反方向，其概率是 5×10^{-8}。由于大多数植物的基因组比这种单一十聚体的扩增基因组的 DNA 大，因此只能扩增出少量条带，经证实，这在确定大量多态性上十分成功（Williams et al.，1990）。因此，这种方法可以在其他基因组信息未知的情况下使用，因为有数以百计的随机十聚体引物可用。这一技术也有缺陷。首先，尽管在同一实验室内可获得重复性结果，但有报道指出该技术缺乏不同实验室之间的重复性和耐热酶资源（Jones et al.，1997）。目前已开发出新版本的 *Taq* 酶，但其在低温下也无法运行，所以也无法用于产生 RAPD 条带，因为引物的长度要求退火温度约为 40 ℃。其次，对分子图谱来说 RAPD 条带的分配是基因组特异的，因此对于每一个体而言，由于多态性片段不能跨基因组关联，所以需要绘制出每个独特的 RAPD 图谱。最后，RAPD 是显性标记，意味着不论片段是纯合或杂合都会产生条带，因此要绘出精确的分子图谱需要更大的分离群体。RAPD 的基础可能是引物序列中单一核苷酸的改变（因为 1 个碱基改变相当于引物序列中 10%的错配，而这将导致退火了的引物在稳定性上有 10℃的改变）或引物位点之间的插入或缺失。因此导致了 RFLP 相同的基因组变异也是 RAPD 变化的原因，除非单碱基改变并不需要限制性内切酶识别位点。

5.2.3 扩增片段长度多态性

这些标记基本上是通过 PCR 扩增检测的 RFLPs。PCR 反应所需的已知特定引物可以避免，方法是将已知接头连接到限制性片段的末端。这一方法要求首先用两个不同的限制性内切酶消化基因组 DNA。之后将接头加到限制性片段的末端，然后这些接头将作为引物，用于 PCR 反应。接下来，会观察到大量的扩增片段，在所有可以扩增的、有各种大小可能的限制性片段的背景下，识别多态性片段。利用凝胶或 DNA 自动测序仪将扩增条带的复杂混合物分离。多态性条带可被克隆和测序，产生序列标记位点（STS）标记，该标记使用实际基因组序列进行后续的特征描述。利用对甲基化敏感或不敏感的限制性内切酶同裂酶可研究基因组中高甲基化或低甲基化区域产生的潜在后生效应。与 RAPD 一样，AFLP 也是显性标记。

5.2.4　微卫星与 SSRs

微卫星或 SSRs 是源自短串联重复序列（通常小于 6 bp），如（GA）$_n$、（AAT）$_n$ 和（GT）$_n$ 的遗传标记。微卫星和 SSR 这两个术语一般可替换，尽管微卫星的单元一般比 SSRs 长 2～3 bp。这种序列分布广泛，大多数动植物基因组中都有其分布，它的多态性是由在给定的位点上重复次数的变异性导致的（图 5-2）。如图 5-2 所示，长度多态性由重复次数的变化所产生。图 5-2 所示的 3 类重复：a 为 5 次重复，b 为 2 次重复，c 为 10 次重复。若 SSR 为二核苷酸重复，则 a 和 b 的实际大小差别是 6 bp，a 和 c 的是 10 bp，b 和 c 的是 16 bp。因此，分离扩增产物需要高分辨力。因而早期 ABI 自动测序仪因其在描述 SSR 长度多态性特征上的重要作用，重新焕发生命力。

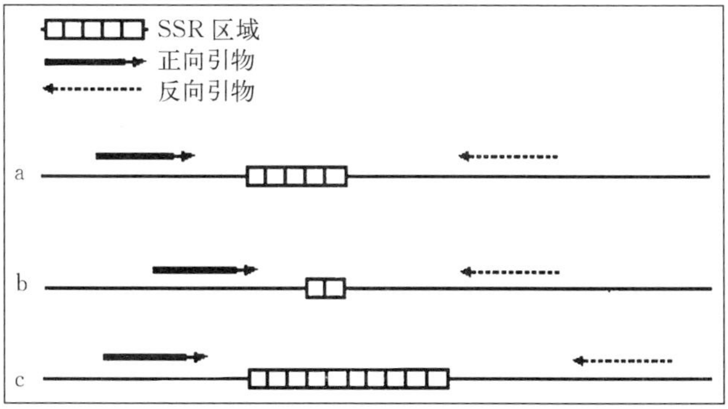

图 5-2　位点上 SSR 重复次数的不同导致的长度多态性

与 RFLPs 一样，针对每一物种的每个 SSR 引物都需分离。现今采用的是第二代测序技术，使得它们从基因组文库或富集基因组文库中的分离更加快速（Panaud et al.，1995；Santana et al.，2009）或通过对 cDNA 序列进行分析而生成。富集技术包括生物素标记 SSR 片段的使用，它被用于将 PCR 扩增的基因组片段的互补序列进行物理删除。该技术早期使用 AFLP 引物来提供基因组的表达。最初，对所有回收的片段都需进行克隆和测序。不过，新一代的 454 测序方法能对整组的片段进行测序和组合。之后使用 SSRIT 程序分析重叠群（http：//www.gramene.org/db/markers/ssrtool）来识别出含 SSRs 的重叠群。在 SSR 区域周围有充足序列的地方，设计引物，并筛选目标种质以发现多态性引物。虽然现今较之从前能更快识别 SSRs 区域，但每一 SSR 都要被筛选以发现多态性的 SSRs，因而也十分耗时。

SSRs的第二个来源是表达序列标签（EST）测序数据。mRNAs常发生三联体重复，因为在蛋白质结构和顺序不变的情况下，一个额外的氨基酸将会插入到该蛋白质中。三联体重复也在mRNA的5′和3′非翻译区存在。识别这些SSRs明显需要大型的EST序列数据库，因此其只局限在有大型序列数据库的物种中。源自EST的SSRs的多态性水平也可能低于基因组富集方法确定的多态性水平。

5.2.5 单核苷酸多态性

单核苷酸多态性（SNP）是指基因组序列中一个单核苷酸（A、T、C或G）在两个个体的DNA样品之间不同，产生了DNA序列变异。例如，SNPs可将DNA序列从AGGATTCA变成AGGATTTA。SNPs既可发生在基因组的编码（基因）区，又可发生在非编码区。如上文详述，SNPs指的是限制性位点的缺失和获得导致的这类RFLPs。许多SNPs对细胞功能没有影响，因为不改变蛋白质结构（事实上，任何发生在氨基酸密码子第三位置的SNP如果不改变所得蛋白质的氨基酸序列，将没有任何效果）。它们的高频率（也许在植物DNA中高达2%～3%）意味着在连锁作图中特别有用（Kristensen et al.，2001；Lai，2001）。然而，SNPs的识别依赖于获得可用的序列信息。而序列信息的获得要么来自不同品种，要么来自远系繁殖的个体，这样杂合性才会存在。信息学工具可用于比较序列并识别变异，但以跟踪文件为形式的原始数据在决定多态性的真实性上至关重要。举例说明，若ESTs构建成unigene集的部分，任何差异可能在共有序列的形成中被消除，并需要检索。由于没有区分真SNP和测序错误的先验方法，因此每一潜在的SNP都须验证。如果每个多态性可独立地遗传，即使频率为1%，这些多态性都会产生极大量的单倍型。然而，观察到的单倍型相对较少，这表明SNP产生的速率可能类似于生成单倍型块的基因组区域中发生重组的速率。因此，SNPs可能最有用的是用来定义单倍型，而不是自己单独的信息，所以SNPs的利用可能涉及使用单倍型进行连锁不平衡研究，而不是作为单个分子标记的特定的SNPs。

5.3 作图群体

这些群体通常来源于人工控制的杂交，杂交的亲本选择、杂交方案和标记系统取决于作图实验的目标。所选择的亲本需要在所用的分子标记上和表型性状上有足够大的差异性。分子图谱的商业价值通常在于标记和表型性状之间的关联，因此在作图群体中应分离尽可能多的表型和数量性状。特别需要注意的一点是当涉及大量杂交种时，染色体重排有可能会导致重组抑制，标记之间的遗传距离和物理距离可能显著不同。

5.3.1 作图种群类型

5.3.1.1 F_2 种群

这个群体是通过两个亲本杂交得到 F_1 植株，然后 F_1 自交或同胞交配而产生。因此，这一群体是两种亲本基因组之间单个减数分裂事件的结果。这一群体最易构建，很适合用作初步研究。然而，由于它是原始亲本基因组之间的随机重组事件所导致的一系列个体，因而无法被重构。因为 F_2 群体是仅一个周期减数分裂的结果，所以无法精细作图，除非对大量个体进行计分。此外，由于每个个体是不同的基因组所导致的，因而无法对数量性状位点（QTL）作图，也无法解决环境和生长（G×E）的相互作用问题。F_2 个体通过自交可产生 F_3 群体，这使得能确定表达显性性状的 F_2 个体的遗传状态。继续自交将最终产生重组近交系。表 5－2 给出了随着自交和回交后代数量增加，该位点纯合的比率。

表 5－2 随着自交或回交后代增多，近亲繁殖的水平

（数值不考虑连锁累赘，所以每一代的实际纯合性水平较低）

自交	每个位点纯合系的百分比（%）	回交	父母基因组复发的百分比（%）
F_3	75.0	BC_2	75.0
F_4	87.5	BC_3	87.5
F_5	92.25	BC_4	92.25
F_6	96.875	BC_5	96.875
F_7	98.437 5	BC_6	98.437 5
F_8	99.218 75	BC_7	99.218 75

5.3.1.2 回交

F_1 植物与亲本一方杂交形成回交群体。一般目标性状是隐性的亲本作为回交亲本。本文中，重组仅发生在亲本的一方，即 F_1。数轮回交导致轮回亲本基因组的再生，如果与特定性状的选择关联起来，可用于开发近等基因系(NILs)。这些系有不同的等位基因，在相同遗传背景下影响某一特定性状，因此可确定性状和背景之间的相互作用。此外，如果开发出不同基因的 NILs，之后这些 NILs 可组合在一起以了解不同环境中不同基因之间的相互作用，免除了不同遗传背景的影响。

5.3.1.3 双单倍体（DH）

双单倍体通常来自花药培养。不过，在一些特别的系统，如大麦属，双单倍体可来自种间杂交之后的染色体消减。衍生自 F_1 个体花药培养的 DH 植株都将不同，因为它们是 F_1 减数分裂的独立结果。它们在每个位点上也都是纯合的，因而 DH 群体是作图的永久性群体。每一 DH 个体都将真实繁育，因而在多种环境中性状的重复是可能的，可用来检测 QTL 和基因型与环境（G×E）的影响。该方法的一个主要缺点是并不是所有植物都适于花药培养，甚至那些可进行花药培养的植物也需要监测培养诱导的变异问题。

5.3.1.4 重组近交系（RILs）

重组近交系由 F_1 代连续自交或同胞交配而成。每一后代将在越来越多的位点上纯合，直到每一系最终完全纯合。这些系相当于双单倍系，但需要更长时间来获得，可源自任何植物物种。通常这些系由单粒传法而得。所选植物的一粒种子在每一代生长。这些 RILs 完成分离的时间取决于该植物的生命周期，如果首次开花耗费数年，那形成 RILs 要耗费数百年。当这些系开发出来，就永久存在，并可被分发给其他研究人员，以便所有人可将标记加到相同的图谱上。它们也可以不断增加，与其他作图的永久性群体一样，用于多种环境中的性状的重复，可用于检测 QTL 和基因型与环境（G×E）的影响。重组间隔能否解决取决于被筛选的 RILs 数量。

5.3.1.5 近等基因系（NILs）

近等基因系是由 F_1 的重复自交或回交而成。每一代通常都选择某一特定性状的两种形式，以便 RILs 除了在目标位点上不同外，几乎在所有位点上都相同。比如，豌豆表型为圆粒/皱粒的 NILs 的开发。该例子中，圆粒是显性，而皱粒则是隐性。将自交系圆粒亲本和自交系皱粒亲本杂交，得到的 F_1 全部是圆粒的。当这个 F_1 与皱粒亲本回交，BC_1（回交 1 代）分离出的圆皱比为 1∶1。圆粒种子种下，再次与皱粒亲本回交。后代也分离出 1∶1 的圆粒和皱粒种子。经过几轮的回交，植物最终基本上纯合，除了决定圆和皱的位点不同，其他与皱粒亲本基本相同。最后一代是圆粒种子个体本身（在该位点上仍然是杂合）和挑选后的圆粒和皱粒的家族。圆粒种子个体自交一次，以确保它在该位点上是纯合的（不分离皱粒种子），两系可倍增，在除了决定圆和皱的基因型的位点外，在每一位点上相同，产生 NILs 需要耗费大量的时间，并且只对挑选下的性状有用。不过，NILs 对与性状相关联的多效性效应的功能基因组学的研究起到重要作用。

5.3.1.6 测交

测交相当于与隐性亲本进行回交，使得数据聚焦于亲本一方的重组。两点测交指两亲本杂交，亲本一方在某一特定标记上杂合，另一方为空。因此，在

测交后的 F_1 后代中，该标记将按 1∶1 比例分离。这一方法可更迅速地用于性状作图。

5.4　集团分离分析法

除 NILs 外，集团分离分析法适用于前文所述所有群体。此方法以同源系原则为基础。例如，F_2 的个体按照某一性状，如疾病抗性或敏感性分类，根据它们的表型分组。从每一组选择 10 个个体来提取 DNA，并混合成 DNA 池。采用一系列在亲本间具有多态性的标记对两个 DNA 池进行筛选。若标记未与某一性状关联，则每个 DNA 池扩增后都将显示出相同的带型。若标记与性状紧密相连，则两个 DNA 池的带型将与原始亲本的带型相似。在这个池中的 10 个个体可被确定足以作为第一次筛选。

5.5　使用的作图软件

许多类型的软件可用来处理杂交的原始数据以绘制遗传图谱。最早开发的程序现在仍在使用，只是进行了一些修改，使得速度更快，能处理更大的数据集。还有一些程序不仅用来绘制连锁图谱，还可处理更复杂的数据，特别是识别数量性状位点。

5.5.1　Linkage-1

该程序开发于 1983 年，用来促进对连锁位点的分析。这是对用于绘制图谱的标记生成连锁计分的基本程序。该程序可分析从 F_2 代、回交种和位点间（相同分离群体中显示不同遗传模式的位点）单因子分离比例的不同组合得来的数据。该程序可分析显性和共显性基因与标记，利用卡方分析来测试成对的分离位点之间的独立分配。将每个标记基因型的比率与在该位点上的预定分离比率比较，以得到预期分离模式。当 P 值显示偏离独立分配的预期频率，之后计算每对位点的重组频率。程序的早期版本得出了 2-locus 表、卡方与 P 值和重组率及标准误差（图 5－3）。之后可利用其他程序或手动将这些数据编译成一张图。想使用最初的 Linkage-1 程序，请访问洛克菲勒大学网站：ftp：//linkage. rockefeller. edu/software/linkage。该软件已得到改进，通过并行运行多个算法以容纳更大的数据集，现在以 FASTLINK 为名，可在 NCBI 的基因分析软件站点上找到（网址：http：// www. ncbi. nlm. nih. gov/CBBresearch/Schaffer/genetic _ analysis. html）。相同的 Linkage 程序正被使用，但当前的版本能更快得出结果，特别是对于大型数据集来说。

A Total number of individuals compared = 40 out of 40

The two loci being compared are PGM3 and DOM1

DOM1	PGM3 BB	BC	CC
A-	2	17	10
aa	11	0	0

Chi-square	Df	P	r +/- SE
3.15119E+01	2	<0.01	5.0000E-02 +/ 3.5811 E-02

B Total number of individuals compared = 40 out of 40

The two loci being compared are PGM3 and PG11

PG11	PGM3 BB	BC	CC
AA	12	8	0
AB	1	9	10

Chi-square	Df	P	r +/- SE
1.93665E+01	2	<0.01	4.0000E-02 +/ 4.3818 E-02

C Total number of individuals compared = 40 out of 40

The two loci being compared are PGM2 and PG11

PG11	PGM2 AB	BB
AA	10	10
AB	7	13

Chi-square	Df	P	r +/- SE
9.20716E-01	2	0.5 - 0.25	None

This valuer is not significant

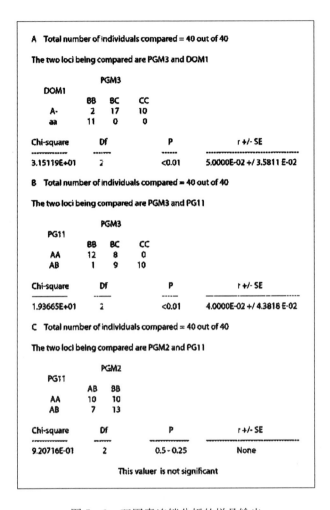

图 5-3 双因素连锁分析的样品输出

［显示了 3 种不同的家族类型（A、B、C），每个基因的预期比率和表 5-1 中的相同］

（Total number of individuals compared＝40 out of 40：比较个体总数为全部 40 个）

A. 两个比较的位点为 PGM3 和 DOM1

B. 两个比较的位点为 PGM3 和 PG11

C. 两个比较的位点为 PGM2 和 PG11

（来自 Suiter，1983）

5.5.2 MapMaker

MapMaker 套装软件包括原始的 MapMaker、MapMaker/EXP 和 MapMaker/QTL。MapMaker 可分析 3 代系谱，利用 MapMaker/EXP，可将

通过孟德尔标记的杂交得来的试验数据构建图谱。MapMaker/EXP 可分析的数据包括 F_2 代、回交种、RILs 和 F_3 -自交群体。EXP 分析数据的方法类似于 Linkage-1，计算每一对标记的最大似然距离和连锁似然值（LOD）。当 LOD 值大于某一阈值（缺省值通常为 3.0，表示发生的可能性为 10^{-3}），而距离小于某一阈值（30cM），这些标记被认为是连锁的。之后将位点分配到连锁群，基于这些数据构建图谱（图 5-4）。MapMaker/QTL 利用遗传图谱对涉及数量性状的多基因作图，并可利用 MapMaker/EXP 生成的图谱进行作图。MapMaker/EXP 和 QTL 的软件网址：http：// linkage. rockefeller. edu/soft/list. html。这个网站也有详细的在线教程。

```
15> map
=====================================================================
Map:
  Markers              Distance
    4  T24              14.8 cM
   11  C15               6.4 cM
    8  T125             18.9 cM
   12  T71              24.0 cM
    9  T83              18.1 cM
    6  T209             28.6 cM
   10  T17             ----------
                       110.8 cM    7 markers    log-likelihood= -688.99
=====================================================================
```

图 5-4　从 MapMaker/EXP 中输出

（Marker：标记；Distance：距离；log-likehood：似然对数）

（Lincoln et al.，1992）

5.5.3　JoinMap

开发这一作图软件是为了将独立试验的图谱数据整合到一张综合图谱中（图 5-5）。不同类型的数据，如 F_2 代、回交种、RILs 和成对的重组率可集成，只要这些数据集有一些共同的标记。这一程序也可利用单一试验的数据来创建图谱。信息文件的编码类似于 MapMaker，这样 MapMaker 的数据也可轻松编入。JoinMap 用以下方式合并作图数据。

读取数据，对于所有数据集共有的标记对，所估计的重组频率被加权，并被单个值取代。然后通过重组频率来计算 LOD 值。基于 LOD 值建立共有数据的连锁群，随后通过与已定位在图谱上的 LOD 值进行比较，构建包含其余数据的图谱。JoinMap 可进行评估，可从 Kyazma 上购买，网址是 http：//www. kyazma. nl/。该网站也提供关于作图和 QTL 分析的课程信息。

相较于 Linkage-1 和 MapMaker 套装软件，JoinMap 更易安装，拥有更友好的用户界面。

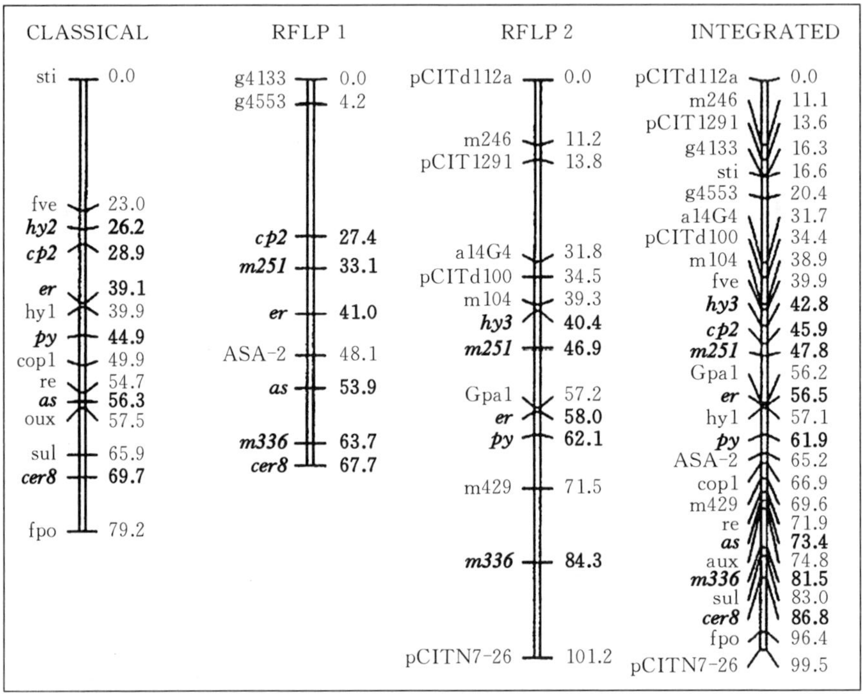

图 5－5　3 张独立的连锁图谱和拟南芥 2 号染色体的综合图谱

（粗体标注的标记不止出现在一张个体图中。为保证清晰，这里只显示出一部分标记）

（来自 Stam，1993）

5.5.4　其他软件

最近的文献中提到的其他作图程序有 Carthagene，像 JoinMap 一样，既可整合数据，又可创建新图，类似 Linkage 用卡方分析的 Surveyor，还有 CPROP、GMendel、CriMap、GeneHunter、Vitesse、Allegro、Merlin、Solar、SimWalk2 和 Superlink。其中一些程序和它们最适用的数据已得到审核（Dudbridge，2003）。洛克菲勒大学网站还列出了许多其他程序。截至 2010 年 8 月，共列出 580 项程序用于遗传杂交数据的基本操作，包括动植物育种与 QTL 分析、群体遗传学和人类系谱分析。程序按字母顺序列出，附上了每个程序的特点和应用的简短介绍，并链接了软件和参考文献。

5.6　分子遗传图谱的发展

前述部分已经独立地描述了作图的几个方面。遗传和物理作图通常是相关科学和农业各界共同参与的综合活动。几乎所有的植物改良都依赖于杂交和从这些杂交种中选取特定的产出。可能的例外包括识别有用的体细胞无性系变异和诱变育种，虽然在后一种情况下利用遗传图谱可取得更成功的选育结果。

第一性状通常是表型性状，它们容易被计分。然而，许多表型性状的表达可能干扰其他性状的计分，所以在某个特定的杂交中只有少数性状可遵循。因此，早期的遗传图谱由来自不同杂交的关联的数据汇集而成。分子标记的出现基本上使无限个位点可被放置在单个杂交的连锁图上。

一张初始的基因图谱可能仅包含少量的标记，理想地分布在所有染色体上。这使性状和另外的标记可定位于某一特定染色体上。然而，对于密度更高的作图，需要添加更多的标记，并开始利用边际报酬递减定律。若应用随机标记，则区间越小，越需要筛选更多的标记以填补连锁群中的特定空白。因此某一形式的区域锚定将更为有用。

第一代分子遗传连锁图谱通常基于分离距离约为20 cM（厘摩）（两位点间的重组频率约为1%时，它们的图距是1 cM）的标记。因为覆盖率是这一水平，所以任何新标记或性状不会距离已知位置超过10 cM。因此，该标记和目标基因或性状共遗传的概率至少为90%。随着对准确性的要求越来越高，越来越多的标记放置在遗传图谱上。要确定一系列紧密连锁标记的顺序，有必要扩大作图群体。如果标记之间的距离是20 cM，在有100个个体的一个分离群体中，平均发生20次重组。但是，要对距离是1 cM的标记进行作图，同样的群体就不够大了。

至此讨论已涉及遗传图谱的方方面面。物理图谱是染色体中DNA序列的线性顺序和碱基对中已知位置之间的距离。遗传和物理图谱的调和有时很难做到，因为染色体区域可能降低重组频率，所以1 cM的遗传距离没有完全对等的碱基对。它可在染色体中和染色体之间产生差异，甚至会因其所测量的个体的性别而产生差异。整合多张图谱也会产生问题。若作图群体不同，那么染色体重排将导致特定的标记具有不同位置的问题。例如，如果两组利用不同的起始群体各自绘制相应的分子图谱，一个群体发生易位，另一个群体未发生，那么一些标记甚至都不会绘制到相同的染色体上。解决这个问题的一种方法是开发分子细胞遗传学图谱。

原位杂交的发展使物理序列定位在染色体上成为可能。物理图谱的发展可以标记低拷贝DNA序列的大片段并将这些片段杂交到染色体上，然后可视化杂交的位点。如果这些染色体能被复染和识别，那么每一片段都可物理定位在特定的染色体位点上。

5.7　结论

遗传作图是植物改良的必然之路。对于其他分子信息少的植物，它们可以利用的一整套标记类型经过不断发展推动了现代育种技术的应用。然而，许多这样的调查是以长期杂交项目为基础，而现今这些项目通常难以得到支持。植物材料需要耗费相当长时间来开发，基础设施若不能保证，改良项目必将受阻。

缩略语

AFLP：扩增片段长度多态性
DH：双单倍体
EST：表达序列标签
G×E：基因型与环境交互作用
mRNAs：信使 RNA
NILs：近等基因系
PCR：聚合酶链式反应
PIC：多态性信息含量
QTL：数量性状位点
RAPD：随机扩增多态性 DNA
RFLP：限制性片段长度多态性
RILs：重组近交系
SNP：单核苷酸多态性
SSR：简单重复序列

参考文献

Botstein D，White R L，Skolnick M，et al，1980. Construction of a genetic linkage map in man using restriction fragment length polymorphisms [J]. Am J Hum Genet，32：314-331.

Dudbridge F，2003. A survey of current software for linkage analysis [J]. Human Genomics，1：63-65.

Jones C J，Edwards K J，Castaglione S，et al，1997. Reproducibility testing of RAPD，AFLP and SSR markers in plants by a network of European laboratories [J]. Mol Breed，3：381-390.

Kristensen V N，Kelefiotis D，Kristensen T，et al，2001. High-throughput methods for detection of genetic variation [J]. BioTechniques，30：318-322.

Lai E，2001. Application of SNP technologies in medicine：lessons learned and future challenges [J]. Genome res，11：927-929.

Lincoln S E，Daly M J，Lander E S，1992. Mapping genes controlling quantitative traits：using MapMaker/QTL version 1.1，a tutorial and reference manual [R]. 2nd ed. A Whitehead Institute for Biomedical, Research Technical Report.

Panaud O, Chen X, McCouch S R, 1995. Frequency of microsatellite sequences in rice (*Oryza sativa* L.) [J]. Genome, 38: 1170-1176.

Santana Q, Coetzee M, Steenkamp E, et al, 2009. Microsatellite discovery by deep sequencing of enriched genomic libraries [J]. BioTechniques, 46: 217-223.

Senior M L, Heun M, 1993. Mapping maize microsatellites and polymerase chain reaction confirmation of the targeted repeats using a CT primer [J]. Genome, 36: 884-889.

Stam P, 1993. Construction of integrated genetic linkage maps by means of a new computer package: JoinMap [J]. Plant J, 3: 739-744.

Sturtevant A H, 1913. The linear arrangement of six sex-linked factors in *Drosophila*, as shown by their mode of association [J]. J Expt Zool, 14: 43-59.

Suiter K A, Wendel J F, Case J S, 1983. Linkage-1: a PASCAL computer program for the detection and analysis of genetic linkage [J]. J Hered, 74: 203-204.

Vos P, Hogers R, Bleeker M, et al, 1995. AFLP: a new technique for DNA fingerprinting [J]. Nucl Acid Res, 23: 4407-4414.

Williams J G, Kubelik A R, Livak K J, et al, 1990. DNA polymorphisms amplified by arbitrary primers are useful as genetic markers [J]. Nucleic Acids Research, 18: 6531-6535.

第 6 章　芭蕉属植物简单遗传性状的作图和标记

Rodomiro Ortiz[①]

摘　要： 香蕉和大蕉由于缺乏传统遗传学上的研究，可用的遗传标记很少。芭蕉属植物遗传分析上的主要障碍：常见的品种多为三倍体，生育力很低，体外胚胎萌发后可回收的杂交种很少，导致种群规模小。尽管如此，关于芭蕉属植物的传递遗传学研究仍旧开展起来了，尤其是在 20 世纪 90 年代。本文探讨了包括白化病、顶端优势、侏儒症和果实单性结实在内的一些目标性状的重要性。本文还对作图基因进行了讨论。

关键词： 芭蕉属植物的目标性状，性状作图

6.1　简介

尽管芭蕉属植物是重要的农作物，但因为缺乏在传统遗传学上的研究，香蕉和大蕉可用的遗传标记很少（Ortiz and Vuylsteke，1994a）。芭蕉属植物遗传分析上的主要障碍：常见的品种多为三倍体，生育力很低，体外胚胎萌发后可回收的杂交种很少，导致种群规模小。尽管如此，关于芭蕉属植物的传递遗传学研究仍旧开展起来了，尤其是在 20 世纪 90 年代（Ortiz，2000）。芭蕉属植物遗传研究的最早结论之一是关于果实单性结实和花序轴雄性部分苞片及花的持续性。有互补基因作用的 3 个独立的显性 P_i 基因控制前者，而互补基因控制后者（Simmonds，1952）。芭蕉属植物基因组联盟（http：// www.musagenomics.org）的成员培育出了分离种群，来鉴定以下的遗传标记，包括寄主植物对病原体和害虫的抗性、果实单性结实、连锁不平衡分析和与水稻进行比较遗传分析的标记（http：// www.musagenomics.org/index.php? id=49）。作图群体是 F_1（如果亲本一方是显性杂合而另一方是隐性的，可用测交分离；如果亲本双方是共享一个双等位基因的基因型，可分离成为 F_2）和 F_2

---

① Department of Plant Breeding and Biotechnology，Swedish University of Agricultural Sciences，Box 101，SE-230 53，Alnarp，Sweden；e-mail：rodomiroortiz@gmail.com.

世代。杂交种包括二倍体种，如 *M. acuminata*、*M. balbisiana* 和某些三倍体栽培种。第一张芭蕉属植物遗传图谱是基于限制性片段长度多态性（RFLP）标记和二倍体群体中的同工酶（Faure et al.，1993），并加入了微卫星（或简单重复序列，即 SSR）和扩增片段长度多态性（AFLP）标记。

6.2 目标性状的重要性

白化病（a_i；Ortiz and Vuylsteke，1994a）：杂合胚胎体外萌发后在小植株上观察到的缺乏叶绿素的幼苗。在杂交后代上观察到的分离比率表明白化病至少是由两个独立的隐性等位基因（a_1 和 a_2）控制。白化病可能是研究作物遗传负荷的有趣性状，通过监测这种有害的隐性等位基因的频率能了解二倍体繁殖群体通过轮回选择带来的好处。

顶端优势（*ad*；Ortiz and Vuylsteke，1994b）：因顶芽释放的物质导致的侧芽生长受到抑制已被认为是大蕉持续生育力的限制因素。大蕉-香蕉育种群体的分离比率表明顶端优势的遗传是由一个主要的隐性基因 *ad* 控制。通过测量大蕉-香蕉杂交种最高吸芽的高度得知，显性 *Ad* 等位基因改善了作物的分蘖。*Ad* 基因可调节赤霉素产生，但这一等位基因显示出不完全显性和可变表达。

侏儒症（*dw*；Ortiz and Vuylsteke，1995a）：假茎的高度，即土壤到最高叶子叶柄的距离，被用于大蕉栽培品种亚组的分类。据此将其分成大、中、小3个高度。节间短的芭蕉属无性系被称为"矮小"品种（Ortiz and Vuylsteke，1998）。这种侏儒症由单一隐性基因 *dw* 控制，这一基因的位置似乎邻近着丝粒，在四倍体水平表现出剂量效应。矮蕉品种较少遭受风害，由于种植密度高和吸芽快速发展，产量有所增加。

果实单性结实（P_i；Simmonds，1952）：芭蕉属植物栽培品种的果实发育依靠营养性单性结实，这表现在非授粉子房自动长成果实，果实的小腔内充满可食用的果肉。自主刺激引起的单性结实至少由3个独立、互补、显性的基因 P_1、P_2 和 P_3 相互作用导致。P_1 的位点似乎远离着丝粒，在第一个交叉之后。Ortiz 和 Vuylsteke（1995b）报道果实大小和重量与 P_1 的位点上等位基因替代效应部分相关。

寄主植物抗性：获得更多关于寄主植物抗性的遗传学知识将促进抗病品种的开发，这被认为是获得稳定产量的最佳手段。利用在二倍体和多倍体水平分离的群体可以阐释以下几种抗性的遗传控制：抗香蕉条纹病毒（Ortiz，1996）、香蕉叶斑病（bs_1+2bsr_i；Ortiz and Vuylsteke，1994c）和穴居线虫（二基因系统；Dochez et al.，2009）。上述病原体和害虫显著影响果穗重量，从而导致芭蕉属作物的减产，如由 *Mycosphaerella fijiensis* 所致的香蕉叶斑病是危害全球大蕉和香蕉生产的主要元凶。

二倍体繁殖群体中寄主植物抗香蕉叶斑病的分离比例符合一个遗传模型，这个模型拥有一个重要的、位于着丝粒和第一个交叉之间的隐性抗性等位基因（bs_1），还拥有产生加性效应的两个独立的等位基因（bs_2 和 bs_3）。一个相似的模型在四倍体水平上解释了这个结果，假设有利抗性等位基因的 4 个拷贝存在于各自的位点（bs_i^4）时，它们有剂量效应。在二倍体水平上纯合的 bs_1 基因型，单独或只与 bsr_i 基因型组合在一起不会导致部分抗性，而只是导致较不敏感的反应。只有所有 3 个有利等位基因纯合的二倍体才会产生部分抗性的基因型。由于剂量效应，有纯合 bs_1 基因型的四倍体将出现部分抗性。相似地，即使存在不利的 Bs_1 易感等位基因，任何 bsr_i 位点的纯合基因型有较不敏感的表型，这意味着在四倍体水平上随着有利的等位基因数量的增加，可以战胜寄主植物的易感等位基因和病原体的毒性基因作用。进一步的遗传研究（Craenen and Ortiz，1997）表明，bs_1 位点上位点内相互作用能显著调节叶面症状的表观，而 bs_1 位点的加性效应和位点内的相互作用影响寄主植物的病害发展。因此，bs_1 位点上的基因活动通过减缓病害在寄主植物上的发展，以维持对香蕉叶斑病的持久抗性。

6.3 利用集团分离分析法来标记基因

根据分离群体中的个体对特定性状表型的表达对其进行分组，并测试其在各种群集团之间的等位基因的差异频率，即集团分离分析法（BSA；Michelmore et al.，1991）。用于绘制遗传图谱的分子标记可以用于 BSA。亲本间的分子多态性和各自的与特定性状主要调控基因紧密连锁的标记将与该基因共分离。大多数种群按照它们的性状表达分组并用多态性标记测试这样的极端组群后，两个集团内的每个标记的两个标记等位基因的频率都应当显著偏离预期的 1∶1 比例。当已知许多分子标记在染色体的位置时，则无需知道分离种群中每个个体的基因型就能推断出紧密连锁的 DNA 标记与性状的图谱位置。

香蕉条纹病毒（BSV）的表达：Lheureux 等（2003）对 F_1 三倍体群体中香蕉条纹病（BSD）的发生率进行了分析，这个群体来自无病毒和病原体的二倍体 *M. balbisiana* 和四倍体 *M. acuminata* 亲本的种间杂交，其后代 50%表达 BSV 粒子。它们的目的在于阐明在这些杂交种中无论是否发生 BSD，都是由于激活了 BSV-O1 内源性逆转录病毒序列（EPVR）整合到芭蕉属植物基因组中的结果，而不是外源的感染，并且在育种过程中的杂交可能是参与触发 BSV 成分游离表达的因素之一。这个研究使用聚合酶链式反应（PCR）扩增来确定是否存在 BSV-O1 EPRV。研究人员认为这个内源性的序列对 *M. balbisiana* 的基因组是特异的，并且发生在纯合状态。这一研究能够通过 BSA 来鉴定，不论 *M. balbisiana* 是否感染 BSV，10 个 AFLP 标记都能共分离，而 *M. acuminata* 的基因组中则没有这个现象。存在 BSV 粒子时，7 个 AFLPs 分离，而不存在 BSV 粒子时，3 个 AFLPs 分离。这些结果表明，BSV 表观的遗传控制来自单基因等位基因系统，承担载体作用的是 *M. balbisiana*。

寄主植物对镰孢枯萎病的抗性：将对镰孢枯萎病有抗性和易感的幼苗和品种与随机扩增多态性 DNA（RAPD）技术结合在一起，用来鉴定寄主植物对这种病原体的抗性关联的标记。在野生 *M. acuminata* ssp. *malaccensis* 的 4 个幼苗群体中，对 *F. oxysporum* f. sp. *cubense* 4 号生理小种的分离比例调整到 1∶3之后，4 个幼苗群体中的 3 个出现单隐性因子，另外一个幼苗群体的寄主植物抗性受单显性因子影响（Javed et al.，2004）。13∶3 这一比例表明有两个位点控制寄主植物对镰孢枯萎病的抗性，这一点也在第五个幼苗群体中观察到。5 个幼苗群体中的这个结果表明，在遗传学上，它们的野生亲本对镰孢枯萎病 4 号小种的寄主植物的抗性位点在遗传上明显不同，这也许是病原体与其宿主之间共同进化的结果。一些多态性 RAPD 标记在有抗性或易感幼苗中表现出特定的带型。所有易感幼苗都出现 03 号引物的 400 bp 的主要条带，而有抗性的后代则没有，另外，24 号引物的 300 bp 的条带只在多数易感幼苗中观察到，而 21 号引物的 1 kb 的条带只出现在对镰孢枯萎病有抗性的幼苗中。

Zambrano 等（2007）利用 10 种对镰孢枯萎病有抗性的和 8 种易感的有不同倍性水平的栽培种和野生种来筛选与该病原体寄主植物抗性关联的 RAPD 标记。引物 OPK-03 检测出的主要片段为 485 bp，是唯一一个在易感品种中出现条带，而在抗性品种和野生种中未出现条带的引物。

侏儒症：Engelborghs 等（2000）使用 AFLP 和这一 DNA 标记技术的几个衍生技术来描述 4 对离体型侏儒大蕉 Curare Enano 和它的正常大小植株。三酶切 AFLP（TE-AFLP）产生更少和更短的片段，从而导致矮小植株品种和其正常大小植株之间更少的多态性。cDNA-AFLP 揭示了一个正常的特异片段，而 cDNA TE-AFLP 能识别矮小植株的特异片段，这显示出 AFLP 技术及其衍生技术在区分相似度高的芭蕉属植物基因型上的巨大价值。

6.4 基因作图

香蕉条纹病毒（BSV）的表达：考虑到测交策略，在 F_1 杂交群体中，AFLP 标记的分离允许构建一个包含了与 BSV 感染相关联位点的连锁群的母系遗传图谱（Lheureux et al.，2003）。为了进行 BSA 分析，使用了分别由 5

个健康和 5 个患病的 F_1 后代组成的两个 AFLP 模板库。两个模板库采用 AFLP 的指纹技术，利用了 *Eco*R Ⅰ 和 3/*Mse* Ⅰ 的 64 个组合，外加 3 个引物延伸。标记跨度为 69.8 cM，标记之间的遗传距离为 0.6～39.2 cM。BSV 表达位点（BEL）距离最近的标记（ACACAC）1.2 cM。

寄主植物抗香蕉叶斑病：为了将各自的连锁群锚定到芭蕉属植物核心图上，Carreel 等（1999）使用来自 *M. acuminata* spp. *burmannicoides*（Calcutta 4）和 *M. acuminata* spp. *banksii*（Madang）杂交的含有 153 株植物的 F_2 分离群体绘制出一张 110 个 AFLP 标记连同 39 个显性 RFLP 和 SSR 标记的连锁图。其中一个 RFLP 与寄主植物对香蕉叶斑病的抗性密切相关，与这一抗性有关的第二位点重要性低一些，被绘制在另一张连锁图中。

缩略语

AFLP：扩增片段长度多态性
BSA：集团分离分析法
BSD：香蕉条斑病
BSV：香蕉条纹病毒
EPVR：内源性准逆转录病毒序列
RAPD：随机扩增多态性 DNA
RFLP：限制性片段长度多态性
SSR：简单重复序列

参考文献

Carreel F, Abadie C, Carlier J, et al, 1999. Genome mapping andgenetic analysis of the black leaf streak resistance in bananas [J]. Infomusa, 8: 11 - 111.

Craenen K, Ortiz R, 1997. Effect of the *bs₁* gene in plantain-banana hybrids on response to black sigatoka [J]. Theor Appl Genet, 95: 497 - 505.

Dochez C, Tenkouano A, Ortiz R, et al, 2009. Host plant resistance to *Radopholus similis* in a diploid banana hybrid population [J]. Nematology, 11: 329 - 335.

Engelborghs I, Swennen R, Sagi L, 2000. Fluorescent AFLP analysis on azacytidine and gibberellin treated banana (*Musa* spp.) plants to assess differences in cytosine methylation and the mechanism of dwarfism [J]. Med Fac Landbouww Univ Gent, 65/3b: 387 - 396.

Faure S, Noyer J L, Horry J P, et al, 1993. A molecular marker-based linkage map of diploid bananas (*Musa acuminata*) [J]. Theor Appl Genet, 87: 517 - 526.

Javed M A, Chai M, Othman Y R, 2004. Study of resistance of *Musa acuminata* to *Fusarium* oxysporum using RAPD markers [J]. Biologia Plantarum, 48: 93 - 99.

Lheureux F，Carreel F，Jenny C，et al，2003. Identification of genetic markers linked to banana streak disease expression in inter-specific *Musa* hybrids [J]. Theor Appl Genet，106：594－598.

Michelmore R W，Paran I，Kesseli R V，1991. Identification of markers linked to disease-resistance genes by bulked segregant analysis：a rapid method to detect markers in specific genomic regions by using segregating populations [J]. Proc Natl Acad Sci USA，88：9828－9832.

Ortiz R，1996. The potential of AMMI analysis for field assessment of *Musa* genotypes to virus infection [J]. HortScience，31：742－762.

Ortiz R，2000. Understanding the *Musa* genome：an update [J]. Acta Hort，540：157－168.

Ortiz R，Vuylsteke D，1994a. Genetic analysis of apical dominance and improvement of suckering behaviour in plantain [J]. HortScience，119：1050－1053.

Ortiz R，Vuylsteke D，1994b. Inheritance of albinism in banana and plantain（*Musa* spp.）[J]. HortScience，29：903－905.

Ortiz R，Vuylsteke D，1994c. Inheritance of black sigatoka disease resistance in plantain-banana（*Musa* spp.）hybrids [J]. Theor Appl Genet，89：146－152.

Ortiz R，Vuylsteke D，1995a. Effect of the parthenocarpy gene P_1 and ploidy on fruit and bunch traits of plantain-banana hybrids [J]. Heredity，75：460－465.

Ortiz R，Vuylsteke D，1995b. Inheritance of dwarfism in plantain（*Musa* spp.，AAB group）[J]. Plant Breeding，114：466－468.

Ortiz R，Vuylsteke D，1998. Quantitative variation and phenotypic correlations in banana and plantain [J]. Sci Hort，72：239－253.

Simmonds N W，1952. Segregations in some diploid bananas [J]. J Genet，51：458－469.

Zambrano A Y，Martinez G，Gutierrez Z，et al，2007. Marcador RAPD asociado a la resistencia a *Fusarium oxysporum* en *Musa* [J]. Interciencia，32：775－779.

第7章 复杂性状的分子作图

Rodomiro Ortiz①

摘 要：孟德尔遗传学在阐释芭蕉属植物质量性状变异上贡献很大。高通量“组学”技术的最新进展将提供对复杂性状作图和深刻理解的新方法，使遗传信息能够翻译成生物功能。芭蕉属植物基因组学联盟提供的作图群体中最重要的目标性状包括寄主植物对害虫及病原体的抗性和果实单性结实。尽管在开发分子标记及其在芭蕉属植物遗传图谱中进一步的使用上取得了一定的进展，但是关于香蕉和大蕉的数量性状位点（QTL）作图的报道仍很缺乏。本章将讨论芭蕉属植物数量性状作图上所取得的进展。

关键词：数量性状，作图群体

7.1 引言

孟德尔遗传学在阐释芭蕉属植物质量性状变异上贡献很大（Ortiz，2000）。同样的，生物统计学方法促进了对以下知识的了解：寄主植物对病原体和害虫的抗性遗传会影响作物产量，倍性或基因作用会影响果穗和果实的性状。数量性状变异也会受到环境及基因型与环境互作的影响（Ortiz and Tenkouano，2011）。高通量“组学”技术的最新进展将提供对复杂性状作图和深刻理解的新方法，使遗传信息能够翻译成生物功能。

7.2 目标性状和群体

芭蕉属植物基因组学联盟提供的作图群体中最重要的目标性状包括寄主植物对害虫及病原体的抗性和果实单性结实（http：//www. musagenomics. org/index. php? id=49）。大多数 F_1 或 F_2 分离群体都源于二倍体种之间的杂交，如

① Department of Plant Breeding and Biotechnology，Swedish University of Agricultural Sciences，Box 101，SE-230 53，Alnarp，Sweden；e-mail：rodomiroortiz@gmail. com.

M. acuminata ssp. *malaccencis*，spp. *banksii* 或 *M. balbisiana*。已经使用的分子标记有限制性片段长度多态性（RFLP）、微卫星或简单重复序列（SSR）、随机扩增多态性 DNA（RAPD）、扩增片段长度多态性（AFLP）或由巴西（UCB，EMBRAPA）、喀麦隆（CRBP）、法国（CIRAD）、印度（IIHR）、尼日利亚（IITA）和乌干达（IITA 和 NARO）的研究团队开发的抗性基因类似物（RGA）。

虽然细菌、真菌、昆虫、线虫和病毒显著影响了香蕉和大蕉的产量及整个环境的稳定性，芭蕉属植物育种家还有其他的目标性状（图 7-1），即侏儒症、早熟、短周期、吸芽行为、根系改良、重果穗、大果和在二倍体和多倍体育种群体上都符合最终用户需求的高质量。在发展中国家的一些地区，特别是非洲撒哈拉以南的热带地区，收获后的损失也是阻碍生产扩大的重要问题（Vuylsteke et al.，1997）。

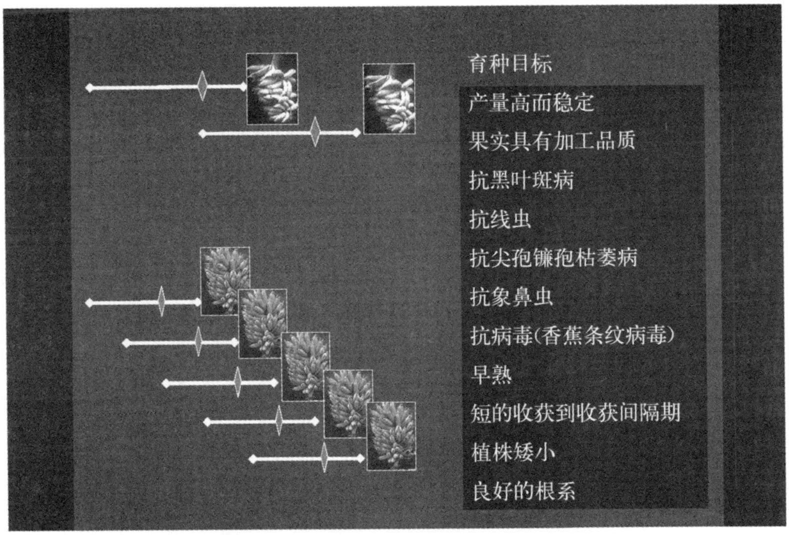

图 7-1 大蕉和香蕉育种的育种目标
（H-H：收获到收获，R=寄主植物抗性）
（图片来自 Abdou Tenkouano 博士，AVRDC，坦桑尼亚）

7.3 框架图和标记

第一张芭蕉属植物遗传图谱是基于 92 个二倍体个体的 F_2 群体中分离的 58 个 RFLP、28 个 RAPD 和 4 个同工酶标记。研究人员总共检测了 90 个位点，其中 77 个在 15 个连锁群上，而 13 个独立分离。36%的位点上出现偏分离，大多数偏向父本，这可能是由于染色体结构重排导致的（Fauré et al.，

1993)。第一张饱和的图谱利用了生育力强的二倍体香蕉品种 M53 自交产生的 89 个二倍体个体，它们包含 300 多个标记，分散在代表 *M. acuminata* 基因组的 11 个连锁群中，其中 100 个是共显性 RFLP 标记，30 个是 SSR 标记，而其他的是显性 AFLP 标记。一些标记显现出了由 *M. acuminata* 亚种特有的易位所导致的高偏分离率（Noyer et al.，1997）。Carreel 等（1999）利用 Calcutta 4×Madang 杂交产生的 153 个二倍体个体，绘制出了含有 110 个 AFLP 标记的连锁图，该图谱与 39 个 RFLP 和 SSR 标记相联系，可将连锁群锚定在香蕉核心遗传图谱上。研究人员观察到在 F_2 的后代上有 58 个标记出现重要的等位基因偏分离。最近，绘制成功了由 167 个 SSR 和 322 个 DArT 标记构成的，覆盖范围为 1 197 cM 的两张亲本图，该图谱来自 *M. acuminata* 两个遗传上远缘的种质 Borneo 和 Pisang Lilin（P. Lilin）杂交的 180 个 F_1 后代（Hippolyteet al.，2010）。基于每个亲本的配子重组率，高比率的标记（21.7%）偏离（$P \leqslant 0.05$）了预期的孟德尔比率。这些偏离的标记分布在每个亲本的不同连锁群。第一张饱和图谱作为"芭蕉属植物的参考图谱"，用于进一步的分析（Hippolyte et al.，2010）。

微卫星和其他 DNA 标记系统已被用于描述芭蕉属植物育种群体的特征。例如，SSR 标记提供一些见解，包括三倍体大蕉中 $2n$ 卵细胞的分离和四倍体杂交种的杂合性水平及用于其 $3x \times 2x$ 杂交的亲本。来自母系的大蕉基因型（Obino l'Ewai）的等位基因中有约 75%发生了分离，这证实了在三倍体可育大蕉的 $2n$ 大孢子形成期间发生了重组（Crouch et al.，1998）。对野生二倍体种质 Calcutta 4 及其自交后代的分析表明这一种质高度杂合。数目可变串联重复序列（VNTR）标记也用于评估二倍体和四倍体大蕉-香蕉杂交种及其亲本的变异（Crouch et al.，1999a）。一般情况下，相比它们的二倍体全同胞，$4x$ 杂交种与三倍体大蕉亲本联系更紧密。VNTR 分析能识别在次生三倍体杂交种的后续育种中的最有用的杂交种。杂交种与三倍体大蕉亲本基于 VNTR 的遗传相似性和基于 8 个农艺描述特征的表型距离指数之间观察到负相关，但是杂交种与二倍体野生亲本的分子遗传相似性和各自的表型距离指数之间没有关联。出乎意料的是，许多 VNTR 从二倍体和三倍体亲本及其后代中产生大量的扩增产物，这表明在 A 和 B 基因组中都存在高频率的位点重复。Crouch 等（1999b）的进一步研究表明 AFLP、RAPD 和 VNTR 试验检测出亲本基因型之间和子代群体内的高多态性。不过，AFLP 试验的复用率最高，而 VNTR 分析法检测的多态性水平最高。全同胞四倍体杂交群体的 AFLP 分析证实了之前基于 VNTR 分析的报告，即在三倍体大蕉地方品种的 $2n$ 配子形成过程中存在高频率的重组。全同胞三倍体杂交群体的 VNTR 和 RAPD 分析都表明，在四倍体杂交种的 n 配子形成过程中存在高频率的部分同源重组。用不同类型标记的遗传相似性估计值之间低度相关。微卫星也用于 DNA 指纹识别，来区分次生的三倍体全同胞杂交种（Ortiz et al.，1998）。这些研究结果表明芭蕉属植物繁殖群体遗传特性的重要性，以及微卫星和用于芭蕉属植物标记辅助基因研究的其他 DNA 标记的价值。

尽管在分子标记的开发及分子标记在芭蕉属植物遗传图谱、遗传学和育种研究的进一步应用上取得了一定的成果，但是关于香蕉和大蕉的数量性状位点（QTL）作图的报告还很缺乏（第 6 章）。因此，本章不含关于 QTL 作图软件的参考或性状方面 QTL 和染色体位置的细节。

7.4　QTL 的孟德尔定律

20 世纪 90 年代出现了一些研究，尝试利用倍性多态性和为数不多的可用遗传标记来解释某些性状的数量性状变异（QTV），如株高（Ortiz and Vuylsteke，1995a）、寄主植物对香蕉叶斑病的抗性（Craenen and Ortiz，1997）和果穗重量与果实体积（Craenen and Ortiz，1996；Ortiz et al.，1997）。由倍性和遗传标记解释的 QTV 数量因检查标记的特性和杂交的不同而不同。表 7-1 总结了关于果穗和果实特性的 QTV 遗传模型，这些模型分别由控制以下性状的主要位点解释：寄主植物抗香蕉叶斑病（bs_1）、果实单性结实（P_1）的基因之一和大蕉改良最成功的杂交种中它们的相互上位作用。

表 7-1　基于倍性（x）的多元回归模型，对培育最成功的大蕉-香蕉杂交种（Obino l'Ewai×Calcutta 4）的果穗和果实性状的数量性状变异进行解释的标记位点，有利于产生理想的四倍体后代

(After Ortiz et al.，1997)

性　状	回归模型	标准错误	R^2（%）
穗重（BW，kg/植株）	$BW=4.4x+3.5A_{P_1}+4.2D_{bs_1}-5.5A_{bs_1}+6.3$	2.4	88
果指数（H）	$H=3.61D_{P_1}+3.1$	1.2	75
果实数（F）	$F=55.3D_{P_1}+44$	31.3	82
平均果重（AFW，g）	$AFW=58x+33.2A_{P_1}+49.4D_{P_1}-7.3A_{bs_1}-65.5$	22.4	94
平均果长（FL，cm）	$FL=4.2-10.2A_{P_1}-7D_{bs_1}+6.5A_{bs_1}-0.01(A_{P_1}\times D_{bs_1})+9.3(A_{bs_1}\times D_{P_1})$	2.4	94
平均果宽（FW，cm）	$FC=2.9-8A_{P_1}-4.7D_{bs_1}+4.4A_{bs_1}+0.4(A_{P_1}\times A_{bs_1})-0.2(A_{P_1}\times D_{bs_1})+6.2(A_{bs_1}\times D_{P_1})$	1.2	94

注：A_{P_1}＝P_1 位点等位基因替换效应，A_{bs_1}＝bs_1 位点等位基因替换效应；D_{P_1}＝P_1 基因座优势效应（比如：基因座内相互作用），D_{bs_1}＝在 bs_1 基因座内的基因座内相互作用的净效应。

这些结果并不令人意外，因为倍性、果实单性结实和寄主植物对香蕉叶斑病的反应极大影响着果穗和果实性状的表型。该研究显示通过分离倍性和源自三倍体作物如大蕉的杂交后代的遗传标记来解剖 QTV 具有可行性。

7.5 标记辅助育种

20 世纪 90 年代末研究人员首次尝试将 DNA 标记作为间接选择工具用于大蕉和香蕉育种。例如，基于系谱和分子数据，微卫星揭示了杂交种的表现与遗传相似性指数有关联，但并非密切相关（Tenkouano et al.，1998）。这项研究表明，可用的遗传模型并不适用于具有后代间基因组大小多态性的群体。Tenkouano 等（1999a，b）进一步研究表明，基于系谱的亲子关系估计值与分子数据完全不同。根据芭蕉属植物减数分裂的可用模型，三倍体母系种质对其二倍体后代的标记贡献高于预期；相反，母系种质对其四倍体后代的贡献低于预期。对于亲本不同的克隆体，系谱的相似性最低，而对于全同胞而言，系谱的相似性最高。标记相似性和系谱关系没有关联。DNA 标记可对遗传关联性提供更准确的描述，而系谱分析可用于芭蕉属植物育种中选择合适的杂交种。同样的，利用系谱距离和均等的亲本贡献能最好地预测次生三倍体后代的串重。假设亲本贡献不均等，那利用 DNA 标记数据能最准确地预测果实体积。将多倍体和种群结构考虑在内，由亲子关系系数和一套独立的 DNA 标记来确定的遗传模型能用于芭蕉属作物的进一步关联作图研究。

7.6 结论

芭蕉属植物的数量性状，如果穗质量、果穗组成（果梳和果指数）或果实的重量和大小，取决于遗传学、染色体倍性和该作物生长的地点和时间（植株或宿根苗周期、季节、位置、年份和作物耕种）。虽然芭蕉属的地方品种是可用的，但是杂交种和育种材料在产量和其他数量性状上各不相同，它们的遗传因素也不易检测，这是因为涉及大量的基因和彼此之间的相互作用（即上位性），还有环境影响。但是，也有实例表明，具有复杂性状的其他作物，能确定其中各个基因和它们对特定性状的贡献。虽然这样的复杂性状并不如串重本身一样由许多基因控制，其组成成分性状（果实数或平均果重）可能涉及简单的途径，可以更容易地通过测量表型在 DNA 序列水平上描述。通过确定涉及组件特性的基因，可以发现这些基因中控制性能改良的等位基因，然后通过标记辅助育种将这些等位基因引入高级育种材料。在这点上，关联作图应该可以发现不同表型的基因序列上发生的简单变化之间的关联。一些技术能确定每个基因的一个或两个变化，通过完全测序来展开分析，这些基因横跨芭蕉属植物育种项目中可用遗传资源的大的多样性框架。对所有天然存在的等位基因进行精确测序，将能确定改善多基因性状的等位基因的功能。要在这方面取得成功，需要对不同环境下的等位基因的影响进行高质量的表型分析。

缩略语

AFLP：扩增片段长度多态性
CRBP：香蕉和大蕉地区研究中心
CIRAD：法国国际农业研究中心
EMBRAPA：巴西农牧研究院
IIHR：印度园艺研究所
IITA：国际热带农业研究所
NARO：国家农业研究组织
QTV：数量性状变异
RAPD：随机扩增多态性DNA
RFLP：限制性片段长度多态性
RGA：抗性基因类似物
SSR：简单重复序列
UCB：巴西利亚天主教大学
VNTR：数目可变串联重复序列

参考文献

Carreel F，Abadie C，Carlier J，et al，1999. Genome mapping andgenetic analysis of the black leaf streak resistance in bananas [J]. InfoMusa (ProMusa)，8：Ⅲ-Ⅳ.

Craenen K，Ortiz R，1996. Effect of the black sigatoka resistance gene *bs*$_1$ and ploidy level in fruit and bunch traits of plantain-banana hybrids [J]. Euphytica，87：97－101.

Craenen K，Ortiz R，1997. Effect of the bs1 gene in plantain-banana hybrids on response to black sigatoka [J]. Theor Appl Genet，95：497－505.

Crouch H K，Crouch J H，Jarret R L，et al，1998. Segregation at microsatellite loci in haploid and diploid gametes of *Musa* [J]. Crop Sci，38：211－217.

Crouch J H，Crouch H K，Tenkouano A，et al，1999a. VNTR-based diversity analysis of $2x$ and $4x$ full-sib *Musa* hybrids [J]. Elec J Biotechnol，2：99－108.

Crouch J H，Crouch H K，Constand H，et al，1999b. Comparison of PCR-based molecular marker analyses of *Musa* breeding populations [J]. Mol Breed，5：233－244.

Fauré S，Noyer J L，Horry J P，et al，1993. A molecular marker-based linkage map of diploid bananas (*Musa acuminata*) [J]. Theor Appl Genet, 87: 517-526.

Hippolyte I, Bakry F, Seguin M, et al, 2010. A saturated SSR/DArT linkage map of *Musa acuminata* addressing genome rearrangements among bananas [J]. BMC Plant Biol, 10: 65-65.

Noyer J L, Dambier D, Lanaud C, et al, 1997. The saturated map of diploid banana (*Musa acuminata*) [C]// Plant and animal genomes V conference abstracts. 12-16 January San Diego, USA.

Ortiz R, 2000. Understanding the *Musa* genome: an update [J]. Acta Hort, 540: 157-168.

Ortiz R, Vuylsteke D, 1995a. Inheritance of dwarfi sm in AAB plantains [J]. Plant Breeding, 114: 466-468.

Ortiz R, Vuylsteke D, 1995b. Effect of the parthenocarpy gene P_1 and ploidy in bunch and fruit traits of plantain and banana hybrids [J]. Heredity, 75: 460-465.

Ortiz R, Tenkouano A, 2011. Genotype by environment interaction and *Musa* improvement [M]// Pillay M, Tenkouano A. Banana breeding: progress and challenges. Boca Raton, FL, USA: CRC Press: 235-247.

Ortiz R, Craenen K, Vuylsteke D, 1997. Ploidy manipulations and genetic markers as tools for analysis of quantitative trait variation in progeny derived from triploid plantains [J]. Hereditas, 126: 255-259.

Ortiz R, Vuylsteke D, Crouch H K, et al, 1998. TM3x: triploid black sigatoka resistant *Musa* hybrid germplasm [J]. HortScience, 33: 362-365.

Tenkouano A, Crouch J H, Crouch H K, et al, 1998. Genetic diversity, hybrid performance and combining ability for yield in *Musa* germplasm [J]. Euphytica, 102: 281-288.

Tenkouano A, Crouch J H, Crouch H K, et al, 1999a. A comparison of DNA marker and pedigree methods for genetic analysis in plantain and banana (*Musa* spp.) clones. Ⅰ. Estimation of genetic relationships [J]. Theor Appl Genet, 98: 62-68.

Tenkouano A, Crouch J H, Crouch H K, et al, 1999b. A comparison of DNA marker and pedigree methods for genetic analysis in plantain and banana (*Musa* spp.) clones. Ⅱ. Predicting hybrid performance [J]. Theor Appl Genet, 98: 69-75.

Vuylsteke D, Ortiz R, Ferris R S B, et al, 1997. Plantain improvement [J]. Plant Breed Rev, 14: 267-320.

第 8 章　芭蕉属的图位克隆

Andrew James[①,*]　Rodomiro Ortiz[②]　Robert Miller[③]

摘　要：图位克隆（MBC）是一种正向遗传学方法，目的是确定诱导或自然发生的变异产物的特定性状的核苷酸序列。数量性状位点（QTL）是许多重要农艺性状的基础，这些位点由位置来定义，除非能识别相应的突变体，否则 MBC 或关联作图是获得数量性状控制的基因序列的唯一方法。数量性状位点（QTL）受到许多位点和环境因素的限制，因而表型无法提供可靠的基因型信息。然而 MBC 的 QTL 克隆已在模式植物中成功运用。

MBC 最常见的问题是作图数据的不一致。来自芭蕉属植物种间的、跨倍性的杂交种的分离后代由于种群规模小，一直被视作遗传分析的主要障碍。然而，研究人员已在开发芭蕉属植物的遗传标记和数量性状变异上取得一定进展。可用于芭蕉属植物图位克隆的基因组资源包括抗性基因、抗性基因类似物（RGA）及 RGA 的遗传标记。细菌人工染色体（BAC）资源也正被用于芭蕉属植物的 RGA 物理作图。

相比其他经济重要性相等或小的作物，香蕉和大蕉图位克隆的平台（即高分辨率的遗传图谱、作图种群、表达序列标签和全基因组序列）发展仍相对落后。不过，最近一张相对高密度的 *M. acuminata* 的遗传图谱面世了。由于这张遗传图谱包含许多 SSR 标记，因此对于无法采用高昂的高通量技术，如微阵列技术的研究人员来说将十分有用。

关键词：图位克隆，遗传标记，抗性基因类似物，基因组文库

---

① Centro de Investigacion Cientifi ca de Yucatan，Calle 43 No. 130，Col. Churburna de Hidalgo，Merida，C. P. 97200，Yucatan，Mexico；e-mail：andyj007@cicy. mx.

② Department of Plant Breeding and Biotechnology，Swedish University of Agricultural Sciences，Box 101，SE-230 53，Alnarp，Sweden；e-mail：rodomiroortiz@gmail. com.

③ Laboratório de Microbiologia：Interação Planta-Praga，Instituto de Ciências Biológicas，Departamento de Biologia Celular，Universidade de Brasília，Campus Universitário Darcy Ribeiro，Asa Norte，70910-900，Brasília，D. F. ，Brasil；and Universidade Católica de Brasília，SGAN Qd. 916 Módulo B，Av. W5 Norte，CEP：70. 790-160，Brasília/DF，Brazil；e-mail：rmiller@pos. ucb. br or bruna@tba. com. br.

*　通信作者。

8.1 引言

许多具有农业重要性的性状由基因控制，而这些基因的产物、功能和在基因组中的顺序和位置是未知的。为了分离这些基因，采用的方法应该无需事先知道基因产物的信息。其方法之一为图位克隆（MBC），也被称为定位克隆。图位克隆（MBC）是一种正向遗传学方法，目的是确定诱导或自然发生的变异产物的特定性状的核苷酸序列。数量性状位点（QTL）是许多重要农艺性状的基础，这些位点由位置来定义，除非能识别相应的突变体，否则 MBC 或关联作图是获得数量性状控制的基因序列的唯一方法。MBC 对于理解拟南芥等模式物种生态型的表型的遗传基础也十分重要（Jander et al.，2002）。

MBC 的基础是重组，与物理距离直接相关；标记/基因之间的物理距离越大，同源染色体之间发生的重组事件（RE）就越多。在模式生物中，用来鉴定含有性状位点的染色体区域的连锁分析依赖于遵循系谱中性状的遗传原则，旨在寻找共遗传的分离性状和遗传标记。性状位点被推断位于标记位点的附近。对于 MBC 必须满足两个条件：①性状的分离必须基于遗传差异，②有关的基因可以被定位到与分子标记连锁的染色体位置上。植物的 MBC 要产生最高的效率，应该具有以下可用的资源：包括植株兼性自花授粉；各种环境下可用的高质量表型数据；可用的大的 F_2 分离群体或染色体缺陷系；高密度的分子标记遗传图谱，最好是来源于等基因系的（因为这使标记和性状位点之间的连锁不平衡最大化）；增代时间短；由酵母或细菌人工染色体（BAC）载体维持的大插入片段的基因组文库；几个良好表征的表达序列标签（EST）或 cDNA 文库；一张理想的物理图谱（例如，从基于荧光的高信息含量指纹 BAC 文库组装的连续克隆）或完整的基因组序列。

要检测 DNA 多态性的最大数量，两个紧密关联的基因组的序列是必需的。例如 *Oryza sativa* 和 *O. indica* 的两个基因组可通过 80 127 个多态位点（国际水稻基因组测序项目 2005）进行区分，而 *Arabidopsis* 的两种生态型，Col-0 和 Ler 可通过 56 670 个多态位点进行区分（Jander et al.，2002）。

MBC 遵循如下步骤：在为所需表型分离的大的 F_2 作图群体中鉴定感兴趣的表型/突变体（如抗病性）。分辨率主要由作图群体的大小来确定；拟南芥中 10～40 kb 的分辨率需要 1 000 株植物或 2 000 条染色体（Lukowitz et al.，2000）。

① 识别 F_2 作图群体中与目标位点共分离的分子标记。标记应是共显性，以便对植物的两个同源染色体进行基因分型，从而使信息量最大化。

② 鉴定与标记探针杂交的大的插入基因组克隆；其可能在 COSMID、FOSMID、细菌（BAC）、酵母（YAC）或可转化的（TAC 或 BIBAC）人工染色体载体中。如果存在植物转化方案，那么 TAC 或 BIBAC 克隆可以直接用于互补试验。

③ 一旦发现与性状共分离的标记，则通过重叠克隆之间的杂交，从标记启动“染色体步移”；它会持续进行，直到“步移”遇到已知基因反向的标记。该项技术在植物中面临的主要困难是大量的重复序列。如果步移通过这样的区域，步移可能会转向不同的染色体，这样就无法发现那个基因。这种费力、费时的方法现已被“染色体登录”取代了（Tanksley et al.，1995）。这项技术的目的是通过鉴定与靶基因的物理距离小于用于克隆分离的基因组文库的插入片段的平均距离的分子标记来减少“步移”。这可通过使用更高密度的分子标记图谱和/或使用近缘种/家族（比较图）衍生的直系同源探针来实现。然后用 DNA 标记来筛选文库并“登录”在含有该基因的克隆体上。

④ 为了证明候选基因功能的遗传互补（转化），克隆侧翼标记之间的 DNA 片段并导入缺乏该基因的植物中。

⑤ 对候选基因测序，并确定该功能是否已知。

8.2　集团分离分析法

集团分离分析法（BSA）是一种快速作图策略，适用于单基因质量性状，以鉴定与突变遗传关联的标记（Michelmore et al.，1991）。BSA 比较源自单次杂交的 F_2 或 RI 分离群体的个体的两个批量的 DNA 样品。每个库中，个体在目标性状或基因上相同，但在其他基因上是任意的。分析一个性状对比鲜明的两个群来鉴定区分它们的标记。在某一性状上不同的两个分离群体库仅在该性状的基因座上不同。两库之间多态性的标记将与确定用于构建这两库的性状的基因座遗传连锁。因为对 DNA 库而不是个体进行检查，所以相较于为了识别与表型相关标记而进行的遗传图谱构建，PCR 反应或 DArT 等标记阵列的数目更少。随着连锁图谱的随机标记越来越多，作图的效率越来越低；因此 BSA 提供一种方法，集中于感兴趣的区域、标记稀少的区域，或用于定位在最初用于生成遗传图谱的群体中不分离的基因。

8.3　数量性状位点（QTL）的图位克隆

QTL 受数目不详的位点和环境因素的控制，因而个体表型无法提供可靠的基因型信息；但是 MBC 已成功对番茄（*Fw2.2*、*Brix9－2－5*）、拟南芥

（*Frigida*）和水稻（*Hd1*、*Hd6*）进行QTL克隆；经发现全部由单基因控制（综述见 Remington et al.，2001）。它们中的4个基因具有重大影响，而第五个产生的影响很小；因此对这些影响小的基因进行克隆是可能的。使用由 F_1 对亲本的重复回交而产生的近等基因系（NILs），将残留的遗传和环境变化最小化，从而将这些QTL转换成单一的孟德尔因子。随后对 F_2 或 F_3 子代进行精密标度的QTL作图。如果植物自交不亲和，有近交衰退，世代周期长，并且如果没有可用的互补的转化方案，则会出现许多问题。为了克服这些问题，果蝇中使用了染色体缺失系；然而，对于大多数植物来说缺失系还不适用。另一种可能性是利用关联作图，尽管存在种群混合，连锁不平衡（LD）和物理距离不一致，LD衰减率可能太慢或太快的问题。如果基因组被良好描述，例如玉米，则可以在QTL区域中鉴定候选基因，尽管一个QTL中可能含有数百个基因。

8.4 MBC在模式植物中的应用

基于位置克隆的第一个植物基因之一是番茄的抗病基因 *Pto*（Martin et al.，1993），尽管这耗费了数年的努力。同样的，通过染色体步移在拟南芥中克隆第一个基因也耗费了数年时间。然而，随着两个拟南芥全基因组测序的完成及对于单核苷酸多态性（SNP）和插入缺失检测的高通量方法的使用，现已有超过50 000个多态性标记，假定在12个月内可完成5个生命周期，则对任一有突变表型的拟南芥基因克隆只需要不到1人/年（Lukowitz et al.，2000；Jander et al.，2002）。对拟南芥基因功能的了解也有助于鉴定其他植物的同源基因。

在玉米中，作图群体很容易生成，因为雄花和雌花是分离的。对玉米进行MBC的第一步是识别突变所在的染色体臂。有两种方法可用：①使用易位系（需要1个或2个杂交）；②BSA分析法（更高效）。染色体被分成几个“箱子”，突变体被定位到箱子上（10个正常植株DNA和10个突变体DNA，加上每一染色体臂的两个标记），从个体制备DNA，从而将突变映射到限定的区间。如在拟南芥中，映射到单个BAC需要1 000株植物，尽管玉米基因组大小是它的20倍，但它可能具有相似数目的基因，并且重组大多发生在基因内（Bortiri et al.，2006）。2006年的玉米遗传图谱包括3 149个基于基因的插入缺失多态性（IDPs），来自两项作图项目的可用多态性标记数目总共为5 179个。为了确保基因锚定到物理图谱上，用来源于EST的基因特异性寡核苷酸的（“overgo”）探针来搜索BAC重叠群。当overgo探针检测基因家族的成员，或存在不准确的BAC位置时，可能会出现问题；而通过消除在许多重叠群中检测到的阳性 BAC 可以避免这些问题发生。标记的另一个来源是 BAC 末端序列（BES），一个玉米的 BAC 文库已生成了 154 293 个 BES。

由于谷物基因组之间的基因顺序和数目的保守型，利用 Gramene 网站可将已经开发了标记的玉米序列用来 BLAST 水稻序列。一旦识别了同线型的水稻区域，就可以通过 BLAST 或 Gramene 的对比图功能，利用水稻基因来寻找直系同源的玉米基因。玉米遗传图谱的物理连锁使用基于荧光的高信息量指纹识别技术将 BAC 克隆组装到重叠群。两个标记之间的作图间隔缩短了，因而跨越了几个玉米的 BAC（一个或两个 BAC 克隆可能容纳 0.1 cM）和直系同源的水稻序列中的几个基因。然后，搜索水稻序列的可能候选基因以确定候选的玉米类似基因。若发现几个候选基因与某一标记相连锁，则有必要比较它们的 RNA 水平。通过重组、对比、对等位基因测序和转化来验证候选基因。玉米中首次发表的定位克隆论文是着眼于区别类蜀黍和玉米的 QTL 区域，所采用的识别方法是紧密连锁到 BAC 重叠群的分子标记、侧翼标记和水稻基因组的同线区域。然后通过使用 3 106 个个体的重组将 QTL 变窄为 1 042 bp 的基因（Doebly and Stec，1993）。

8.4.1　MBC 可能遇到的问题（例子来自拟南芥）

（1）最常见的问题是作图数据不一致，因此，重要的是使用不止一个的作图群体。表型计分可能出错，因为许多表型不完全表现或受生长条件影响；这些都必须在 F_3 或 F_4 群体中重新检查。

（2）当给定的性状受多于一个基因座影响时，对天然存在的变异的分析会出现问题。对这些基因座的精细定位需要降低作图群体的遗传复杂性；例如，通过创建其中仅一个基因座保持多态性的重组近交系。如，拟南芥中的性状：昼夜节律、开花时间、种子大小、休眠、次生代谢和毛状体密度。

（3）表观遗传突变。这些是不是由于 DNA 序列变化导致的基因表达和功能上的可遗传变化，例如，拟南芥属 SUPERMAN 的花发育基因，可能是由于胞嘧啶甲基化的作用，因此无法定位这些位点。

（4）遗传距离与物理距离的比率因在染色体上的位置不同而不同。例如，*GURKE* 基因显示很少的重组，因为它的位置十分靠近着丝粒，并且存在对重组有很大影响的重复 DNA；1%的重组率对应于 100～400 kb，但靠近着丝粒的 1%重组率对应的是 1 000～2 500 kb。如果基因位于端粒附近，也会出现类似的问题。

（5）存在 DNA 重排。例如，两个种质之间 500 kb 的倒位意味着倒位中所有的重组事件将产生无法作图也无法存活的减数分裂产物。辐照和 T-DNA 插入也可引起难以进行作图的 DNA 重排。

8.5 芭蕉属植物改良的性状、目标基因和遗传知识

8.5.1 芭蕉属植物遗传学的现状

尽管香蕉和大蕉是热带地区的一种主食，但直到20世纪90年代才出现对香蕉和大蕉性状遗传的可用研究成果（概述请见 Ortiz 1995 年发表的论文和其中的参考文献）。尽管芭蕉属作物的繁育体系、倍性种类和无性繁殖使对香蕉和大蕉进行基因分析困难重重，但关于香蕉和大蕉的遗传研究贯穿了整个20世纪90年代，研究主要集中在西非的国际热带农业研究所的高降雨站（IITA，奥纳，尼日利亚）和东非的 IITA 东部和南部非洲区域中心（ESARC，坎帕拉，乌干达）。也出现了其他遗传研究成果（尤其是由分子生物学分析推动的）（Crouch et al.，1998，1999a，b），主要来自法国国际农业研究中心（CIRAD，法国）和巴西农牧研究院（EMBRAPA，巴西）。

来自芭蕉属种间的跨倍性杂交的分离后代的群体规模小，一直被视作遗传分析的主要障碍，这解释了关于高度不育三倍体大蕉和香蕉的遗传研究空白的原因。然而，IITA 的研究成果已经揭示了一些性状的遗传，许多遗传标记也得以面世（表 8-1）。此外，源自三倍体杂合大蕉和二倍体纯种野香蕉（Ortiz et al.，1998）杂交种和二倍体-二倍体的大蕉与香蕉杂交种的分离后代测交的可行性为之后进一步的芭蕉属植物遗传研究铺平了道路（Ortiz and Vuylsteke，1996；Vuylsteke et al.，1997；Ortiz，2000）。通过这样的方式揭露了之前无法通过遗传分析了解到的三倍体芭蕉属植物基因组的相关情况。

表 8-1 形态学、生殖和农艺性状的基因作用类型

性状	基因作用类型	参考文献
白化症	两个具有互补作用的独立的隐性等位基因（a_i）	Ortiz and Vuylsteke，1994a
2*n* 花粉	至少一个显性基因	Ortiz，1997a
侏儒症	一个主要的隐性（*dw*）的短假节间修饰	Ortiz and Vuylsteke，1995a
	影响植株高度	Ortiz，2000
顶端优势和吸芽行为	一个主要的隐性基因（*ad*）	Ortiz and Vuylsteke，1994b
	显性基因（*Ad*）表现不完全显性和可变表达	Ortiz，2000
雄蕾苞片持续性和两性花	显性互补基因的两个独立的2点的系统	Simmonds，1952；Ortiz，1996a

（续）

性　状	基因作用类型	参考文献
果实单性结实	3 个独立互补显性基因（P_i），在大蕉 Calcutta 4 后代中具有至少 1 个等位基因（P_1）分离基因座	Simmonds，1952；Ortiz and Vuylsteke，1995b
果穗方向	由至少 3 个显性位点的上位性效应调节的寡基因系统	Ortiz and Vuylsteke，1998
假茎蜡质	一个隐性基因（wx）	Ortiz et al.，1995b

8.5.2　性状遗传

20 世纪 90 年代以前，二倍体香蕉中的果实单性结实（Simmonds，1952）是最著名的关于重要描述符的遗传研究。20 世纪 90 年代的首批研究成果中发现在二倍体大蕉-香蕉的杂种幼苗中出现了白化病，即任何植物组织中完全缺乏叶绿素（Ortiz and Vuylsteke，1994a），这为重新研究几乎难以涉足的芭蕉属植物基因组提供了一种方法。在先前的芭蕉属植物遗传研究的 40 多年后，这份报告确定白化病是由两个独立的隐性等位基因的遗传控制。这一发现还显示，有害的隐性等位基因存在于法国大蕉及二倍体香蕉的栽培品种和育种群体中。后者表明用于农艺性状的表型轮回选择可以消除这类有害基因。

在大蕉-香蕉后代中的进一步遗传研究导致孟德尔方式遗传的等位基因的发现，不论其是在单基因或多基因上位系统，控制 $2n$ 花粉、侏儒症、顶端优势、雄蕾苞片持续性及两性花、果穗方向和假茎蜡质（表 8－1）。此外，阐明了宿主植物对黑叶斑病、香蕉象鼻虫、香蕉条纹病毒和穴居线虫的抗性基础，并添加到以前关于黄叶斑病、马可病和巴拿马病的抗性的遗传学知识中（表 8－2）。

表 8－2　宿主植物对主要芭蕉属植物病虫害的抗性遗传

病虫害	宿主植物抗性遗传	参考文献
黑叶斑病	一个主要的隐性抗病基因（bs_1）和两个独立的等位基因的加性效应（bsr_2 和 bsr_3）与有利基因剂量效应	Ortiz and Vuylsteke，1994c
	bs_1 位点的座位内相互作用显著调节叶表面症状的出现，而 bs_1 位点的加性效应和座位内相互作用影响病害在宿主植物中的发展	Craenen and Ortiz，1997
黄叶斑病	在 *M. acuminata* spp. *microcarpa* 和 spp. *errans* 中具有剂量效应的多基因	Vakili，1968
	在 *M. acuminata* spp. *malaccensis* 中的显性基因	Rowe，1984
	在 *M. acuminata* spp. *burmannica* 中的隐性基因	Shepherd，1990

（续）

病虫害	宿主植物抗性遗传	参考文献
香蕉条纹病毒病（BSV）	由特异杂交产生的上位效应可能诱导 BSV 的易感性和症状的表达	Ortiz，1996b
	涉及 BSV 出现的遗传机制，和单基因等位基因系统赋予载体对 *M. balbisiana* 亲本的作用	Lheureux et al.，2003
穴居线虫	两个具有加性效应和交互影响的显性基因，即 1 个位点的隐性纯合子基因型抑制其他位点显性等位基因	Dochez et al.，2009
香蕉象鼻虫	对抗性亲本不完全或部分显性的基因	Ortiz et al.，1995a
镰孢枯萎病（或巴拿马病）	1 号小种的 1 个主要显性基因	Vakili，1965a
细菌性枯萎病（或马可病）	几个隐性基因	Vakili，1965b

8.5.3 寄主植物抗性的表型分析

精确的表型分析仍是对数量性状变异，特别是对害虫的部分抗性进行恰当评估的主要因素。田间筛查应在特定病原体或害虫普遍存在的地区。总是对已知的易感性检查进行对比。在这方面，基于生物统计学对寄主植物抗性进行评级的标度是避免错误分组的强有力的工具（表 8－3）。可以通过使用已知的易感宿主的“扩散行”来增强病原体或害虫的均匀感染。由于不均匀变化引起的空间调整，应在对“扩散行”进行方差显著性分析之后进行（表 8－4）。最佳批量大小（或每一批的植物数量）及适当的复制数量可用于大蕉和香蕉杂交种、甜蕉和东非高地蕉中寄主植物抗黑叶斑病的田间筛选（Ortiz and Vuylsteke，1994a；Okoro et al.，1997；Nokoe and Ortiz，1998），也可用于筛选根部健康的大蕉。同样，也有宿主植物抗香蕉象鼻虫（Ortiz et al.，1995a；Kassim et al.，2010）和穴居线虫（Dochez et al.，2009）的筛选方案。

表 8－3 比较对易感性和抗性参考品种的已知宿主反应，将宿主对某一病原体或害虫的反应进行分组

与已知易感宿主比较	与已知抗性宿主比较	杂交种的宿主反应
明显不同	不是明显不同	抗性
不是明显不同	明显不同	易感
明显不同	明显不同	部分抗性
不是明显不同	不是明显不同	不确定

注：根据新复极差法检验。

表 8-4　宿主对病原体反应的方差分析

变异来源	自由度	F 检验中的 P 值
Blocks（B）	b−1	不显著=均匀扩散
Reps/B	（r−1）b	不显著=均匀扩散
剩余*	（p−1）br	
合计	pbr−1	

* 变异来源最小化以获得低的变异系数。b 代表分组，r 代表复制，p 代表植物。

8.5.4 遗传标记和数量性状变异

在大蕉-香蕉杂交种中，主要的隐性基因控制顶端优势、黑叶斑病抗性和侏儒症（表 8-1 和表 8-2）。然而，在它们的分离后代上观察到连续变异。此外，控制果实单性结实的显性基因（P_1），控制黑叶斑病抗性（bs_1）和倍性的隐性基因显著影响在大蕉-香蕉四倍体、三倍体和二倍体杂交种中观察到的数量变异（表 8-5；Vandenhout et al.，1995）。这些结果表明在多基因系统，观察到的大量的数量性状变异可以由在几个位点具有大的表型效应的等位基因来解释。这些结果也证明了黑叶斑病对大蕉-香蕉杂交种产量的影响。例如，四倍体香蕉杂交种的果穗重量似乎与该疾病的发育时间显著相关（Craenen and Ortiz，1998）。

表 8-5　基因对大蕉和香蕉杂交种中寄主植物抗黑叶斑病、株高、串重和果实大小的影响

基　因	表型效应	参考文献
大蕉 Calcutta 4 杂交种 Bobby Tannap 中的 dw	纯合 dw 杂种显示更短的假结节，与正常植株相比更矮小	Ortiz and Vuylsteke，1995a
大蕉 Calcutta 4 杂交种中的 P_1	更高的倍性和改变 P_1 位点的隐性基因为显性基因将提高果穗和果实的重量，尽管这两个因素不能说明这些性状的所有数量性状变异	Ortiz and Vuylsteke，1995b
大蕉 Calcutta 4 杂交种中的 bs_1	部分控制果穗重量和果实大小的数量性状变异的 bs_1 和倍性的组合效应：部分抗性和不易感染的杂种显示比它们更易感的全同胞更高的产量	Craenen and Ortiz，1996
大蕉 Calcutta 4 杂交种中的 P_1 和 bs_1	两个位点占据了大蕉产量数量变异的 80%	Ortiz et al.，1997

上位性可以进一步提高高产大蕉-香蕉杂交种的果实大小。bs_1 和 P_1 位点上的加性×加性效应和加性×座位内相互作用均可增加果实重量。这一发现并不足为奇，因为非加性基因作用控制无性繁殖作物的产量（Peloquin and Ortiz，1992）。通过比较源自法国大蕉和野生香蕉 Calcutta 4 的半同胞杂交后代（Ortiz，1997c），这类基因的作用得到进一步的证实。源自 Obino l'Ewai 的杂交后代的平均果穗重高于源自 Bobby Tannap 的杂交后代，而 Bobby Tannap 本身的果穗重却高于 Obino l'Ewai。这一结果可由控制大蕉果穗重的非加性基因作用（杂合度和上位性）来解释。因此，潜在的果串重无法通过亲本的表现来预测，而应使用具体的配合力鉴定来预测（Tenkouano et al.，1998a）。

杂交种的表现也与基于谱系和分子数据的遗传相似指数有关，但不显著相关（Tenkouano et al.，1998b）。这样的研究表明，假设遗传模型要么是二体生物或多体遗传，那遗传模型未必适合具有代际基因组大小多态性的群体。在这方面，Tenkouano 等（1999b）指出，对后代果串重的最佳预测是基于家谱距离和平等的亲本贡献。当利用 DNA 标记数据，并且对不平等的亲本贡献做出假设时，对果实大小预测是最准确的。亲本杂合性会对所有性状产生大量残留物。Tenkouano 等（1999a）进一步指出，DNA 标记可以提供更精确的基因亲缘描述，但以系谱为基础的分析对于芭蕉属植物育种中预期的亲代组合是有用的。

8.5.5 数量遗传学

方差分量的计算基于基因型、环境和基因型与环境（G×E）相互作用的每个循环绘图均值计算的，其允许估计重复性、广义遗传性和组内相关系数（Ortiz，1995）。表 8-6 提供了对形态学和农艺性状及用于测量宿主植物对黑叶斑病抗性的一些估值。低遗传力值是因为这些性状的遗传变异缺失或显著的 G×E 相互作用。相反地，高性状遗传力表明不显著的 G×E 相互作用或高度的遗传变异。低重复性表示环境的显著影响。组内相关的幅度暗示着亲本来源、倍性或杂交种之间产生的变异占总遗传变异的比例。这种遗传信息使得可以使用不同的育种方案。例如，由于加性遗传变异而导致的在杂交组内关联的具有高遗传力的性状表型轮回选择；以及对由于杂交种之间非加性遗传变异导致的组内低关联的性状进行后代测验的轮回选择；或者是对两种类型的遗传变异都显著的性状的相互轮回选择。

表 8-6 芭蕉属植物杂交种（PB）及其他种质（M）中的广义遗传力（H^2），同一杂交中相同倍性水平杂交种之间的组内联系（r_x）和宿主植物对黑叶斑病的抗性（R）、农艺性状、产量及组成成分

性　状	H^2	r_x	R	参考文献
植株高度（cm）	0.75 M		1.75 M	Ortiz，1997b
		0.59 PB		Ortiz，1997c
植株周长（cm）	0.86 M		3.08 M	Ortiz，1997b

（续）

性　　状	H^2	r_x	R	参考文献
开花时伸展的叶数	0.64 PB	0.78 PB	0.56 PB	Craenen and Ortiz，1997
开花时具有黑叶斑病斑点的最幼嫩叶	0.57 PB	0.69 PB	0.59 PB	Craenen and Ortiz，1997
开花时具有黑叶斑病症的最幼嫩叶	0.57 PB	0.72 PB	0.49 PB	Craenen and Ortiz，1997
开花时被黑叶斑病影响的总叶片区域（%）	0.45 PB	0.58 PB	0.33 PB	Craenen and Ortiz，1997
开花时由黑叶斑病造成的具有坏死斑点的叶指数	0.37 PB	0.47 PB	0.29 PB	Craenen and Ortiz，1997
果穗采收时最高吸芽的高度（cm）	0.82 M		2.20	
		0.57 PB		Ortiz，1997c
果串重量（kg）	0.68 M		1.05 M	Ortiz，1997b
	0.88 PB		2.35 PB	Ortiz and Vuylsteke，1995b
		0.57 PB		Ortiz，1997c
每串果梳数（#）	0.78 M		1.70 M	Ortiz，1997b
	0.75 PB		0.33 PB	Ortiz and Vuylsteke，1995b
		0.43 PB		Ortiz，1997c
每串果实数（#）	0.80 M		1.99 M	Ortiz，1997b
	0.82 PB		0.33 PB	Ortiz and Vuylsteke，1995b
		0.54 PB		Ortiz，1997c
果实填满的时间（d）	0.14 PB		0.01 PB	Ortiz and Vuylsteke，1995b
		0.01 PB		Ortiz，1997c
果实重量（g）	0.94 PB		1.32 PB	Ortiz and Vuylsteke，1995b
		0.54 PB		Ortiz，1997c
果实长度（cm）	0.90～0.93 M		4.87～6.42 M	Ortiz，1997b
	0.94 PB		7.84 PB	Ortiz and Vuylsteke，1995b
		0.65 PB		Ortiz，1997c
果实周长（cm）	0.86～0.90 M		3.48～4.24 M	Ortiz，1997b
	0.94 PB		2.02 PB	Ortiz and Vuylsteke，1995b
		0.67 PB		Ortiz，1997c

8.5.6 育种群体中标记分析推动的遗传知识发展

虽然没有关于芭蕉属植物连锁不平衡的信息，但是从其他研究报告中可以推断出它对香蕉和大蕉的起源或对源自 $2n$ 配子（或有孢子体染色体数的配子）的二者的育种都十分重要。例如，Crouch 等（1998，1999b）利用微卫星分析了四倍体大蕉-香蕉杂交种的遗传构成以表征它们衍生的配子的性质。在分析四倍体杂交群体时，观察到由母系大蕉基因型供给的 75%的等位基因发生分离。这些数据表明，在三倍体大蕉形成 $2n$ 大孢子时发生了重组。

Crouch 等（1999a）对具有亲本基因型的 $2x$ 和 $4x$ 全同胞大蕉-香蕉杂交种进行比较分析，结果表明，四倍体杂交种通常与它们的三倍体大蕉亲本的联系比与二倍体全同胞更加紧密。杂交种与其大蕉亲本的基于可变数目串联重复（VNTR）的遗传相似性和基于 8 个农艺描述符的表型距离指数之间呈显著负相关。然而，杂交种与其二倍体野生香蕉亲本分子遗传相似性和各自的表型距离指数之间没有关联。许多微卫星标记从野生香蕉和大蕉基因型及其后代产生了特别高数量的扩增产物，这表明在 A 和 B 基因组中可能都存在高频率的位点重复，这也为利用 DNA 标记来确定多倍体芭蕉属植物的连锁不平衡程度制造了巨大障碍。

随着有性加倍和 $2n$ 配子的分离，三体遗传可能会影响芭蕉属植物的连锁不平衡分析。在这方面，Ortiz 和 Vuylsteke（1994d）报道了三倍体大蕉-二倍体香蕉杂交后的二倍体杂交种中一些遗传标记的三体分离模式。他们的结果表明，所谓的“A 基因组”的同源染色体之间不存在优先配对，但是在第一次减数分裂后期Ⅰ期间细胞极的染色体配对会发生随机分布，这可能是可育二倍体杂交种比例高的原因，因为 AB 杂交种应该不育。这一发现挑战了关于香蕉和大蕉的祖先 *M. acuminata* 和 *M. balbisiana*，即所谓“B 基因组”供体之间的基因组分化假说。

8.6 可用于芭蕉属植物图位克隆的遗传资源

8.6.1 植物抗性基因

已知植物“基因对基因”小种特异性抗性是通过显性等位基因特异性病原体无毒（Avr）基因产物或配体激活遗传优势的等位基因特异性组成型抗病（R）基因蛋白质产物或受体（Flor，1971）。相反，Avr 蛋白质在没有其对应的 R 蛋白质的情况下时，可以充当毒力因子（Dangl，1994；Chisholm et al.，2006）。病原体识别又称为不相容，激活了大量可诱导的防御机制，从而产生不同的防御反应。许多关键的信号分子和反应在植物中常见，包括在感染部位发生快速超敏细胞死亡（HR）；限制病原体传播，下游产生信号分子，如活性氧（ROS）、一氧化氮（NO）和酚类物质积累，组织与细胞壁加强（Dixon et al.，1994）；维管束阻塞；病程相关（PR）蛋白合成，植物激素信号分子水杨酸（SA）、茉莉酸（JA）和乙烯积累；以及植物抗毒素等抗菌蛋白质形成（Hammond-Kosack and Jones，1996）。ROS 的细胞释放（Lamb and Dixon，1997）可提高超氧阴离子自由基（$\cdot O_2^-$）、羟自由基（·OH）和过氧化氢（H_2O_2）的水平。后一种成分被认为在病原体氧化、植物抗毒素生物合成、防御基因的激活和触发系统中获得抗性（SAR），即触发病原体攻击位点的植物组织来抵抗病原体的袭击中发挥重要作用（Dong，2001；Métraux，2001）。

来自一个或几个抗性基因的特定等位基因很可能决定病原体识别（称为单基因或寡基因的），引发不亲和性。尽管已经假定 R 蛋白质及其相应的 Avr 蛋白质之间的直接相互作用，但是这种一对一的相互作用的发生很少得到证实。在保卫假说中，解释了为什么 Pto 蛋白激酶需要 NBS-LRR 蛋白质 Prf 来激活 Avrpto 识别之后的防御（Van der Biezen and Jones，1998），*R* 基因可能与被称为 guardee 的植物蛋白质相互作用，而这通常是由 Avr 蛋白质靶向的。因此，通过 *R* 基因监测攻击性毒力蛋白质的修饰蛋白质，使得 R 蛋白质修饰的 guardee-Avr 蛋白质的复合物诱导信号转导和防御反应。保卫假说的证据已通过拟南芥 R 蛋白质被观察到（Mackey et al.，2002）。鉴于 *R* 基因可提供对病原体的抗性，在植物的适合度下降方面代价最小，因而这种来源的抗性渗入在常规育种项目中十分常见（Pink，2002）。尽管这种品种特异性抗性可以限制病原体的繁殖，限制接种潜力的增加，并减缓疫情的发展，但抗性通常持续时间短暂，如单一种植中常见的那样。这可能由病原体 *Avr* 基因突变导致，其可能在适合度下降的情况下发生。

已知在每个物种中存在大量的 *R* 基因（Michelmore and Meyers，1998），现已对模式植物和不同作物种类的 40 多个 *R* 基因进行了描述（Martin et al.，2003）。尽管识别出的病原体范围广泛，但是 *R* 基因在氨基酸序列和结构蛋白基序上具有显著的同源性，这表明作为受体系统的蛋白质与蛋白质之间相互作用有相同机制，并且在植物防御中的病原体信号诱导上发挥相同作用（Ellis et al.，2000）。基于蛋白质的保守结构域，研究人员已经确定许多抗性基因的结构类别。

第一个 R 蛋白类包含 4 个亚家族，全部具有胞质核苷酸结合位点（NBS）和富含亮氨酸重复序列（LRR）结构域；第二个 R 蛋白类中，细胞外 LRR 可以被短跨膜结构域（受体样蛋白质）锚定；而在第三类中可以连接到胞质丝氨酸-苏氨酸激酶结构域；第四类只包含丝氨酸-苏氨酸激酶，没有伴随的 LRR；而第五类将 R 蛋白和锚定在膜上的卷曲螺旋结构域归类在一起；第六类包含各种各样的蛋白质。

最普遍的一类 *R* 基因编码具有 NBS 和 LRR 结构域的蛋白质（Meyers et al.，1999）。迄今为止，它们在植物中唯一已知的功能是赋予植物抗病性（Nimchuk et al.，2003），有超过 20 个已得到证实的 *R* 基因源自众多的单子叶植物和双子叶植物。在拟南芥中，编码抗病性蛋白质特征的 207 个基因中（TAGI，2000），149 个具有 NBS-LRR 结构域（Meyers et al.，2003）。这一类的基因已被证明能够抵抗多种病原体和害虫，包括真菌、细菌、病毒、线虫和蚜虫（Dangl and Jones，2001）。在包含大量 LRR 重复的 LRR 结构域的快速进化的 C 末端部分，特异性被编码，至少编码部分。随后的信号被认为涉及 NBS 结构域的 N 末端区域（Belkhadir et al.，2004）。

8.6.2 抗性基因类似物

抗性基因类似物（RGA）已被鉴定横跨多种植物分类群，例如拟南芥（Aarts et al.，1998）、水稻（Leister and Katagiri，2000）和其他（详情请阅 Chelkowski and Koczyk，2003）。对于芭蕉属植物，最近在野生和栽培的 *M. acuminata* 和 *M. balbisiana* 种中鉴定出 non-TIR NBS-LRR RGA（Pei et al.，2007）。Miller 等（2008）对作为抗病供体物种用于香蕉育种项目中的 *M. acuminata* Calcutta 4 中 non-TIR NBS-LRR RGA 多样性进行了大规模的分析。52 个不同的 NBS 编码蛋白质序列重叠群中，33 个包含连续的开放阅读框

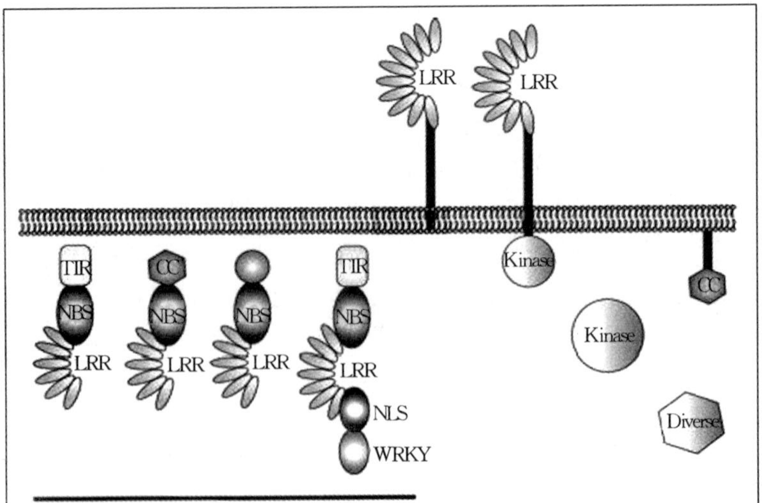

图 8-1　已知 R 蛋白类的表征

（LRR：亮氨酸重复结构；Kinase：激酶；NBS：编码包含核苷酸结合位点）

(ORF)。拟南芥基因组中编码 NBS 型抗性蛋白质的 161 个推定基因中有 30%属于 non-TIR 类，这是相当多的（TAGI，2000）。non-TIR NBS-LRR RGA 的表征也已在 *M. acuminata* ssp. *malaccensis* 中得到描述（Peraza-Echeverria et al.，2008），连同不同的 A 和 B 基因组品种和芭蕉属种 *M. ornata*、*M. schizocarpa*、*M. textilis*、*M. velutina*（Mohamad and Heslop-Harrison，2008）。考虑到水稻中报道过这种基因序列的丰度，对 *M. acuminata* NBS-LRR 序列多样性的详尽分析有可能需要进一步的输入（Goff et al.，2002；Zhou et al.，2004）。还报道了来自其他 *R* 基因家族的 RGA 表征，其中有在 Zebrina GF 分离的类受体蛋白质 Cf 直系同源物（Wiame et al.，2000），以及在 *M. acuminata* cv. Tuu Gia 中的 *Pto* 型受体丝氨酸-苏氨酸激酶（Peraza-Echeverria et al.，2007）。

8.6.3 RGA 遗传标记

不同的植物物种中的 RGA 已被绘制到抗病性的数量性状位点（QTL）的基因组区域上（Irigoyen et al.，2004；Xu et al.，2005；Xiao et al.，2007）。这样的共定位促成了一个假设，这些基因涉及抗性 QTL 的效应（Pflieger et al.，2001）。Miller 及其同事（2008）报道了 30 个 RGA，显示出来自 *M. acuminata* 亲本的限制性酶切 DNA 上出现单个位点或多个位点的多态性，而该亲本材料用于芭蕉属植物正在绘制的参考遗传图谱（表 8－7，图 8－2）（Hippolyte et al.，2008）。水稻上也报道 RGA 标记中的这种多态性（Ramalingam et al.，2003）。鉴定与在抗病性分离群体中的抗性 QTL 共分离的 RGA 标记，可以提供潜在的标记辅助性状选择。现已有计划将芭蕉属植物 RGA 标记放置在巴西农牧研究院将来绘制出的图谱上，该图谱来自在叶斑病抗性上差异巨大的杂交种。RGA 标记在高分辨率遗传图和随后的图位克隆上的应用也将推动对芭蕉属植物 *R* 基因的发现。

一个例子是对 *M. acuminata* spp. *microcarpa* 遗传图谱亲本 Borneo（B）和 Pisang Lilin（PL）的基因组 DNA 限制性酶切，以 RGA MaRGA43 为探针，进行 Southern 杂交，观察到多态性。R 代表 Raoul 标记大小（Appligene，Illkirch，France；引自 Miller et al.，2008）。

表 8－7　*M. acuminata* RFLP-RGA 标记的多态性

（引自 Miller et al.，2008）

RGA 重叠群	Genebank 检索号	在 *M. acuminata* 亲本 Borneo 和 Pisang Lilin 观察到的多态性		
		Dra Ⅰ	*Hind* Ⅲ	*Eco*R Ⅴ
MaRGA03	ER935975	单态的	单态的	多态的（多位点）
MaRGA04	ER935976	多态的（多位点）	多态的（多位点）	多态的（多位点）

（续）

RGA 重叠群	Genebank 检索号	在 *M. acuminata* 亲本 Borneo 和 Pisang Lilin 观察到的多态性		
		Dra Ⅰ	*Hind* Ⅲ	*Eco*R Ⅴ
MaRGA07	ER935979	多态的（多位点）	多态的（多位点）	多态的（多位点）
MaRGA08	ER935980	多态的（多位点）	多态的（多位点）	多态的（多位点）
MaRGA09	ER935981	多态的（多位点）	多态的（多位点）	多态的（多位点）
MaRGA10	ER935982	多态的（多位点）	多态的（多位点）	多态的（多位点）
MaRGA11	ER935983	多态的（多位点）	多态的（多位点）	多态的（多位点）
MaRGA12	ER935984	多态的（多位点）	多态的（多位点）	多态的（多位点）
MaRGA13	ER935985	单态的	多态的（多位点）	单态的
MaRGA14	ER935986	多态的（多位点）	多态的（多位点）	多态的（多位点）
MaRGA15	ER935987	多态的（多位点）	多态的（多位点）	多态的（多位点）
MaRGA16	ER935988	多态的（多位点）	单态的	单态的
MaRGA17	ER935989	多态的（多位点）	多态的（多位点）	多态的（多位点）
MaRGA18	ER935990	单态的	多态的（多位点）	多态的（多位点）
MaRGA19	ER935991	多态的（多位点）	多态的（多位点）	多态的（多位点）
MaRGA20	ER935992	多态的（多位点）	多态的（多位点）	多态的（多位点）
MaRGA22	ER935994	多态的（多位点）	多态的（多位点）	多态的（多位点）
MaRGA23	ER935995	多态的（多位点）	多态的（多位点）	多态的（多位点）
MaRGA24	ER935996	单态的	多态的（多位点）	多态的（多位点）
MaRGA25	ER935997	单态的	多态的（多位点）	多态的（多位点）
MaRGA27	ER935999	多态的（多位点）	多态的（多位点）	多态的（多位点）
MaRGA28	ER936000	多态的（多位点）	多态的（多位点）	多态的（多位点）
MaRGA29	ER936001	多态的（多位点）	多态的（多位点）	多态的（多位点）
MaRGA37	ER936009	多态的（多位点）	多态的（多位点）	多态的（多位点）
MaRGA38	ER936010	单态的	单态的	单态的
MaRGA39	ER936011	单态的	单态的	单态的
MaRGA40	ER936012	多态的（多位点）	多态的（多位点）	多态的（多位点）
MaRGA41	ER936013	多态的（多位点）	单态的	单态的
MaRGA42	ER936014	无杂交	无杂交	无杂交
MaRGA43	ER936015	多态的（多位点）	多态的（多位点）	多态的（多位点）
MaRGA44	ER936016	多态的（多位点）	多态的（多位点）	多态的（多位点）
MaRGA45	ER936017	多态的（多位点）	多态的（多位点）	多态的（多位点）
MaRGA46	ER936018	单态的	多态的（多位点）	单态的

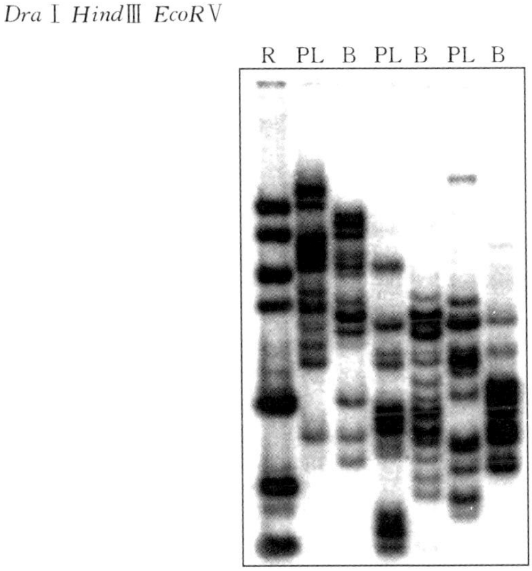

图 8－2　利用一个 RGA 遗传标记在 *M. acuminata* 亲本上观察到的多态性

8.6.4　利用 BAC 文库的 RGA 物理图谱

迄今为止，利用两个不同的 ^{32}P 标记的 RGA 探针，在 *M. acuminata* Calcutta 4（AA）、*M. acuminata* Grande Naine（AAA）和 *M. balbisiana* Pisang Klutuk Wulung（PKW）（BB）中共鉴定出 86 个独特的 RGA 阳性 BAC 克隆（Miller et al.，2008）。鉴于这些大基因家族在多个基因位点间频繁出现，如此大量的点可以被预测为 *R* 基因。高分辨率荧光原位杂交（FISH）也被用于将一个这样的 RGA BAC 克隆（MBP _ 32/N20）物理绘制到粗线染色体位置上去（de Capdeville et al.，2008），因为这一技术既适用于细胞遗传学图谱的开发，也适用于遗传图谱上标记位置的验证。

Miller 及其同事（2008）也经常报道个体 BAC 克隆包含 RGA 的串联重复。这样的 *R* 基因家族和 RGA 簇在植物基因组中是普遍的（Hulbert et al.，2001），它是旁系同源串联重复的结果（Meyers et al.，2003）。有了这样的簇，*R* 基因可能将抗性赋予病原体的不同菌株或不同的分类群（van der Vossen et al.，2000）。聚类网络也可能代表 *R* 基因进化的变异库。由于 Miller 等（2008）描述了 33 个连续的芭蕉属植物 RGA，继续应用探针为鉴定芭蕉属植物中独特抗性位点的推定提供了有效方法。包含 RGA 位点的芭蕉属植物 BAC 克隆的测序和分析正在进行（Baurens et al.，2007），未来将阐明控制这些基因区域进化的网络和机制。这些数据推动的遗传标记的开发将能对物理和遗传图谱之间进行比较。

8.7 源自 cDNA 和 EST 文库的遗传标记

在巴西利亚天主教大学（UCB）、巴西农牧研究院（EMBRAPA）、法国国际农业研究中心（CIRAD）和日本国家农业科学研究所（NIAS）进行的合作研究中，进行了大规模的鉴定以便分析体外接种过 *Mycosphaerella fijiensis* 的有抗性和易感的 *M. acuminata* 栽培品种的叶片的基因表达（Passos et al.，2007）。迄今为止，已对每一文库近 8 000 个随机选择的克隆进行单向的 5′末端测序，而色谱数据存放在芭蕉属植物基因组数据库 Data *Musa*（http://genoma. embrapa. br/musa/index. html/DATA _ musa. html）。利用内部 EST 处理流程，除了发现基因和表达分析之外，数据也可被用来创建分子标记。在至少有跨越超过 10 个碱基的两个重复单元的情况下，可用 consensi 的计算搜索来定位 SSR。由于遗传图谱可以富集基因区域，因此从 EST 中衍生出的这些标记具有优势。迄今为止，已经设计了超过 300 种 SSR 的引物，在很多被验证抗叶斑病上具有显著差异的亲本二倍体 *M. acuminata* 品种上具有多态性。

8.8 遗传作图和创建分离群体

现已生成了几个分离群体来鉴定目标性状的位置，然而，唯一同行评审过的一个作图群体的出版物来自 CIRAD（Fauré et al.，1993）。表 8－8 给出了正在进行的关于分离群体研究的简要总结，而 CIRAD 的工作详情见下文和表 8－9。

表 8－8　芭蕉属种及品种的分离群体

亲　本	研究单位	目　的
M. acuminata burmannicoides Calcutta 4（AA）×*M. acuminata banksii* Madang（AA）	法国国际农业研究中心	对 *M. fijiensis* 的抗性/易感性
M. acuminata truncata（AA）×*M. acuminata* Madang（AA）		
M. acuminata Pahang（AA）×Pisang Jari（cv. AA）		线虫抗性（www. musagenomics. org）
M. a. Pahang（AA）×Yangambi Km5（AAA）		

（续）

亲　　本	研究单位	目　　的
M. acuminata×野生 *M. balbisiana*	印度园艺研究所	使用分子标记作图（Pers. comm. R. Kundapura，2007）
AA×AA BB×BB		
对比品种/野生种质	印度园艺研究所	研究枯萎病和线虫抗性（www.musagenomics.org）
Calcutta 4 和 Pisang Lillin	乌干达国家农业研究组织	单性结实的群体分离（www.musagenomics.org）
SF247（AA）×*M. acuminata microcarpa* Borneo（AA）	国际热带农业研究所	单性结实和黑叶斑病抗性的群体分离（Vroh-Bi，2008）
6142（AA）×8075（AA）		
M. acuminata ssp. *malaccencis* *Foc*-（*Fusarium oxysporum* f. sp. *cubense*）Tropical Race 4 抗（父本）×*Foc*-TR4 易感（母本）	马来亚大学	对 *Foc*-TR4 抗性的群体分离（Kayat et al.，2009）

Fauré 等（1993）对作图群体进行了研究，第一张图谱包括野生二倍体 *M. acuminata* spp. *banksii* 和单性结实基因分离的二倍体 AA 栽培类型（SF265）之间的杂交种。该图谱以一个有 92 个个体的 F_2 群体为基础，使用 58 个限制性片段长度多态性（RFLP）、4 个同工酶和 28 个随机扩增多态性 DNA（RAPD）标记；检测出 90 个位点，其中 77 个在 15 个连锁群上，13 个独立分离。该图谱的总长度为 606 cM。36%的总位点显示偏分离。对香蕉基因组作图的首次尝试强调了易位是芭蕉属植物遗传学的一个主要参与事件。第二张遗传图谱的绘制使用了两种二倍体 AA 品种（M53 和 SFB5）的自交后代，显示了 11 个连锁群，推测对应于香蕉的 11 对染色体（Noyer et al.，1997；Lagoda et al.，1998）。该图以 100 个共显性 RFLP、30 个微卫星和 170 个 AFLP 为基础，而其中的 17 个被认为是共显性。该图谱的总长度为 1 200 cM。所有标记的 36%显示出偏分离（表 8－9）。

CIRAD 研究人员在 *M. acuminata* 中已经鉴定出了 7 组易位，包括具有相同染色体结构的种质（Standard、North Malaysia、Malaysia Mountains、North A、North B、Indonesia 和 East Africa）。对分离的 F_2 群体（AFCAM；表 8－9）和两张之前遗传图谱进行图谱比照，以便进一步研究香蕉中的易位

表 8-9 为了使用分子标记进行遗传作图而创建的芭蕉属植物分离群体

种/品种	群体大小	F_1/F_2	基因座数目	连锁群	总长度（cM）	标记间长度（cM）	LOD值	偏分离（%）	标记	非伴性基因座	表型性状	来源
M. acuminata spp. *banksii* ×AA 栽培种（SF265） 群体名称：SFB5	92	F_2	90	15	606	10	3.0	36	58 个 RFLP；4 个同工酶；28 个 RAPD	13	单性结实	Faure et al.，1993
自交 M53（*M. acuminata malaccencis* × *M. acuminata banksii* 和 *M. acuminata banksii*×AA cv. Paka）	89	F_1	185	18			4.0		120 个 AFLP；40 个 RFLP；24 个 SSR；1 个同工酶	9	束顶病	Noyer et al.，1997
M53 和 SFB5 复合图	181	F_1	373	11	1 227	3	4.75	36	100 个共显性；RFLP；30 个 SSR；170 个 AFLP	9		Lagoda et al.，1998
AAAA IDN 110×*Musa balbisiana* Pisang Klutuk Wulung（BB）	249	F_1	n/a	n/a			3.0	n/a	使用 AFLP 进行集团分离分析	n/a	香蕉条纹病释放因子	Lheureux et al.，2003
M. acuminata AA Calcutta 4（trans. group North）× *M. acuminata* AA Madang（trans. group Standard） 群体名称：AFCAM	244	F_2	244	13	n/a	n/a	n/a	n/a	19 个 RFLP；47 个 SSR；178 个 AFLP	n/a	黑叶斑病抗性/易感性	Vilarinhos et al.，2006
M. acuminata microcarpa Borneo× *M. acuminata* Pisang Lilin 群体名称：BORLI	180	F_1	489	11	1 197	2.8	5.0	n/a	167 个 SSR；322 个 DArT	n/a		Hippolyte et al.，2010

事件（Vilarinhos et al.，2004，2006）。该比较证实连锁群是严格保守的，但易位事件使配子形成和分离变得复杂。此外，对于生成遗传图谱的这些尝试揭示了芭蕉属植物中配子分离的复杂性，即使在二倍体水平上。已报道了关于多倍体供体的研究（Crouch et al.，1998，1999），显示出高水平的偏分离率。用于遗传作图的群体也由于易位富集材料后代的衰弱而减小。

鉴于大多作图方案都面临主要由易位导致的大型群体生产和维持的问题，最近的一个尝试是，利用两种野生二倍体 *M. acuminata* 种质，*M. acuminata* spp. *microcarpa* Borneo 和 *M. acuminata* spp. *malaccensis* Pisang Lilin 之间的杂交体构建香蕉遗传图谱。据报道，该杂交只发生了一次易位事件。该图谱含有饱和的 SSR 标记和 DArT（表 8－9）（Hippolyte et al.，2010）。在现今香蕉基因组测序项目的框架下，CIRAD 通过胚胎拯救开发出一种从野生 *M. acuminata* spp. *malaccencis* Pahang 自交衍生的新的作图群体。其目的是获得含有 SSR 和 DArT 标记的图谱，每兆碱基对包括两个标记。这一作图项目将帮助组装 Pahang 的基因组序列，并将作为微卫星标记核心集的基础，用于未来的香蕉作图项目。

Lheureux 等（2003）已经证明了集团分离分析法在定位香蕉特定表型性状中的应用，他们调查了 AAAA 四倍体和 BB 二倍体种间杂交产生的同源三倍体作图群体（AAB）中香蕉条纹内源性拟逆转录病毒（BSV）的激活（表 8－9）。该分析使用了 5 个患病和 5 个健康个体的 52 个池，使用了 64 个 AFLP 引物组合；然后使用每个条件特定性的标记来测试其余的后代，可以构建 BSV 表达位点（BEL）的遗传图谱。最近的标记与 BEL 的距离为 1.7 cM。尽管该位点未被克隆，但是这项工作展现了 BSA 和遗传作图在芭蕉属植物图位克隆上的潜力。

8.9　大型插入基因组文库的发展

表 8－10 总结了芭蕉属植物基因组学联盟可用的大型插入 BAC 和 BIBAC 基因组文库（www.musagenomics.org）。Calcutta *Hind* Ⅲ 和 *M. acuminata* 文库都可以用作三维库，这减少了在 BAC 中寻找候选基因的成本和时间。野生二倍体 Calcutta 4 已被用于育种计划，主要用于遗传分析和产生抗黑叶斑病的新品种，并被认为可进行全基因组测序。然而，分子分析表明该克隆高度杂合，因此已从 *M. acuminata* spp. *malaccencis* Pahang 生成了一个双单倍体。Pahang 的 BAC 文库现在正在为法国 Genoscope 开展的全基因组测序做出贡献。

表 8-10 芭蕉属植物基因组学联盟可用的大的插入基因组的文库

种	克隆名称	克隆载体	克隆位点	平均插入大小	克隆数目	覆盖基因组	参考文献
M. acuminata	Calcutta 4	pECBAC-1	*Bam*H1	110	17 280	3*x*	INIBAP Annual Report，2003（www. musagenomics. org）
M. acuminata	Calcutta 4	pIndigoBAC-5	*Hind*Ⅲ	100	55 152	9*x*	Vilarinhos et al.，2003
M. acuminata	Grand Naine	pIndigoBAC-5	*Hind*Ⅲ	145	55 296	4. 5*x*	Vilarinhos et al.（www. musagenomics. org）
M. acuminata	Tuu Gia	pCLD04541	*Hind*Ⅲ	100	30 700	5. 1*x*	Ortiz-Vàzquez et al.，2005
M. acuminata	Pahang，双单倍体	pCC1BAC	*Bam*H1	140	23 040	5*x*	www. musagenomics. org
M. acuminata	Pahang，双单倍体	pIndigoBAC-5	*Hind*Ⅲ	110	31 104	5. 7*x*	www. musagenomics. org
M. balbisiana	Pisang Klutuk Wulung	pIndigoBAC-5	*Hind*Ⅲ	135	36 864	9*x*	Safàr et al.，2004

8.10 芭蕉属植物中 MBC 的展望

显而易见的是，相较之其他经济重要性相等或较低的作物，香蕉和大蕉图位克隆的平台（即高分辨率的遗传图谱、作图群体、EST 和全基因组序列）的发展仍相对落后。然而，最近公布了一张 *M. acuminata* 相对高密度的参考遗传图谱（Hippolyte et al.，2010）。由于这张遗传图谱包含了许多 SSR 标记，对于无法采用昂贵的高通量技术（例如微阵列技术）的研究人员来说将十分有用。更重要的是，Genoscope 和 CIRAD 目前正在对双单倍体 *M. acuminata malaccensis* Pahang 的基因组进行测序（由法国国家研究署资助），并将其提供给公众。该序列连同一张更高密度的遗传图谱将开启芭蕉属植物的定位克隆，它也将作为对 A 基因组重排和对 B 基因组测序的模板，这有助于大量单核苷酸多态性的发现，以及促进芭蕉属植物基因组的遗传作图、定位克隆和多样性分析。

缩略语

AFLP：扩增片段长度多态性
BAC：细菌人工染色体
BES：BAC 末端序列
BIBAC：二元细菌人工染色体
BSA：集团分离分析法
cDNA：互补 DNA

CIRAD：法国国际农业研究中心
cM：厘摩
DArT：多样性阵列技术
EMBRAPA：巴西农牧研究院
ESARC：东部和南部非洲区域中心
EST：表达序列标签
FISH：荧光原位杂交
HR：过敏反应
IDP：插入缺失多态性
JA：茉莉酸
LD：连锁不平衡
MBC：图位克隆
NBS-LRR：核苷酸结合区富含亮氨酸重复
NIAS：日本国家农业科学研究所
NIL：近等基因系
NO：一氧化氮
PCR：聚合酶链式反应
PR：病程相关
QTL：数量性状位点
RAPD：随机扩增多态性 DNA
RE：重组事件
RFLPs：限制性片段长度多态性
RGA：抗性基因类似物
ROS：活性氧物种
SA：水杨酸
SAR：获得性抗性
SNP：单核苷酸多态性
TAC：可转化的
UCB：巴西利亚天主教大学
YAC：酵母人工染色体

参考文献

Aarts M G，Lintel T E，Hekkert B，et al，1998. Identification of *R*-gene homologous DNA fragments genetically linked to disease resistance loci in *Arabidopsis thaliana* [J]. Mol

Plant Microbe Interact，11：251－258.

Baurens F C，Miller R N G，Lescot M，et al，2007. Comparative genomics of *Musa* and rice genomes：insight into an RGA cluster［C］. 6th Plant Genomics European Meeting（Plant GEM），October 3－6，Tenerife.

Belkhadir Y，Subramaniam R，Dangl J L，2004. Plant disease resistance protein signalling：NBS-LRR proteins and their partners［J］. Curr Opin Plant Biol，7：391－399.

Blomme G，Ortiz R，1996. Field techniques for root health assessment in plantains［J］. MusAfrica，9：6－7.

Bortiri E，Dave J，Sarah H，2006. Advances in maize genomics：the emergence of positional cloning［J］. Curr Opin Plant Biol，9（2）：164－71.

Bradeen J M，Simon P W，1998. Conversion of an AFLP fragment linked to the carrot Y_2 locus to a simple，codominant，PCR-based marker form［J］. Theor Appl Genet 97：960－967.

Chelkowski J，Koczyk G，2003. Resistance gene analogues of *Arabidopsis thaliana*：recognitionby structure［J］. J Appl Genet，44：311－321.

Cheung F，Town C D，2007. A BAC end view of the *Musa accuminata* genome［J］. BMC Plant Biol，7：29.

Chisholm S T，Coaker G，Day B，et al，2006. Host-microbe interactions：shaping the evolution of the plant immune response［J］. Cell，124：803－814.

Craenen K，Ortiz R，1996. Effect of the black sigatoka resistance gene bs_1 and ploidy level in fruit and bunch traits of plantain-banana hybrids［J］. Euphytica，87：97－101.

Craenen K，Ortiz R，1997. Effect of the bs_1 gene in plantain-banana hybrids on response to black sigatoka［J］. Theor Appl Genet，95：497－505.

Craenen K，Ortiz R，1998. Influence of black sigatoka disease on the growth and yield of diploid and tetraploid hybrid plantains［J］. Crop Protec，17：13－18.

Crouch H K，Crouch J H，Jarret R L，et al，1998. Segregation at microsatellite loci in haploid and diploid gametes of *Musa*［J］. Crop Sci，38：211－217.

Crouch J H，Crouch H K，Constandt H，et al，1999a. Comparison of PCR-based molecular marker analyses of *Musa* breeding populations［J］. Mol Breed，5：233－244.

Crouch J H，Vuylsteke D，Ortiz R，1998b. Perspectives on the application of biotechnology to assist the genetic enhancement of plantain and banana（*Musa* spp.）［J］. Elec J Biotechnol，1（1）：1－12 http://www.ejb.org/content/vol1/issue1/full/2/.

Crouch J H，Crouch H K，Tenkouano A，et al，2003. VNTR-based diversity analysis of $2x$ and $4x$ full-sib *Musa* hybrids［J］. Electronic J Biotechnol，2：130－139.

De Capdeville G，Souza Jr. MY，Szinay D，et al，2008. The potential of high-resolution BAC-FISH in banana breeding［J］. Euphytica，166：431－443.

Dangl J L，1994. The enigmatic avirulence genes of phytopathogenic bacteria［J］. Curr Topics Microbiol Immunol，192：99－118.

Dangl L，Jones J，2001. Plant pathogens and integrated defence responses to infection［J］.

Nature, 411: 826 - 833.

Dixon R A, Harrison M J, Lamb C J, 1994. Early events in the activation of plant defense responses [J]. Annu Rev Phytopathol, 32: 479 - 501.

Dochez C, Tenkouano A, Ortiz R, et al, 2009. Host plant resistance to radopholus similis in a diploid banana hybrid population [J]. Nematology, 6: 215 - 221.

Doebly J, Stec A, 1993. Inheritance of the morphological differences between maize and teosinte: comparison of the results for two F_2 populations [J]. Genetics, 134: 559 - 570.

Dong X, 2001. Genetic dissection of systemic acquired resistance [J]. Curr Opin Plant Biol, 4: 309 - 314.

Ellis J, Dodds P, Pryor T, 2000. Structure, function and evolution of plant disease resistance genes [J]. Curr Opin Plant Biol, 3: 278 - 284.

Faure S, Noyer J L, Horry J P, et al, 1993. A molecularmarker-based linkage map of diploid bananas (*Musa acuminata*) [J]. Theor Appl Genet, 87: 517 - 526.

Flor H H, 1971. Current status of the gene-for-gene concept [J]. Annu Rev Phytopathol, 9: 275 - 276.

Goff S A, Ricke D, Lan T H, et al, 2002. A draft sequence of the rice genome (*Oryza sativa* L. ssp. *japonica*) [J]. Science, 296: 92 - 100.

Hammond-Kosack K E, Jones J D G, 1996. Inducible plant defense mechanisms and resistance gene function [J]. Plant Cell, 8: 1773 - 1791.

Hippolyte I, Seguin M, Bakry F, et al, 2008. Musa genetic mapping [C]. XVI International Plant and Animal genome conference (PAG), January, SanDiego, CA, USA.

Hippolyte I, Bakry F, Seguin M, et al, 2010. A saturated SSR/DArT linkage map of *Musa acuminata* addressing genome rearrangements among bananas [J]. BMC Plant Biol, 10: 65 - 65.

Hulbert S H, Webb C A, Smith C M, et al, 2001. Resistance gene complexes: evolution andutilization [J]. Annu Rev Phytopathol, 39: 285 - 312.

INIBAP Annual Report, 2003. The *Musa* genomics consortium [R]. International Network for Improvement of Banana and Plantain, Montpellier, France.

Irigoyen M L, Loarce Y, Fominaya A, et al, 2004. Isolation and mapping of resistance gene analogs from the *Avena strigosa* genome [J]. Theor Appl Genet, 109: 713 - 724.

Jander G, Norris S R, Rounsley S D, et al, 2002. Arabidopsis map-based cloning in the post-genome era [J]. Plant Physiol, 129: 440 - 450.

Kassim S, Nyine M, Pillay M, 2010. A screening method for banana weevil (*Cosmopolites sordidus* Germar) resistance using reference genotypes [J]. Afr J Biotechnol, 9: 4725 - 4750.

Kayat F, Bonar N, Waugh R, et al, 2009. Development of a genetic linkage map for genes associated with resistance and susceptibility to *Fusarium oxysporum* f. sp. *cubense* from an F_1 hybrid population of *Musa* acuminata ssp. *malaccensis* [J]. Acta Hort, 828: 333 - 340.

Lagoda P J L, Noyer J L, Baurens F C, et al, 1998. Results of a *Musa* mapping project

[Z]//Use of novel DNA fingerprinting techniques for the detection and characterization of genetic variation in vegetatively propagated crops. IAEA Technical Doc, 1047: 35 - 43.

Lamb C, Dixon R A, 1997. The oxidative burst in plant disease resistance [J]. Annu Rev Plant Physiol Plant Mol Biol, 48: 251 - 275.

Leister R T, Katagiri F, 2000. A resistance gene product of the nucleotide binding site—leucine rich repeats class can form a complex with bacterial avirulence proteins in vivo [J]. Plant J, 22: 345 - 354.

Lheureux F, Carreel F, Jenny C, et al, 2003. Identification of genetic markers linked to banana streak disease expression in inter-specific *Musa* hybrids [J]. Theor Appl Genet, 106: 594 - 598.

Lukowitz W, Gillmor C S, Scheible W R, 2000. Positional cloning in Arabidopsis: why it feels good to have a genome initiative working for you [J]. Plant Physiol, 123: 795 - 805.

Mackey D, Holt B F, Wiig A, et al, 2002. RIN4 interacts with Pseudomonas syringae type Ⅲ effector molecules and is required for RPM1-mediated resistance in Arabidopsis [J]. Cell, 108: 743 - 754.

Martin G B, Brommonschenkel S H, Chunwongse J, 1993. Map-based cloning of a protein kinase gene conferring disease resistance in tomato [J]. Science, 262: 1432 - 1436.

Martin G B, Bogdanove A J, Sessa G, 2003. Understanding the functions of plant disease resistance proteins [J]. Annu Rev Plant Biol, 54: 23 - 61.

Meyers B C, Dickerman A W, Michelmore R W, et al, 1999. Plant disease resistance genes encode members of an ancient and diverse protein family within the nucleotide-binding superfamily [J]. Plant J, 20: 317 - 332.

Meyers B C, Kozik A, Griego A, et al, 2003. Genome-wide analysis of NBS-LRR-encoding genes in Arabidopsis [J]. Plant Cell, 15: 809 - 834.

Métraux J P, 2001. Systemic acquired resistance and salicylic acid: current state of knowledge [J]. Eur J Plant Pathol, 107: 13 - 18.

Michelmore R W, Meyers B C, 1998. Clusters of resistance genes in plants evolve by divergent selection and a birth-and-death process [J]. Genome Res, 8: 1113 - 1130.

Michelmore R W, Paran I, Kesseli R V, 1991. Identification of markers linked to disease-resistance genes by bulked segregant analysis: a rapid method to detect markers in specific genomic regions by using segregating populations [J]. Proc Natl Acad Sci USA, 88: 9828 - 9832.

Miller R N G, Bertioli D J, Baurens F C, et al, 2008. Analysis of non-TIR NBS-LRR resistance gene analogs in *Musa acuminata* Colla: isolation, RFLP marker development, and physical mapping [J]. BMC Plant Biol, 8: 15.

Miller R N G, Passos M A N, Menezes N N P, et al, 2010. Characterization of novel microsatellite markers in *Musa acuminata* subsp. *burmannicoides*, var. Calcutta 4 [J]. BMC Res. Notes, 3: 149.

Mohamad A, Heslop-Harrison J S, 2008. Genomes, diversity and resistance gene analogs in

Musa species [J]. Cytogenet. Genome Res，121：59－66.

Nimchuk Z，Eulgem T，Holt B F，et al，2003. Recognition and response in the plant immune system [J]. Annu Rev Genet，37：579－609.

Nokoe S，Ortiz R，1998. Optimum plot sizes for banana trials [J]. HortScience，33：130－132.

Noyer J L，Dambier D，Lanaud C，et al，1997. The saturated map of diploid banana (*Musa acuminata*) [C]//Plant and Animal Genomes V Conference，January 12－16，SanDiego，CA，USA.

Okoro J，Ortiz R，Vuylsteke D，1997. Optimum plot size for black sigatoka evaluation in East African highland bananas [J]. Tropicultura，15：186－189.

Ortiz R，1995. *Musa* genetics [M]//Gowen S. Bananas and plantains. London，UK：Chapman and Hall：84－109.

Ortiz R，1996a. Segregation for persistent neutral (hermaphrodite) flowers and male bracts in plantain-banana hybrids [J]. InfoMusa，5 (2)：19－20.

Ortiz R，1996b. The potential of AMMI analysis for field assessment of *Musa* genotypes tovirus infection [J]. HortScience，31：829－832.

Ortiz R，1997a. Occurrence and inheritance of 2*n* pollen in *Musa* [J]. Ann Bot，79：449－453.

Ortiz R，1997b. Morphological variation in *Musa* germplasm [J]. Genet esour Crop Evol，44：393－404.

Ortiz R，1997c. Genetic and phenotypic correlations in plantain-banana euploid hybrids [J]. Plant Breed，116：487－491.

Ortiz R，2000. Understanding the *Musa* genome：an update [J]. Acta Hort，540：157－168.

Ortiz R，Vuylsteke D，1994a. Inheritance of albinism in banana and plantain (*Musa* spp.) [J]. HortScience，29：903－905.

Ortiz R，Vuylsteke D，1994b. Genetic analysis of apical dominance and improvement of suckering behaviour in plantain [J]. HortScience，119：1050－1053.

Ortiz R，Vuylsteke D，1994c. Inheritance of black sigatoka disease resistance in plantain-banana (*Musa* spp.) hybrids [J]. Theor Appl Genet，89：146－152.

Ortiz R，Vuylsteke D，1994d. Trisomic segregation ratios and genome differentiation in AAB plantains [J]. InfoMusa，3 (1)：21.

Ortiz R，Vuylsteke D，1995a. Inheritance of dwarfi sm in AAB plantains [J]. Plant Breed，114：466－468.

Ortiz R，Vuylsteke D，1995b. Effect of the parthenocarpy gene P_1 and ploidy in bunch and fruit traits of plantain and banana hybrids [J]. Heredity，75：460－465.

Ortiz R，Vuylsteke D，1996. Recent advances in *Musa* genetics，breeding and biotechnology [J]. Plant Breed Abst，66：1355－1363.

Ortiz R，Vuylsteke D，1998. Segregation for bunch orientation in banana and plantain hybrids [J]. Euphytica，101：79－82.

Ortiz R，Vuylsteke D，Dumpe B，et al，1995a. Banana weevil resistance and corm hardness in *Musa* germplasm [J]. Euphytica, 86: 95 - 102.

Ortiz R, Vuylsteke D, Ogburia N M, 1995b. Inheritance of waxiness in the pseudostem of banana and plantain [J]. J Hered, 86: 297 - 299.

Ortiz R, Craenen K, Vuylsteke D, 1997. Ploidy manipulations and genetic markers as tools for analysis of quantitative trait variation in progeny derived from triploid plantains [J]. Hereditas, 126: 255 - 259.

Ortiz R, Vuylsteke D, Crouch J H, 1998. *Musa* genetics, 'Calcutta-4', and scientific ethics: reply to Shepherd's letter [J]. InfoMusa, 7 (2): 31 - 32.

Ortiz-Vàzquez E, Kaemmer D, Zhang H B, et al, 2005. Construction and characterization of a plant transformation-competent BIBAC library of the black Sigatoka-resistant banana Musa acuminata cv. 'Tuu Gia' (AA) [J]. Theor Appl Genet, 110 (4): 706 - 713.

Passos M A N, Emediato F L, Nunes F A C, et al, 2007. Analysis of ESTs in the interaction between *Musa acuminata* and *Mycosphaerella fijiensis* [C]// 53rd Brazilian Genetics Congress. Águas de Lindóia /SP, Sociedade Brasileira de Genética.

Pei X, Li S, Jiang Y, et al, 2007. Isolation, characterization and phylogenetic analysis of the resistance gene analogues (RGAs) in banana (*Musa* spp.) [J]. Plant Sci, 172: 1166 - 1174.

Peloquin S J, Ortiz R, 1992. Techniques for introgressing unadapted germplasm to breeding populations [M]// Stalker H T, Murphy J P. Plant breeding in the 1990s. Wallingford, UK: CAB International: 485 - 507.

Peraza-Echeverria S, James-Kay A, Canto-Canche B, et al, 2007. Structural and phylogenetic analysis of Pto-type disease resistance gene candidates in banana [J]. Mol Genet Genom, 278: 443 - 453.

Peraza-Echeverria S, Dale J L, Harding R M, et al, 2008. Characterization of disease resistance gene candidates of the nucleotide binding site (NBS) type from banana and correlation of a transcriptional polymorphism with resistance to *Fusarium oxysporum* f. sp. *cubense* race 4 [J]. Mol Breed, 22: 565 - 579.

Picq C, Vézina A, 2004. First International Congress on *Musa*: Harnessing Research for Improved Livelihoods [C]. Penang, Malaysia.

Pink D A C, 2002. Strategies using genes for non-durable resistance [J]. Euphytica, 1: 227 - 236.

Pflieger S, Palloix A, Caranta C, et al, 2001. Defense response genes co-localize with quantitative disease resistance loci in pepper [J]. Theor Appl Genet, 103: 920 - 929.

Ramalingam J, Vera-Cruz C M, Kukreja K, et al, 2003. Candidate defense genes from rice, barley and maize and their association with qualitative and quantitative resistance in rice [J]. Mol Plant Microbe Interact, 16: 14 - 24.

Remington D L, Ungerer M C, Purugganan M D, 2001. Map-based cloning of quantitative trait loci: progress and prospects [J]. Genet Res Camb, 78: 213 - 218.

Rowe P, 1984. Breeding bananas and plantains [J]. Plant Breed Rev, 2: 135 - 155.

Safár J, Noa-Carrazana J C, Vrána J, et al, 2004. Creation of a BAC resource to study the structure and evolution of the banana (*Musa balbisiana*) genome [J]. Genome, 47 (6): 1182 - 91.

Shepherd K, 1990. Genetic improvement of bananas in Brazil: aspects related to resistance to the genus Mycosphaerella [C]//Fullerton R A, Stover R H. Sigatoka leaf spot diseases of bananas. Proceedings of International Workshop, San José, Costa Rica, 28 March - 1 April 1989. INIBAP, Montpellier, France: 237 - 242.

Simmonds N W, 1952. Segregations in some diploid bananas [J]. J Genet, 51: 458 - 469.

Tanksley S D, Ganal M W, Martin G B, 1995. Trends in Genetics, 11 (2): 63 - 68.

TAGI: The Arabidopsis Genome Initiative, 2000. Analysis of the genome sequence of the-flowering plant Arabidopsis thaliana [J]. Nature, 6814: 796 - 815.

Tenkouano A, Ortiz R, Vuylsteke D, 1998a. Combining ability for yield and plant phenology in plantain-derived populations [J]. Euphytica, 104: 151 - 158.

Tenkouano A, Crouch J H, Crouch H K, et al, 1998b. Genetic diversity, hybrid performance and combining ability for yield in *Musa* germplasm [J]. Euphytica, 102: 281 - 288.

Tenkouano A, Crouch J H, Crouch H K, et al, 1999a. A comparison of DNA marker and pedigree methods for genetic analysis in plantain and banana (*Musa* spp.) clones. Ⅰ. Estimation of genetic relationships [J]. Theor Appl Genet, 98: 62 - 68.

Tenkouano A, Crouch J H, Crouch H K, et al, 1999b. A comparison of DNA marker and pedigree methods for genetic analysis in plantain and banana (*Musa* spp.) clones. Ⅱ. Predicting hybrid performance [J]. Theor Appl Genet, 98: 69 - 75.

Vakili N G, 1965a. Fusarium wilt resistance in seedlings and mature plants of *Musa* species [J]. Phytopathology, 55: 135 - 140.

Vakili N G, 1965b. Inheritance of resistance in *M. acuminata* in bacterial wilt caused by the tomato race of *Pseudomonas solanacearum* [J]. Phytopathology, 55: 1206 - 1209.

Vakili N G, 1968. Responses of *Musa acuminata* species and edible cultivars to infection by *Mycosphaerella musicola* [J]. Trop Agric (Trinidad), 45: 13 - 22.

Van der Biezen E A, Jones J D G, 1998. Plant disease-resistance proteins and the gene-for-gene concept [J]. Trends Biochem Sci, 12: 454 - 456.

Van der Vossen E A, van der Voort J N, Kanyuka K, et al, 2000. Homologues of a single resistance-gene cluster in potato confer resistance to distinct pathogens: a virus and a nematode [J]. Plant J, 23: 567 - 576.

Vandenhout H, Ortiz R, Vuylsteke D, et al, 1995. Effect of ploidy on stomatal and other quantitative traits in plantain and banana hybrids [J]. Euphytica, 83: 117 - 122.

Vilarinhos A D, Piffanelli P, Lagoda P, et al, 2003. Construction and characterization of a bacterial artificial chromosome library of banana (*Musa acuminata* Colla) [J]. Theor Appl Genet, 106: 1102 - 1106.

Vilarinhos A, Carreel F, Rodier M, et al, 2006. Characterization of translocations in banana by FISH of BAC clonesanchored to a genetic map [C]// Abstracts of the international conference plant and animal genome XIV. January, San Diego, CA, USA.

Vos R, Hogers R, Bleeker M, et al, 1995. AFLP: a new technique for DNA fingerprinting [J]. Nucl AcidsRes, 23 (21): 4407 - 4414.

Vroh-Bi I, 2008. Mapping agronomic traits in bananas and plantains (*Musa* sp.) [C]// Abstract W47, *Musa* genomics. Plant and Animal Genomes Conference, San Diego, California, USA.

Vuylsteke D, Ortiz R, Ferris R S B, et al, 1997. Plantain improvement [J]. Plant Breed Rev, 14: 267 - 320.

Wiame I, Swennen R, Sági L, 2000. PCR-based cloning of candidate disease resistance genes from banana (*Musa acuminata*) [J]. Acta Hort, 521: 51 - 57.

Xiao W, Zhao J, Fan S, et al, 2007. Mapping of genome-wide resistance gene analogs (RGAs) in maize (*Zea mays* L.) [J]. Appl Theor Genet, 115: 501 - 8.

Xu Q, Wen X, Deng X, 2005. Isolation of TIR and nonTIR NBS-LRR resistance gene analogues and identification of molecular markers linked to a powdery mildew resistance locus in chestnut rose (*Rosa roxburghii* Tratt) [J]. Theor Appl Genet, 111: 819 - 830.

Zhou T, Wang T, Chen J Q, et al, 2004. Genome-wide identifi cation of NBS genes in japonica rice reveals significant expansion of divergent non-TIR NBS-LRR genes [J]. Mol Genet Genom, 271: 402 - 415.

第9章　芭蕉属植物功能基因组学和转录组学

Michael Pillay①,*　Sivalingam Elayabalan②　Kaliyaperumal Ashokkumar③
Silvas Jabekumar Prince Kirubarkaran④

摘　要：多细胞生物体的基因组分析揭示了大量功能未知或不可预测的基因。现有的许多遗传工具可用于调查基因功能。通常将它们分为两大类：正向和反向遗传学。本章回顾了进行基因功能分析的一些技术，特别强调插入突变、通过同源重组靶向基因破坏、RNA 干扰、病毒诱导的基因沉默（VIGS）技术和靶向诱导基因组局部突变技术（TILLING）。此外，本章也回顾了在芭蕉属植物转录组学研究的某些领域所取得的进展。插入突变已被广泛用于模式植物拟南芥中的克隆基因、启动子、增强子和其他调控序列。插入突变提供了一个直接确定基因功能的途径。大多数相关的其他方法，并不一定能证明基因序列与功能之间的因果关系。通过同源重组对基因定向，虽然烦琐，但现在被证明在水稻中是可行的。RNA 干扰（RNAi）是基于序列特异性的 RNA 降解，这一降解发生在与靶基因序列同源的双链 RNA（dsRNA）的形成之后。RNAi 可使一个多基因家族的一个、几个或所有成员沉默，或者使多倍体的同源基因拷贝沉默。病毒诱导的基因沉默（VIGS）是一款功能强大且快速的技术，可以用于分析植物生长和发育中的基因功能。为了克服剔除整个基因的局限性和扩大对活性基因突变的认识，定向诱导基因组局部突变技术（TILLING）应运而生。TILLING 使用化学诱变使几乎所有基因产生了一系列传统等位基因的点突变。基因表达系列分析（SAGE）

① Vaal University of Technology, Private Bag X021, Vanderbijlpark 1900 South Africa; e-mail: mpillay@vut. ac. za.

② Crop Tech Ltd., P. O. Box 1367, Arusha, Tanzania; e-mail: balabiotech@gmail. com.

③ Department of Plant Sciences, University of Saskatchewan, 51 Campus Drive, Saskatoon, SK, S7N 5A8, Canada; e-mail: ashok. kumar@usask. ca.

④ Department of Plant Molecular Biology and Biotechnology, Center for Plant Molecular Biology, Tamil Nadu Agricultural University, Coimbatore-641003-India; e-mail: symsil07@gmail. com.

*　通信作者。

是基于短 cDNA 序列标签的大规模序列分析。本章还总结了细菌人工染色体（BAC）技术、表达序列标签（EST）文库和微阵列技术在芭蕉属植物功能基因组学上的作用。最后，本章概述了比较功能基因组学的最新研究，这有助于研究芭蕉属植物的耐干旱胁迫问题。

关键词：基因剔除，插入突变，RNA 干扰，TILLING，转录组学

9.1 引言

研究人员最重要的发现之一就是通过多细胞生物体的基因组分析揭示了大量功能未知或不可预测的基因（Alonso et al.，2003）。功能基因组学是相对较新的技术，可使用高通量手段一次分析大量基因。它研究个体或基因组的功能。功能基因组学使研究者了解到基因在正常生长和发育过程中的控制作用，以及对环境胁迫的反应。研究人员可以进一步研究遗传和细胞网络中出现的复杂的相互作用，在理想状况下，该技术可以对生物体整个遗传构成进行研究。

现在已经开发了许多遗传工具用于调查基因功能。通常将它们分为两大类：正向和反向遗传学。正向遗传学指通过研究来确定可观察到的表型变异的遗传学基础过程（Tierney and Lamour，2005）。在许多情况下，使用 DNA 损伤剂（诱变剂）诱导了可观察到的变异，但也可能是天然存在的。研究人员最终会对被认为涉及的单个基因或一系列基因进行测序。和正向遗传学从表型到序列的顺序不同，反向遗传学刚好相反，基因序列是已知的，但其确切的功能却是不知道的。反向遗传学中，先破坏或修改特异性基因或基因产物，然后测定表型（Tierney and Lamour，2005）。

现已有许多策略可以用来破坏或改变基因，观察其在生物有机体上的表型效应。这些措施有些针对具体特点的基因，有些带来非定向的随机破坏，它们包括：

（1）靶向基因破坏（Oliver，1996；Kempin et al.，1997）。虽然这种方法在拟南芥中已经使用，但它十分费力，要对每一代生成的数百或数千种转基因植物进行基因测定（Kempin et al.，1997）。

（2）通过正义或反义抑制的基因沉默是过去常用的方法（Baulcombe，1999）。但是，这种方法也需要对每个基因生成几个独立的转基因系。

（3）化学、辐射和转座子诱变被视为用于评估基因功能的最通用方法。转座子和 T-DNA［被转移到植物细胞的 Ti（肿瘤诱导）质粒部分］插入提供分

子标记靶基因的额外优势，而且在许多情况下，通过报道基因和选择标记基因被插入遗传携带（Martienssen，1998）。

转录组学是对基因差异表达的高通量分析，也是用来发现新基因或获得某些生物过程在基因组范围上的其他信息的有力工具（Mochida and Shinozaki，2010）。几乎所有基因的全基因组表达分析已成为可能，即使该分析方法在技术要求上相当苛刻并且成本昂贵。例如，基因表达系列分析(SAGE)(Velculescu et al.，1995）和大规模平行测序技术（MPSS)(Brenner et al.，2000)。

本章回顾了一些进行基因功能分析的技术，特别是插入突变、通过同源重组靶向基因破坏、RNA 干扰、病毒诱导的基因沉默（VIGS）技术和定向诱导基因组局部突变技术（TILLING)。此外，本章也回顾了在芭蕉属植物转录组学研究的某些领域所取得的进展。

9.2　插入突变和基因功能分析

基因抑制（缺失/突变剔除）或过度表达允许该基因序列与一个表型有关，从其中起作用进而推断该基因的功能（Matzke and Matzke，1995）。

基因破坏是获得剔除突变的一种强有力的方式，帮助确定来自基因组序列或表达序列标签的大量未表征的开放阅读框（ORF）的生物学功能(Radhamony et al.，2005)。插入突变已被广泛用于拟南芥研究中的克隆基因、启动子、增强子和其他调节序列（Radhamony et al.，2005)。基因剔除系统提供了一个确定功能的直接途径。大多数其他的剖析基因功能的方法是相互关联的，并不一定证明基因序列与功能之间的因果关系。例如，DNA 芯片提供一种方法来发现基因表达在全基因组范围内受到调节的条件和手段（Becker et al.，2003；Birnbaum et al.，2003)。

然而，除了 mRNA 水平以外，遗传因子也可确定在原位基因产物的活性。因此，表达研究不能证明一个因果关系。相反，无效突变对目标基因的可用性使研究人员可以直接监测这一不足对生物体运作能力的影响（Radhamony et al.，2005)。虽然许多方法，如通过同源重组进行的基因置换在酵母和小鼠中表现良好，但这一技术不易用于植物中的靶向插入。不过水稻上已报道过同源重组靶向有效基因（Terada et al.，2002；Yang et al.，2010；Yu et al.，2010)。

原则上，插入突变的使用为克隆突变基因提供了一种更快速的方法。能够随机地插入染色体内，如转座子（Sundaresan et al.，1995；Martienssen，1998）或 *Agrobacterium tumefacians* 的 T-DNA（Azpiroz-Leehan and Feldmann，1997）的DNA因子能用作诱变剂来诱导植物中的功能突变丧失。转座子是可从一个基因组位置迁移到另一个的可动遗传因子（Hayes，2003）。它们是可将自身插入基因组中的新位置而不必与靶基因座有任何序列关系的DNA序列（Lewin，2004）。基于转座子的签名标记诱变已成功地识别必需基因，以及涉及各种病原体的感染性的基因，现已开发出转座子插入突变策略，并在许多动物和植物模型上使用（Hayes，2003）。基于转座子的方法已被用在拟南芥、玉米和其他植物上（Stemple，2004）。

A. tumefaciens 的Ti质粒部分，又称T-DNA，携带可转化植物细胞的基因，而由T-DNA构成的T-DNA插入突变则是利用T-DNA的这一作用。由于插入的因子的序列已知，所以发生插入的基因可通过多种克隆或PCR策略进行再生。插入突变的一个缺点是突变频率低，因此需要对大量的个体进行筛选，以找出任何给定基因中的突变（Gilchrist and Haughn，2005）。在必需基因中插入通常会造成致死现象，而通过在这些必需基因中产生并不太严重的突变，进而了解基因功能（Till et al.，2003）。

使用T-DNA作为插入突变剂相较于使用转座子的优点是前者在插入之后不换位，并且经过几个世代表现出化学和物理上的稳定性（Radhamony et al.，2005）。T-DNA不仅破坏被它插入的基因的表达，还作为标记用于随后鉴定突变。建立突变表型和插入因子之间的连锁十分重要。尽管可以建立一些联系，但这并不排除该表型是由附近一个独立突变导致的可能性。因此，有必要鉴定插入受损的基因并测试转基因植物中野生型等位基因对突变表型的互补作用。

一个获取测序项目鉴定的基因中突变体的高效程序利用的是植物大型集合的可用性优势，这些植物都通过一个插入因子进行诱变处理。

突变方法已被成功地用于植物生物学中许多性状在遗传和分子基础上的研究。利用定位克隆策略产生突变。这种策略在模式物种如拟南芥上得到发展，因为这些物种有密集遗传图谱，上面有许多可见和分子遗传标记，并且这些物种有完整物理图谱，图谱含有重叠克隆DNA片段集合和总基因组序列。这种方法的限制因素是需要时间和精力来创建作图种群和突变位点的精细作图。

除了模式植物拟南芥，插入突变基因的功能分析已经应用于其他植物，如水稻、番茄和芸薹属植物。玉米Ac/Ds的转座子系统已被证明在水稻的基因标记和功能分析上是有效的（Hiroyuki et al.，1999）。T-DNA标记也已成功地用于水稻的基因发现。番茄中，功能分析的新工具基于插入突变，而在微型品种Micro-Tom背景下的Ac/Ds系统已经建立起来（Meissner et al.，2000）。此外，研究人员也已开发出用萤火虫荧光素酶报告基因进行的启动子诱捕和用*GUS*进行的增强子诱捕。用于分离新的启动子的T-DNA标签策略

已经在 *Brassica napus* 上尝试，其方法是采用无启动子 *gus*：*npt* Ⅱ标签构建体进行下胚轴转化。

利用诱变来发现和研究植物基因的方法被越来越多地用在功能基因组学上。因此，T-DNA 标记与转座子插入、TILLING 等其他基于突变的技术相结合不仅将继续提供拟南芥的有用信息，很可能还会成为对其他植物的功能基因组学研究有用的工具。研究人员正迅速开展对香蕉基因组序列的研究（Cheung and Town，2007），并且这些基于突变的技术可以通过这些研究基因的方法对其基因进行功能分析。Santos 等（2009）进行了全基因组的 T-DNA 标记策略来识别和鉴定新的香蕉启动子。该方法在识别香蕉的新启动子和基因上，以及通过体外发育和低温治疗来监测表达模式上有很大价值。

9.3　同源重组靶向基因破坏

同源重组是在两个携带相同遗传位点的染色体之间发生的 DNA 序列的一种相互交换（Lewin，2004）。正如同源重组已被发现它的发起主要伴随着双链断裂一样，同源重组导致的基因靶向与双链断裂修复相关联。双链断裂修复和合成依赖性链退火模型是用来解释基因靶向的最常用模型（Iida and Terada，2004）。

最近开发出的利用同源重组的反向遗传学系统被用于果蝇研究。这一系统很有前景，不过是一个漫长的过程，并且需要产生特定的转基因果蝇（Stemple，2004）。现在，同源重组导致的可再生基因靶向在水稻上也可行。随着位点特异性重组系统的组合（如 Crelox），同源重组导致的基因靶向作为构建水稻和其他植物基因组的例行程序，前途是光明的（Iida and Terada，2004）。

9.4　RNA 干扰导致的基因沉默

与插入突变相反，RNA 干扰（RNAi）是基于序列特异性 RNA 降解，这一降解是在与靶基因序列同源的双链 RNA（dsRNA）形成之后（Marx，2000；Baulcombe，2004）。通过靶向独一的序列或几个基因共同的序列，RNAi 可使一个多基因家族的一个、几个或所有成员沉默，或者使多倍体的同源基因拷贝沉默（Lawrenceand Pikaard，2003；Miki et al.，2005），并且可能对香蕉这一多倍体作物十分有用。

RNA 干扰（RNAi）是靶基因表达被反义和正义 RNA 抑制的过程。它的工作原理基于双链序列识别并降低与它们互补的序列的能力（Lewin，2004）。

RNA干扰最近已成为沉默基因表达和分析它们的功能丧失表型的强大工具，并且当突变等位基因不可用时它仍可以对基因功能进行分析。

RNAi技术已经成为以下环境中基因靶向的有用方法：真菌（Nakayashiki，2005）、病毒（Baulcombe，2004；Wani et al.，2010）、细菌（Escobar et al.，2001）和植物（Brodersen and Voinnet，2006），因为该技术可对将被测试的成百上千基因的功能进行研究（Godge et al.，2008）。它可以沉默贯穿生物体或在特定组织的基因（Islam et al.，2005），为部分沉默或完全关闭的基因提供多功能性，既在培养细胞中，又在整个生物体中工作，并能在生物体生命周期的特定时期有选择地沉默基因（Milhavet et al.，2003）。植物中RNA沉默的一个有趣特点是，一旦它在某个细胞中被触发，移动信号会产生并在整个植株中扩散，导致整个植物沉默（Dunoyer et al.，2007）。

植物中使用RNAi作为反向遗传技术的最大难点在于，将siRNA传递到目标位点的能力对通量产生限制（Henikoff et al.，2004）。一些将双链RNA（dsRNA）或siRNA传递到不同细胞和组织的方法已经过测试，包括：

① 通过*Agrobacterium*介导转化的方法对与选定基因的dsRNA形成载体进行转化（Waterhouse et al.，2001；Chuang and Meyerowtiz，2000）。

② 通过粒子轰击将*uidA* GUS（β-葡萄糖苷酸酶）和TaGLP2a：GFP（绿色荧光蛋白）报道基因的同源dsRNA传递到单个表皮细胞中（Schweizer et al.，2000）。

③ 通过渗透在植物中引入载体（Liu et al.，2002a）。

④ 通过阳离子寡肽聚精氨酸-siRNA复合体将dsRNA传递到悬浮细胞中。

⑤ 用产生dsRNA的病毒载体感染植物（Dalmay et al.，2000）。

⑥ 将siRNA传递到培养的植物细胞中。

⑦ 通过纳米脉冲激光诱导应力波（LISW）产生基因沉默（Tang et al.，2006）。

RNA介导的基因沉默（RNA沉默）已成为真菌中基因靶向的工具。许多植物病原真菌出现了同源性，而该同源性是以由转基因（共抑制）、反义或dsRNA诱导的基因沉默为基础，这些真菌包括叶霉菌（*Cladosporium fulvum*；Hamada and Spanu，1998）、稻瘟病菌（*Magnaporthae oryzae*；Kadotani et al.，2003；Kim et al.，2009；Chen et al.，2010）、苹果黑星菌（*Venturia inaequalis*；Fitzgerald et al.，2004）、粗糙链孢菌（*Neurospora crassa*；Goldoni et al.，2004）、构巢曲霉（*Aspergillus nidulans*；Hammond and Keller，2005）和禾谷镰孢（*Fusarium graminearum*；Nakayashiki，2005），而真菌病原体中是否适合大规模诱变则有待检验。

抗病毒 RNAi 技术已用于在人类细胞系中的病毒疾病管理（Bitko and Barik，2001；Gitlin et al.，2002；Jacque et al.，2002；Novina et al.，2002）。这些沉默机制（RNAi）已被用来保护和管理植物中的病毒感染（Waterhouse et al.，2001；Ullu et al.，2002）。该项技术产生病毒抗性植物的效力呈现在马铃薯（Waterhouse et al.，1998）、黑吉豆（Pooggin et al.，2003）、木薯（Vanitharani et al.，2003）及烟草上（Qu et al.，2007）。

源自线虫 *Caenorhabditis elegans* 传统研究的 RNAi 技术已显示出有方法能探索这一机制保护植物免受线虫危害的可能性（Wani et al.，2010）。现有两种方法值得提倡。一种依赖于涉及感染过程的靶向基因，另一种针对的是线虫内的必需基因。通过向 *C. elegans* 饲喂 dsRNA，可在 *C. elegans* 中诱导生成 RNAi，因此根据推断，含有寄主植物中必需线虫基因序列的表达 hpRNAs 可能将 dsRNA 传递给饲喂的线虫来使线虫瘫痪或死亡（Wani et al.，2010）。

基因的 RNAi 介导抑制在阻碍线虫发育上起着不可或缺的作用，并可能直接或间接地对发病机制的进展产生不利影响。越来越多的证据表明 RNAi 在植物寄生线虫管理上的有效性，并且囊肿和根结线虫物种中大范围的基因被靶向沉默。

正在进行的研究表明了线虫目标寄生基因在生命周期期间的关联，更重要的是表明了通过 RNAi 技术，农作物在未来可以获得对胞囊线虫的可行的抗性。

基因沉默的 RNAi 和 miRNA 技术是新开发的基因组学工具，因为其更高的沉默效率和更短的筛选时间，相比反义和共抑制有巨大优势（Wani et al.，2010）。

未来发展方向将集中在开发那些在操作时间和空间上能得到控制的微调 RNAi 基因沉默载体。然而，对 RNAi 更好和更全面的了解将使研究人员更有效和更高效地工作，在营养品质上改良作物，并处理作物的各种入侵者（Wani et al.，2010）。

9.5 病毒诱导的基因沉默（VIGS）：植物功能基因组学的一个工具

病毒诱导的基因沉默（VIGS）是一款功能强大、快速的技术，用于分析植物生长和发育中的基因功能（Baulcombe，1999）。VIGS 是保护植物免受外来基因侵袭的 RNA 介导的转录后基因沉默机制。除了能更好地了解植物如何抵抗植物病毒，VIGS 也是一个非常强大的功能基因组学工具，用于敲除某些植物中目标基因的基因表达。如果病毒被用作掺入宿主植物序列的载体，诱导的酶途径将使该宿主基因沉默（基因剔除）。现已利用许多方法来启动这一以同源性为基础的 RNA 降解过程，但所有方法都依赖于目标基因对应的双链 RNA（dsRNA）的活性。它涉及创建设计过的病毒，该病毒携带对应于待沉默的该宿主基因序列的序列。双链 RNA 可被引入到植物中，要么作为稳定发夹 RNA 合成的转基因或通过感染携带植物基因片段的重组病毒（Baulcombe，1999；Wang and Waterhouse，2002）。研究人员第一次注意到这一现象是在烟草 *Nicotiana benthamiana* 感染含有编码八氢番茄红素去饱和酶（胡萝卜素合成必需的一种酶）基因的烟草脆裂病毒（TRV）时，该感染导致光合组织中的光漂白表型。虽然许多植物病毒已被用于开发 VIGS 载体（Ruiz et al.，1998；Ratcliff et al.，2001；Gossele et al.，2002；Turnage et al.，2002），但是以 TRV 为基础的载体在广泛植物种类的应用上也具有潜力，因为 TRV 的宿主范围广阔（Ratcliff et al.，2001）。在效率、应用便捷度和没有疾病症状上，TRV 表现最好。

相比现有的功能基因组学方法，VIGS 具有若干优点，因为它不需要遗传转化来沉默靶植物基因，而且可以在接种病毒的几周内完成对表型的鉴定。该方法可以使多种密切相关的基因沉默，以及剔除会导致胚胎致死的基因沉默。它也更适合于高通量研究，并已成功地用在模型和非模型寄主系统中（Liu et al.，2002c；Lu et al.，2003；Burch-Smith et al.，2004，2006；Constantin et al.，2004；Fofana et al.，2004；Ding et al.，2006）。

至今还没有可用于香蕉和大蕉 VIGS 的研究报告。Renner 等（2009）证明，大麦条纹花叶病毒（BSMV）能感染姜科内的 2 个种，而 BSMV-VIGS 可应用于姜（*Zingiber officinale*）中特定的下调八氢番茄红素去饱和酶。这些结果表明未来 BSMV-VIGS 有可能会延伸到其他单子叶植物，如香蕉。

9.6 TILLING 和 EcoTILLING 技术

传统的反向遗传学方法，如使用转座子来“剔除”特定基因，可以精确地确定表型，但需要费时的转基因方法或复杂的组织培养方法（Colbert et al.，2001）。这种“剔除”方法有局限性，因为整个基因被剔除，所以无法观察到活性基因部分功能丧失的影响。

为了克服剔除整个基因的局限性和扩大对活性基因突变的认识，定向诱导基因组局部突变技术，又称 TILLING 技术的一项新方法应运而生（McCallum et al.，2000）。TILLING 技术使用化学诱变使几乎所有基因产生了一系列传统等位基因的点突变。必需基因含有表型分析需要的亚致死等位基因，

TILLING 技术过程对必需基因来说特别有价值。此外，因为 TILLING 技术不涉及转基因修饰，它不仅对于功能基因组学有吸引力，也对农业应用具有吸引力（Henikof et al.，2004）。TILLING 技术被证实在快速生长的生物体上的可用性提升了它在遗传研究上的价值，这些生物体包括果蝇、拟南芥、斑马鱼、玉米、蒺藜苜蓿和其他生物。关于拟南芥中 TILLING 技术的完整说明请参阅 Henikof 等（2004）。

TILLING 技术中 EMS（甲基磺酸乙酯）的高密度诱变使研究人员开始担忧在表型分析中将背景突变误认为目标基因的突变。不过相较于 TILLING 系，EMS 产生的高密度突变，仍然是遗传学的基本学习工具，背景突变没有出现问题。一方面，对于预计会影响基因控制下的表型性状，如植物高度或大小或叶形的基因，这些基因的突变可能会受到上位相互作用，并且与野生型进行远缘杂交可能是必要的。另一方面，对于预计会影响受少数基因控制的某个表型性状，其突变是不可能产生受背景突变干扰的表型，并且远缘杂交不是进行分析的前提条件（Henikoff and Comai，2003）。

EcoTILLING 技术改编自 TILLING 技术（Colbert et al.，2001；Comai et al.，2004），是一种发现和鉴定单核苷酸多态性（SNP）和小插入/缺失的高通量方法。它已被证明是发现和评估人类、植物和其他生物体核苷酸多样性的一种准确、成本低和高通量的方法（Gilchrist et al.，2006；Till et al.，2006a，b；Nieto et al.，2007；Weil，2009）。

Till 等（2008）首次将 TILLING 技术应用到香蕉。现在正在开发由 EMS 诱变的 Grand Naine 种群。目前也正在研究利用伽马射线照射来评估 TILLING 技术下的不同品种的香蕉，包括 Calcutta 4。目前古巴 INIVIT 正在开发 TILLING 技术诱变的二倍体 Calcutta 4 AA 的诱变种群，该项目为联合国粮农组织/国际原子能机构协作研究项目的一部分，该项目也正在对细胞培养的突变进行调查（Jain et al.，2011）。

Till 等（2010）使用 EcoTILLING 技术来识别香蕉基因库的多态性。他们发现在 80 种芭蕉属种质中有超过 800 个新等位基因。此外，研究人员鉴定出两个单核苷酸多态性，该多态性可能危害对向光性推定重要的基因功能。对杂合多态性与单倍型块的评估揭示了芭蕉属种质中的高水平核苷酸多样性。Till 等（2010）利用这个策略，同时发现二倍体种质中的杂合和纯合多态性，以快速评估相同基因组类型的种质中的核苷酸多样性。他们还得出结论，EcoTILLING 适用于芭蕉属植物多样性的研究，它可用于功能性基因组学研究，并作为传统和诱变育种方法中选择种质的工具。将 TILLING 技术用于香蕉面临着许多挑战。由于广泛食用的三倍体品种为高度不育、不育和/或单性结实，需要进行无性繁殖，所以这使种子诱变无法实行。解决方法是考虑使用体外物质诱发突变。胚胎发生的细胞悬浮液现在可用于香蕉（Strosse et al.，2003），虽然该技术似乎局限于特定品种。要减少嵌合性问题可以通过切除分生组织，使小植株再生，然后进行多轮的分生组织分离（Roux，2004）。用于反向遗传筛选的诱变群体的开发和香蕉中 TILLING 与 EcoTILLING 技术的优化将成为芭蕉属植物研究的有用资源。

9.7 香蕉改良的功能基因组学工具

功能基因组学和系统生物学研究正在推动识别参与控制在农学上有价值性状的遗传变异的基因网络（Varshney and Tuberosa，2007）。此外，将功能基因组学研究的新知识与传统育种方法结合对于推动香蕉改良至关重要。目前，一些功能基因组学工具已被应用于香蕉。

本节的内容包括基因表达系列分析（SAGE）、大规模平行测序技术（MPSS）、细菌人工染色体（BAC）、表达序列标签（EST）、微阵列，以及它们在香蕉改良上的可能应用。

9.7.1 基因表达系列分析（SAGE）

基因表达系列分析（SAGE）是基于短 cDNA 序列标签的大规模序列分析。每个标签是从转录物内的限定位置导出。其大小（14 bp）足以识别相应的基因，而观察到的每个标签的次数提供了对其表达水平的精确测量。SAGE 可以一次对数以千计基因的基因表达进行准确的数量分析（Velculescu et al.，1995）。SAGE 也是一个强大工具，用于识别新型候选基因和鉴定特异代谢或调控途径（Coemans et al.，2005）。SAGE 在植物中主要局限于模式生物，如拟南芥和水稻，因为标签的标注主要取决于 cDNA 文库和 EST 集合的可用性（Matsumura et al.，1999，2003a，b；Fizames et al.，2004）。可以运用多种方式增加标记长度，从而提高标注频率。这些改进包括改进的 SAGE（18 bp 的标签；Ryo et al.，2000）、Long SAGE（21 bp 的标签；Saha et al.，2002）和 SuperSAGE（26 bp 的标签；Matsumura et al.，2003a）。在包含有限的或没有可用的 DNA 和 cDNA/EST 序列的香蕉和大蕉中，产生短标签的传统 SAGE 将无法使用，因为标注测序标签的可能性和可靠性都非常低。不过，非模式植物具有的大量重要性状无法用于模式植物的研究，这强调了需要普遍适用于所有作物的高通量转录分析。这些性状可能包括不同的器官（如肉质果）、特殊发展过程（如无融合生殖或单性结实）和独特的质量性状，如风味、营养或药用物质。产生严重经济后果的植物和病原体的某些相互作用也直接证明了在目标生物体中进行大规模功能分析的必要性。

香蕉中，研究人员利用源自野生二倍体香蕉 *M. acuminata* 的叶材料来产生 SuperSAGE 文库（Coemans et al.，2005）。研究人员分析了对应于 5 292 个表达基因的总共 10 196 个 SuperSAGE 标签。虽然有代表性的最丰富的标签出现了 300 多次，为总转录物的近 3%，但是绝大多数转录只发生 1 次。唯一标签数随着标签拷贝数的上升而急剧下降。例如，对应于所有唯一标签 83.3%的 4 409 个标签只发生 1 次，并构成所分析的组织中存在的转录物的 43.2%。相比，只有代表所有唯一标签 2.1%的 214 个标签的标签拷贝数≥5。然而，这 214 个标签代表所分析的转录池的 40.9%（Coemans et al.，2005）。这些发现类似于对水稻叶片的 SuperSAGE 研究，其中对 12 119 个标签的测序鉴定出 7 546 个（62.3%）标签是唯一的（Matsumura et al.，2003b）。此外，作为 3′RACE 引物的 26bp 的 SuperSAGE 标签被成功地应用，从而可以鉴定未知转录物。因此，SuperSAGE 结合了 3′RACE 和 TAIL-PCR 方法，为支撑序列资源不太丰富的非模式生物体的功能基因组学研究提供了强大的工具（Coemans et al.，2005）。

Sagi 等（2005）进一步的研究报道了含有源自香蕉叶的超过 5 000 个唯一 26 bp 长标签的 SuperSAGE 文库，这使得可以对转录组进行定量分析。除了鉴定丰度极其低的转录物，在感染 *Mycosphaerella fijiensis* 这一香蕉中最重要的真菌病原体之后，对推定差异表达基因进行了鉴定。

将两种类型标记扩展至全长基因的工作正在进行，对这些基因的功能进行分析将有助于对香蕉基因组的理解。

9.7.2 大规模平行测序技术（MPSS）

大规模平行测序技术（MPSS）代表了在 5 μm 直径微珠表面上对数以百万计缀合 32 mer 捕获寡核苷酸标记的 cDNA 的高通量测序，其中每个微珠含有一个特定 cDNA 的约 10 万相同拷贝（“微珠库”），这样避免了单独的 cDNA 分离、模板处理和机器人程序（Brenner et al.，2000）。简言之，细胞中每一 mRNA 的丰度可以通过计算具有相同签名的克隆数来估计。然而，众多的步骤使这一技术在进行表达分析上十分烦琐，这些步骤包括捕获寡核苷酸的组合合成，cDNA 合成和限制，捕获寡核苷酸加到每一 cDNA 的 3′末端，利用 PCR handles 克隆，PCR，产物变性，退火序列的连接，荧光激活细胞分选，克隆模板自由端 16～20 个碱基的测序，流动池管理，适配器连接等。除了这些复杂的技术，也有一些不那么复杂的技术可产生足够的信息，以进行初始转录组分析（例如，随机激活基因表达，RAGE；Harrington，2001）。Mochida 和 Shinozaki（2010）描述了 MPSS 的其他变型。

9.7.3 细菌人工染色体（BAC）和表达序列标签（EST）库

EST 序列是由已转换成 cDNA 的随机挑选的基因转录的部分"一次通过"测序所生成（Adams et al.，1993）。由于不论基因组是否具有复杂性，都可获得 cDNA 和 EST 集合，因而这种方法已不仅应用于模型物种，还应用到一些因为多倍性和/或重复序列数而基因组体积大的物种（Mochida and Shinozaki，2010）。EST 和 cDNA 克隆的快速积累已成为功能基因组学的重要资源。

现今已经公布了几千个 EST（通过 mRNA 分离来分析基因序列），这对于检查基因表达、反应和植物的分化，还有检查多样性十分重要（Santos et al.，2005），而另外数万个 EST 将变得可用。比较 EST 库对于识别胁迫条件下的基因表达差异将十分有价值。Santos 等（2005）构建了生长在低温（5℃）和高温（45℃）条件下的香蕉库，发现库中约 30%的基因已被鉴定在其他物种中涉及对环境胁迫的反应，而两个库之间的表达有显著差异。

M. acuminata 和 *M. balbisiana* 已有几个可用的 BAC 文库（Vilarinhos et al.，2003；Safar et al.，2004；Ortiz-Vázquez et al.，2005；Piffanelli et al.，2008）。对几个 BAC 克隆进行测序，结果显示基因组 DNA 中，除了在有大量转座因子而其中基因较少的区域外，每隔 6.4～6.9 kb 出现一个编码基因（Aert et al.，2004；Cheung and Town，2007）。低深度 454 测序应用程序提供的 DNA 序列数据量最大，并使对其核基因组的重复成分的详细分析成为可能（Hribova et al.，2010）。

9.8 香蕉果实成熟转录组学

香蕉果实成熟过程中经历几个重要的物理化学变化（Godoy et al.，2010）。对成熟过程中出现的控制点的认识和了解可以为开发新的采后技术发展、延长储藏期，并且丰富香蕉产品的营养和感官属性提供科学基础（Lajolo，2004）。

许多研究已经涉及香蕉的果实成熟过程，尤其是涉及产生乙烯的基因。对基因表达的分析将使鉴定参与这一过程的重要基因和调控因子成为可能，从而调控采后香蕉的成熟。编码与成熟过程有关的蛋白的 mRNA 转录子分离是达到这一目的的强有力的工具。

对来自 1 和 3 成熟阶段的香蕉果肉的 cDNA 文库进行差异筛选揭露出，成熟期间果肉中部分蛋白的 cDNA 被下调，包括淀粉合酶、颗粒结合淀粉合酶（GBSS）、几丁质酶、凝集素和 2 型金属硫蛋白（Clendennenand May，

1997)。相反，有些蛋白的 cDNA 显示在早期成熟期间丰度增加，包括内切几丁质酶、β-1,3-葡聚糖酶、类甜蛋白、抗坏血酸过氧化物酶、金属硫蛋白和推定衰老相关蛋白（Clendennen and May，1997）。

在另一项研究中，使用 novel 微量滴定板法对香蕉果肉在不同成熟阶段的 cDNA 文库进行差异筛选。所识别的催熟相关的 cDNA 编码参与乙烯生物合成、呼吸作用、淀粉代谢、细胞壁降解和其他几个关键代谢事件的酶（Medina-Suárez et al.，1997）。Liu 等（1999）研究了与香蕉成熟有关的乙烯生物合成的特性。在自然成熟果实呼吸高峰期的开始，乙烯产量大大增加，伴随着 *MA-ACS1* mRNA 迅速增加，然后剧烈下降。成熟开始，体内 ACC 氧化酶活性大大提高，随后急速下降。编码香蕉 ACC 氧化酶的 *MA-ACO1* 基因表达在呼吸高峰期的前阶段是可检测的，当果实开始成熟，表达增加，之后居高不下，尽管 ACC 氧化酶活性迅速减少。这些结果表明，香蕉果实中的乙烯产生受到 *MAACS1* 的转录调控。

在这些研究中大多数分离的 cDNA 在香蕉中大量表达，反映出差异筛选的局限性。因此，为了分离差异表达水平较低的基因，抑制消减杂交（SSH）的更强大的工具应运而生（Diatchenko et al.，1996）。Manrique-Trujillo 等（2007）报道了从一个果肉 cDNA 消减文库获取的独立基因的首个集合，这些基因在香蕉果实成熟后期差异地表达，对比的是 PCI 1 和 PCI 5 成熟期。将来自消减文库的 EST 聚集起来产生了所述的独立基因集。这除了可以消除冗余，还能够提高序列长度和质量。基因的差异表达组参与和果实成熟相关的过程，例如胁迫、解毒、细胞骨架和挥发性化合物的生物合成。Kesari 等（2007）使用 SSH 方法，识别出香蕉中 37 个在成熟过程中差异表达的 EST 独立基因。这些当中约 50%属于如胁迫、防御和解毒过程。此外，研究人员还鉴定了涉及基因表达调节和其他过程的基因。

从卡文迪什香蕉中分离出 4 个编码 EIN3 类蛋白的香蕉 cDNA（Mbeguié-A-Mbeguié et al.，2008）。当香蕉果实成熟期间观察到的生物化学变化发生时，差异表达基因从果肉分离（Clendennen and May，1997；Medina-Suarez et al.，1997），而在成熟过程开始后，从果皮分离（Drury et al.，1999；Liu et al.，2002b）。

Jin 等（2009）在收获后的 10 d 进行了 cDNA 微阵列分析和关于香蕉果实的 RT-PCR 分析，报道发现 16 个 cDNA 上调，6 个下调。鉴定的 cDNA 参与信号转导、氨基酸代谢、脂类代谢、蛋白水解、柠檬酸的生物合成和代谢，以及钾的吸收和运输。这一信息对帮助了解收获后香蕉成熟过程中乙烯生物合成起始的调控提供了新的线索。果实成熟的转录组学似乎是芭蕉属植物最活跃的研究领域之一。

为了分离在采后香蕉成熟早期的差异表达基因，构建了正向 SSH cDNA 文库（Xu et al.，2007）。利用作为“驱动程序”的来自当日收获香蕉果实的 cDNA 和作为“测试者”的采后（DPH）2 d 香蕉果实的 cDNA 来实施 SSH。研究人员对 SSH 文库中共 289 个克隆进行测序。BLASTX 的结果表明，191 个 cDNA 与 NCBI 数据库的已知序列有显著的序列同源性。191 个 cDNA 中 138 个是单元素集合，而 53 个属于含 2～8 序列的不同集群。识别出的 cDNA 编码参与细胞过程的蛋白，如新陈代谢，蛋白质定向和储存，蛋白质合成，信号转导，运输和细胞内运输，细胞构成、生长和分裂，转录和转录后，感病和防御。为了描述 SSH 库中差异表达 cDNA 的特征，要进行微阵列分析。2-DPH 香蕉果实中发现共有 26 个 cDNA 上调，而这些结果利用逆转录酶 PCR（RT-PCR）得到证实。

Godoy 等（2010）分离 12 个差异表达的 cDNA，并得到圆点印迹法和 Northern 印迹杂交证实。识别的序列中包含与植物水通道蛋白、腺嘌呤核苷酸转运蛋白、亲免素、类似于豆球蛋白的蛋白质、脱氧鸟苷激酶和 ω-3 脂肪酸去饱和酶同源的序列。这些 cDNA 中的一些对应于涉及某些变化的新分离基因，这些变化与呼吸跃变或胁迫-防卫反应有关。这些研究中所产生的信息提供了新线索，有助于了解香蕉成熟机制。

9.9 微阵列技术

作为并行分析大量核酸片段的灵活方法，微阵列分析是在过去的几年里涌现出的新兴技术。它的起源可以追溯到几个不同的学科和技术。微阵列可被看作是分子生物学杂交方法持续发展的产物，如荧光显微镜在细胞生物学中的应用的延伸，以及利用捕获到固体表面诊断测定作为减少所需分析量的方式。随着近年在全基因组和/或大规模的 cDNA 克隆体中被测序的物种数的迅速增加，一些 DNA 微阵列也已经开发用于各种植物物种的转录组分析。在这些领域中使用的思想和原则的趋同，以及在固体支持物上制备核酸小型收集技术的进步，都促成了微阵列和微芯片技术的出现（www. crownscientific. com）。系统级的方法，如微阵列 RNA 分析都非常适合研究生物和非生物胁迫下植物的基因复合体重叠反应。平铺阵列是在特定生物体中跨越整个基因组的高密度寡核苷酸探针，是分析贯穿一个全基因组的表示区域的平台；它是发现新基因并阐明其结构的有效方法。非生物胁迫条件下利用全基因组平铺阵列对拟南芥转录组进行分析，发现了一些被非生物胁迫诱导的反义转录物（Matsui et al.，2008）。类似的分析应该可以在香蕉上进行。

9.10　利用比较功能基因组学来研究芭蕉属植物中的干旱胁迫

芭蕉属植物主要栽种于降水量不稳定，可能存在旱季的热带和亚热带地区。迄今为止，关于芭蕉属植物干旱胁迫耐受性的比较研究还很少，但实地观察表明，相比 A 基因组，B 基因组赋予其更大的耐旱性（Thomas et al.，1998）。香蕉和大蕉显示了在生物和非生物环境下有广泛的重叠反应，而适应不利环境条件（包括干旱）的过程中采用多样的生理、生化和分子策略对于研究分子水平并找到负责胁迫的分子和表达分析带来了挑战（Dunoyer et al.，2007）。然而来自密切相关的，异源物种的市售的高密度寡核苷酸微阵列可以用来探测非模式植物的转录组。

Davey 等（2009）评估了使用市售的、高密度的 Affymetrix 水稻和拟南芥 ATH-1 Gene-Chip® 微阵列来分析香蕉中复杂的植物反应。他们能够快速分析芭蕉属植物转录组对长期干旱胁迫的反应，并确定了一系列结构和调控上的芭蕉属植物基因同系物，而之前发现这些同系物与其他（模式）植物的水分不足反应相关。此外，尽管芭蕉属植物和水稻或拟南芥探针之间有巨大的系统发育差异，公众可用的转录组分析试验比较确定了一系列共同的干旱响应基因，支持了指定的芭蕉属植物基因识别和描述。该方法概述了此策略在描述香蕉和大蕉品种胁迫抗性的特性上具有很大潜力，因为现在可用的序列信息相对较少。将芭蕉属植物的 gDNA 交叉杂交到水稻 Gene-Chip® 基因组阵列之后，约 33 700 个基因特异性探针具有足够高水平的同源性，可用于转录组学分析。在概念验证中，代表芭蕉属栽培品种 Cachaco 的对照和受干旱胁迫叶片的单一生物的 RNA 集被杂交到 Affymetrix 水稻基因组阵列。随后识别出共有 2 910 个在表达水平上有超过两倍差异性的芭蕉基因同源物。这些干旱响应的转录物包括与植物生物和非生物胁迫应答相关的许多功能类别，以及一系列已知参与协调非生物胁迫应答的调节基因。后一组包括 ERF、DREB、MYB、bZIP 和 bHLH 转录因子家族的成员。对干旱敏感的芭蕉属植物转录物种中的 52 个物种与水稻中在抗干旱和寒冷的数量性状位点（QTL）下的潜在基因同源，包括与一个潜在基因相关的两个实例的 QTL。干旱响应转录物名单也收录了公开可用的对比转录组试验中识别的基因（Davey et al.，2009）。

9.11　结论

香蕉是功能基因组学的良好候选者，因为它具有相对小的基因组大小

（600 Mb）和重要的特征，如呼吸高峰果实、单性结实或无性繁殖，而这在模式生物中不存在（Sagi et al.，2005）。如今，遗传学和基因组学的进步已经加强了对植物基因组的结构和功能方面的理解，也加强了对农作物改良方法的基本认识。将功能基因组学研究的新知识与传统育种方法结合对于推动作物改良至关重要。当前的基因组技术可以提供两种方式来改良香蕉和其他作物。首先，更好地理解生物学机制可使改进的筛选方法更有效地选择优良基因型。其次，新知识可以改善决策过程，有助于制定更有效的育种策略（Varsheny et al.，2005）。

缩略语

3′RACE：3′cDNA 末端快速扩增
Ac/Ds：活化子-解离子
BAC：细菌人工染色体
BSMV：大麦条纹花叶病毒
cDNA：互补 DNA
dsRNA：双链 RNA
EMS：甲基磺酸乙酯
EST：表达序列标签
GBSS：颗粒结合淀粉合酶
GUS：β-葡萄糖苷酸酶
MPSS：大规模平行测序技术
PCR：聚合酶链式反应
RNAi：RNA 干扰
RT-PCR：逆转录 PCR
SAGE：基因表达系列分析
SSH：抑制消减杂交
TAIL-PCR：热不对称交错 PCR
T-DNA：肿瘤诱导 DNA
TILLING：定向诱导基因组局部突变技术
TRV：烟草脆裂病毒
VIGS：病毒诱导的基因沉默

参考文献

Adams M，Soares M，Kerlavage A，et al，1993. Rapid cDNA sequencing（expressed sequence tags) from a directionally cloned human infant brain cDNA library [J]. Nat Genet, 4: 373 - 380.

Aert R, Sagi L, Volckaert G, 2004. Gene content and density in banana (*Musa acuminata*) as revealed by genomic sequencing of BAC clones [J]. Theor Appl Genet, 109: 129 - 139.

Alonso J M, Stepanova A N, Leisse T J, et al, 2003. Genome-wide insertional mutagenesis of *Arabidopsis thaliana* [J]. Science 301: 653 - 657.

Azpiroz-Leehan R, Feldmann K A, 1997. T-DNA insertion mutagenesis in Arabidopsis: going back and forth [J]. Trends Genet 13: 152 - 156.

Baulcombe D C, 1999. Fast forward genetics based on virus induced gene silencing [J]. Curr Opin Plant Biol, 2: 109 - 113.

Baulcombe D C, 2004. RNA silencing in Plants [J]. Nature, 431: 356 - 363.

Becker J D, Boavida L C, Carneiro J, et al, 2003. Transcriptional profiling of *Arabidopsis* tissues reveals the unique characteristics of the pollen transcriptome [J]. Plant Physiol, 133: 713 - 725.

Bitko V, Barik S, 2001. Phenotypic silencing of cytoplasmic genes with sequence specific double stranded short interfering RNA and its applications in the reverse genetics of wild type negative strand RNA virus [J]. BMC Microbiol, 1: 34 - 44.

Birnbaum K, Shasha D E, Wang J Y, et al, 2003. A gene expression map of the *Arabidopsis* root [J]. Science, 302: 1956 - 1960.

Brenner S, Johnson M, Bridgham J, et al, 2000. Gene expression analysis by massively parallel signature sequencing (MPSS) on microbead arrays [J]. Nat Biotechnol, 18: 630 - 634.

Brodersen P, Voinnet O, 2006. The Diversity of RNA silencing pathways in plants [J]. Trends Genet, 22: 268 - 280.

Burch-Smith T M, Anderson J C, Martin G B, et al, 2004. Applications and advantages of virus-induced gene silencing for gene function studies in plants [J]. Plant J, 39: 734 - 746.

Burch-Smith T M, Schiff M, Liu Y, et al, 2006. Efficient virus-induced gene silencing in Arabidopsis [J]. Plant Physiol, 142: 21 - 27.

Chen L, Shiotani K, Togashi T, et al, 2010. Analysis of the Rac/Rop small GTPase family in rice: expression, subcellular localization and role in disease resistance [J]. Plant Cell Physiol, 51: 585 - 595.

Cheung F, Town C D, 2007. A BAC end view of the *Musa acuminata* genome [J]. BMC Plant Biol, 7: 29.

Chuang C F, Meyerowtiz E M, 2000. Specific and heritable genetic interference by double-stranded RNA in *Arabidopsis thaliana* [J]. Proc Nat Acad Sci USA, 97: 4985 - 4990.

Clendennen S K, May G D, 1997. Differential gene expression in ripening banana fruit [J]. Plant Physiol, 115: 463 - 469.

Coemans B, Matsumura H, Terauchi R, et al, 2005. Supersage combined with PCR walking allows global gene expression profiling of banana (*Musa acuminata*), a non-model organism [J]. Theor Appl Genet, 111: 1118 - 1126.

Colbert T, Till B J, Tompa R, et al, 2001. High-throughput screening for induced point mutations [J]. Plant Physiol, 126 (2): 480 - 484.

Comai L, Young K, Till B J, et al, 2004. Effi cient discovery of DNA polymorphisms in natural populations by EcoTILLING [J]. Plant J, 37: 778 - 786.

Constantin G D, Krath B N, MacFarlane S A, et al, 2004. Virusinduced gene silencing as a tool for functional genomics in a legume species [J]. Plant J, 40: 622 - 631.

Dalmay T, Hamilton A J, Mueller E, et al, 2000. *Potato virus* X amplicons in *Arabidopsis* mediate genetic and epigenetic gene silencing [J]. The Plant Cell, 12: 369 - 380.

Davey M W, Graham N S, Vanholme B, et al, 2009. Heterologous oligonucleotide microarrays for transcriptomics in a non-model species; a proof-of-concept study of drought stress in *Musa* [J]. BMC Genom, 10: 436.

Diatchenko L, Lau Y F C, Campbell A P, et al, 1996. Suppression subtractive hybridization: a method for generating differentially regulated or tissue-specific cDNA probes and libraries [J]. Proc Natl Acad Sci USA, 93: 6025 - 6030.

Ding X S, Schneider W L, Chaluvadi S R, et al, 2006. Characterization of a *Brome mosaic virus* strain and its use as a vector for gene silencing in monocotyledonous hosts [J]. Mol Plant Microbe Interact, 19: 1229 - 1239.

Drury R, Hortensteiner S, Donnison I, et al, 1999. Chlorophyll catabolism and gene expression in the peel of ripening banana fruit [J]. Physiol Plant, 107: 32 - 38.

Dunoyer P, Himber C, Ruiz-Ferrer V, et al, 2007. Intra-and intercellular RNA interference in *Arabidopsis thaliana* requires components of the microRNA and heterochromatic silencing pathways [J]. Nat Genet, 39: 848 - 856.

Escobar M A, Civerolo E L, Summerfelt K R, et al, 2001. RNAi-mediated oncogene silencing confers resistance to crown gall tumorigenesis [J]. Proc Nat Acad Sci USA, 98: 13437 - 13442.

Fitzgerald A, Van Kha J A, Plummer K M, 2004. Simultaneous silencing of multiple genes inthe apple scab fungus *Venturia inaequalis*, by expression of RNA with chimeric inverted repeats [J]. Fungal Genet Biol, 41: 963 - 971.

Fizames C, Munos S, Cazettes C, et al, 2004. The *Arabidopsis* root transcriptome by serial analysis of gene expression: gene identification using the genome sequence [J]. Plant Physiol, 134: 67 - 80.

Fofana I B, Sangare A, Collier R, et al, 2004. A geminivirus-induced gene silencing system for gene function validation in cassava [J]. Plant Mol Biol, 56: 613 - 624.

Gilchrist E, Haugh E, 2005. TILLING without a plough: a new method with applications forreverse genetics [J]. Curr Opin Plant Biol, 8: 211 - 215.

Gilchrist E J, Haughn G W, Ying C C, et al, 2006. Use of EcoTILLING as an efficient SNP discovery tool to survey genetic variation in wild populations of *Populus trichocarpa* [J]. Mol Ecol, 15: 1367 - 1378.

Gitlin L, Karelsky S, Andino R, 2002. Short interference confers intracellular antiviral immunity in human cells [J]. Nature, 4: 418-430.

Godge M R, Purkayastha A, Dasgupta I, et al, 2008. Virus-induced gene silencing for functional analysis of selected genes [J]. Plant Cell Rep, 27: 209-219.

Godoy A D, Cordenunsi B R, Lajolo F M, et al, 2010. Differential display and suppression subtractive hybridization analysis of the pulp of ripening banana [J]. Sci Hort, 124: 51-56.

Goldoni M, Azzalin G, Macino G, et al, 2004. Efficient gene silencing by expression of double stranded RNA in *Neurospora crassa* [J]. Fungal Genet Biol, 41: 1016-1024.

Gossele V V, Fache I I, Meulewaeter F, et al, 2002. SVISS—a novel transient gene silencing system for gene function discovery and validation in tobacco [J]. Plant J, 32: 859-866.

Hamada W, Spanu P D, 1998. Co-suppression of the hydrophobin gene *Hcf-1* is correlated with antisense RNA biosynthesis in *Cladosporium fulvum* [J]. Mol Gen Genet, 259: 630-638.

Hammond T M, Keller N P, 2005. RNA silencing in *Aspergillus nidulans* is independent of RNA-dependent RNA polymerase [J]. Genetics, 169: 607-617.

Harrington J J, 2001. Creation of genome-wide protein expression libraries using random activation of gene expression [J]. Nat Biotechnol, 19: 440-445.

Hayes F, 2003. Transposon-based strategies for microbial functional genomics and proteomics [J]. Annu Rev Genet, 37: 3-29.

Henikoff S, Comai L, 2003. Single-nucleotide mutations for plant functional genomics [J]. AnnRev Plant Biol, 54: 375-401.

Henikoff S, Till J B, Comai L, 2004. TILLING: traditional mutagenesis meets functional genomics [J]. Plant Physiol, 135: 630-636.

Hiroyuki E, Izawa T, Kawahara M, et al, 1999. Ac as a tool for the functional genomics of rice [J]. Plant J, 19: 605-613.

Hřibová E, Neumann P, Matsumoto T, et al, 2010. Repetitive part of the banana (*Musa acuminata*) genome investigated by low-depth 454 sequencing [J]. BMC Plant Biol, 10: 204 doi: 10.1186/1471-2229-10-204.

Iida S, Terada R, 2004. A tale of two integrations, transgene and T-DNA: gene targeting by homologous recombination in rice [J]. Curr Opin Biotechnol, 15: 132-138.

Islam S M S, Miyazaki T, Tanno F, et al, 2005. Dissection of gene function by RNA silencing [J]. Plant Biotechnol, 22: 443-446.

Jacque J M, Triques K, Stevenson M, 2002. Modulation of HIV-1 replication by RNA interference [J]. Nature, 418: 435-438.

Jain S M, Till B, Suprasanna P, et al, 2011. Mutations and cultivar development of banana [M] // Pillay M, Tenkouano A. Banana breeding: progress and challenges. Boca Raton, FL, USA: CRC Press: 203-217.

Jin Z Q, Xu B Y, Liu J H, et al, 2009. Identification of genes differentially expressed at the onset of the ethylene climacteric in banana [J]. Postharvest Biol Technol, 52: 307-309.

Kadotani N, Nakayashiki H, Tosa Y, et al, 2003. RNA Silencing in the Pathogenic Fungus *Magnaporthe oryzae* [J]. Mol Plant-Microbe Interact, 16: 769 - 776.

Kempin S A, Liljegren S J, Block L M, et al, 1997. Inactivation of the *Arabidopsis* AGL5 MADS-box gene by homologous recombination [J]. Nature, 389: 802 - 803.

Kesaria R, Trivedia P K, Nath P, 2007. Ethylene-induced ripening in banana evokes expressionof defense and stress related genes in fruit tissue [J]. Postharvest Biol Technol, 46: 136 - 143.

Kim J A, Cho K, Singh R, et al, 2009. Rice OsACDR1 (*Oryza sativa* accelerated cell death and resistance 1) is a potential positive regulator of fungal disease resistance [J]. Mol Cells, 30: 431 - 439.

Lajolo F M, 2004. Carbohydrate metabolism during fruit ripening: a functional genomics approach [Z]. www. pesquisaapoiada. fapesp. br/tematico accessed 5 March 2011.

Lawrence R, Pikaard S C, 2003. Transgene-induced RNA interference: a strategy for overcoming gene redundancy in polyploids to generate loss-of-function mutations [J]. Plant J, 36: 114 - 121.

Lewin B, 2004. Genes Ⅷ [M]. USA: Upper Saddle River, Pearson Prentice Hall NJ.

Liu X, Shiomi S, Nakatsuka A, et al, 1999. Characterization of ethylene biosynthesis associated with ripening in banana fruit [J]. Plant Physiol, 121: 1257 - 1265.

Liu H, Cottrell T R, Pierini L M, et al, 2002a. RNA interference in the pathogenic fungus *Cryptococcus neoformans* [J]. Genetics, 160: 463 - 470.

Liu P, Goh C J, Loh C S, et al, 2002b. Differential expression and characterization of three metallothionein-like genes in Cavendish banana (*Musa acuminata*) [J]. Physiol Plant, 114: 241 - 50.

Liu Y L, Schiff M, Dinesh-Kumar S P, 2002c. Virus induced gene silencing in tomato [J]. Plant J, 31: 777 - 786.

Lu R, Martin-Hernandez A M, Peart J R, et al, 2003. Virus-induced gene silencing in plants [J]. Methods, 30: 296 - 302.

Manrique-Trujillo S M, Ramırez-Lopez A C, et al, 2007. Identification of genes differentially expressed during ripening of banana [J]. J Plant Physiol, 164: 1037 - 1050.

Martienssen R A, 1998. Functional genomics: probing plant gene function and expression with transposons [J]. Proc Natl Acad Sci USA, 95: 2021 - 2026.

Marx J, 2000. Interfering with gene expression [J]. Science, 288: 1370 - 1372.

Matsui A, Ishida J, Morosawa T, et al, 2008. *Arabidopsis* transcriptome analysis under drought, cold, high-salinity and ABA treatment conditions using a tiling array [J]. Plant Cell Physiol, 49: 1135 - 1149.

Matsumura H, Nirasawa S, Terauchi R, 1999. Transcript profiling in rice (*Oryza sativa* L.) seedlings using serial analysis of gene expression (SAGE) [J]. Plant J, 20: 719 - 726.

Matsumura H, Reich S, Ito A, et al, 2003a. Gene expression analysis of plant host-pathogen interactions by Super SAGE [J]. Proc Natl Acad Sci USA, 100: 15718 - 15723.

Matsumura H, Nirasawa S, Kiba A, et al, 2003b. Overexpression of Bax inhibitor suppresses the fungal elicitorinduced cell death in rice (*Oryza sativa* L.) cells [J]. Plant J, 33: 425 - 434.

Matzke M A, Matzke A J M, 1995. How and why do plants inactivate homologues (trans) genes [J]? Plant Physiol, 107: 679 - 685.

Mbéguié-A-Mbéguié D, Hubert O, Fils-Lycaon B, et al, 2008. EIN3-like gene expression during fruit ripening of Cavendish banana (*Musa acuminata* cv. Grande Naine)[J]. Physiol Plant, 133: 435 - 448.

McCallum C M, Comai L, Greene E A, et al, 2000. Target induced local lesions in genomes (TILLING) for plant functional genomics [J]. Plant Physiol, 123: 439 - 442.

Medina-Suarez R, Manning K, Fletcher J, et al, 1997. Gene expression in the pulp of ripening bananas [J]. Plant Physiol, 115: 453 - 461.

Meissner R, Chague V, Zhu Q, et al, 2000. A high throughput system for transposon tagging and promoter trapping in tomato [J]. Plant J, 22: 265 - 274.

Milhavet O, Gary D S, Mattson M P, 2003. RNA interference in biology and medicine [J]. Pharmacol Rev, 55: 629 - 648.

Miki D, Itoh R, Shimamoto K, 2005. RNA silencing of single and multiple members in a gene family of rice [J]. Plant Physiol, 138: 1903 - 1913.

Mochida K, Shinozaki K, 2010. Genomics and bioinformatics resources for crop improvement [J]. Plant Cell Physiol, 51: 497 - 523.

Nakayashiki H, 2005. RNA silencing in fungi: mechanisms and applications [J]. Fed Eur Biochem Soc Lett, 579: 5950 - 5970.

Nieto C, Piron F, Dalmais M, et al, 2007. EcoTILLING for the identifi cation of allelic variants of melon eIF4E, a factor that controls virus susceptibility [J]. BMC Plant Biol, 7: 34.

Novina C D, Murray M F, Dykxhoorn D M, et al, 2002. siRNA-directed inhibition of HIV-1 infection [J]. Nat Mediter, 8: 681 - 686.

Oliver S, 1996. A network approach to the systematic analysis of yeast gene function [J]. Trends Genet, 12: 241 - 242.

Ortiz-Vázquez E, Kaemmer D, Zhang H B, et al, 2005. Construction and characterization of a plant transformation-competent BIBAC library of the black sigatoka-resistant banana *Musa acuminata* cv. Tuu Gia (AA) [J]. Theor Appl Genet, 110 (4): 706 - 713.

Piffanelli P, Vilarinhos A D, Safar J, et al, 2008. Construction of bacterial artificial chromosome (BAC) libraries of banana (*Musa acuminata* and *Musa balbisiana*) [J]. Fruits, 63: 375 - 379.

Pooggin M, Shivaprasad P V, Veluthambi K, et al, 2003. RNAi targetting of DNA viruses [J]. Nat Biotechnol, 21: 131 - 132.

Qu J, Ye J, Fang R X, 2007. Artificial microRNA-mediated virus resistance in plants [J]. J Virol, 81: 6690 - 6699.

Radhamony R N, Prasad A M, Srinivasan R, 2005. TDNA insertional mutagenesis in *Arabidopsis*: a tool for functional genomics [J]. Elec J Biotechnol, 8 (1): 82 - 106.

Ratcliff F, Martin-Hernandez A M, Baulcombe D C, 2001. Tobacco rattle virus as a vector for analysis of gene function by silencing [J]. Plant J, 25: 237 - 245.

Renner T, Bragga J, Driscoll H E, et al, 2009. Virus-induced gene silencing in the culinary ginger (*Zingiber officinale*): an effective mechanism for downregulating gene expression in tropical monocots [J]. Mol Plant, 2: 1084 - 1094.

Roux N S, 2004. Mutation induction in *Musa*—review [M]//Jain S M, Swennen R. Banana improvement: cellular, molecular biology, and induced mutations. Enfield, New Hampshire, USA: Science Publishers: 23 - 32.

Ruiz M T, Voinnet O, Baulcombe D C, 1998. Initiation and maintenance of virus-induced gene silencing [J]. Plant Cell, 10: 937 - 946.

Ryo A, Kondoh N, Wakatsuki T, et al, 2000. A modified serial analysis of gene expression that generates longer sequence tags by nonpalindromic cohesive linker ligation [J]. Analytical Biochemistry, 277: 160 - 162.

Safar J, Noa-Carrazana J C, Vrana J, et al, 2004. Creation of a BAC resource to study the structure and evolution of the banana (*Musa balbisiana*) genome [J]. Genome, 47: 1182 - 1191.

Sági L, Remy S, Coemans B, et al, 2005. Functional analysis of the banana genome by gene tagging and SAGE [C]. Plant & Animal Genomes XIII Conference, San Diego, CA, USA.

Saha S, Sparks A B, Rago C, et al, 2002. Using the transcriptome to annotate the genome [J]. Nat Biotechnol, 20: 508 - 512.

Santos C M R, Martins N F, Horberg H M, et al, 2005. Analysis of expressed sequence tags from *Musa acuminata* ssp. *burmannicoides*, var. Calcutta 4 (AA) leaves submitted to temperature stresses [J]. Theor Appl Genet, 110: 1517 - 1522.

Santos E, Remy S, Thiry E, et al, 2009. Characterization and isolation of a T-DNA tagged banana promoter active during *in vitro* culture and low temperature stress [J]. BMC Plant Biol, 9: 77.

Schweizer P, Pokorny J, Schulze-Lefert P, et al, 2002. Double stranded RNA interference with gene functions at the single cell in cereals [J]. Plant J, 24: 895 - 903.

Shinozaki K, Yamaguchi-Shinozaki K, 2007. Gene networks involved in drought stress responseand tolerance [J]. J Exp Bot, 58 (2): 221 - 227.

Stemple D L, 2004. TILLING—a high-throughput harvest for functional genomics [J]. Nat Rev Genet, 5: 145 - 150.

Strosse H, Domergue R, Panis B, et al, 2003. Banana and plantain embryogenic cell suspensions [R]//Vezina A, Picq C. INIBAP Technical Guidelines & The International Network for the Improvement of Banana and Plantain. Montpellier, France: International Plant Genetic Resources Institute.

Sundaresan V, Springer P, Volpe T, et al, 1995. Patterns of gene action in plant development revealed by enhancer trap and gene trap transposable elements [J]. Genes Dev, 9: 1797 - 1810.

Tang W, Weidner D A, Hu B Y, et al, 2006. Efficient delivery of small interfering RNA to plant cells by a nanosecond pulsed laser-induced wave for post transcriptional gene silencing [J]. Plant Sci, 171: 375 - 381.

Terada R, Urawa H, Inagaki Y, et al, 2002. Efficient gene targeting by homologous recombination in rice [J]. Nat Biotechnol, 20: 1030 - 1034.

Thomas D S, Turner D, Eamus D, 1998. Independent effects of the environment on the leaf gas exchange of three banana (*Musa* sp.) cultivars of different genomic constitution [J]. Sci Hort, 75 (1-2): 41 - 57.

Tierney M B, Lamour K H, 2005. An introduction to reverse genetic tools for investigating gene function [J]. The Plant Health Instructor, DOI: 10.1094/PHI-A-2005-1025-01.

Till B J, Reynolds S H, Greene E A, et al, 2003. Large-scale discovery of induced point mutations with high-throughput TILLING [J]. Genome Res, 13: 524 - 530.

Till B J, Zerr T, Bowers E, et al, 2006a. High throughput discovery of rare human nucleotide polymorphisms by EcoTILLING [J]. Nucl Acids Res, 34: e99.

Till B J, Zerr T, Comai L, et al, 2006b. A protocol for TILLING and EcoTILLING in plants and animals [J]. Nat Protoc, 1: 2465 - 2477.

Till B J, Jankowicz-Cieslak J, Nakitandwe J, et al, 2008. TILLING and EcoTILLING in banana [C]. Plant & Animal Genomes XVI Conference. San Diego, CA, USA.

Till B J, Jankowicz-Cieslak J, Sagi L, et al, 2010. Discovery of nucleotide polymorphisms in the *Musa* gene pool by EcoTILLING. Theor Appl Genet, 121: 1381 - 1389.

Turnage M A, Muangsan N, Peele C G, et al, 2002. Geminivirus-based vectors for gene silencing in *Arabidopsis* [J]. Plant J, 30: 107 - 114.

Ullu E, Djikeng A, Shi H, et al, 2002. RNA interference: advances and questions [J]. Phil Trans Roy Soc Lond Brit Biol Sci, 29: 65 - 70.

Vanitharani R, Chellappan P, Fauquet C M, 2003. Short interfering RNA-mediated interference of gene expression and viral DNA accumulation in cultured plant cells [J]. Proc Natl Acad (USA), 100: 9632 - 9636.

Varshney R K, Tuberosa R, 2007. Genomics-assisted crop improvement: an overview [J]. Genomics Approaches and Platforms, 1: 1 - 12.

Varshney R K, Graner A, Sorrells M E, 2005. Genomics-assisted breeding for crop improvement [J]. Trends Plant Sci, 10 (12): 621 - 630.

Velculescu V E, Zhang L, Vogelstein B, et al, 1995. Serial analysis of gene expression [J]. Science, 270: 484 - 487.

Vilarinhos A D, Piffanelli P, Lagoda P, et al, 2003. Construction and characterization of a bacterial artifi cial chromosome library of banana (*Musa acuminata* Colla) [J]. Theor Appl Genet, 106 (6): 1102 - 1106.

Wang M B, Waterhouse P M, 2002. Application of gene silencing in plants [J]. Curr Opin Plant Biol, 5: 146 - 150.

Wani S H, Sanghera G S, 2010. Genetic engineering for viral disease management in plants [J]. Notulae Sci Biol, 2: 20 - 28.

Wani S H, Sanghera G S, Singh N B, 2010. Biotechnology and plant disease control—role of RNA interference [J]. Amer J Plant Sci, 1: 55 - 68.

Waterhouse P M, Graham M W, Wang M B, 1998. Virus resistance and gene silencing in plants can be induced by simultaneous expression of sense and antisense RNA [J]. Proc Nat Acad Sci USA, 95: 13959 - 13964.

Waterhouse P M, Wang M B, Lough T, 2001. Gene silencing as an adaptive defense against viruses [J]. Nature, 411: 834 - 842.

Weil C F, 2009. TILLING in grass species [J]. Plant Physiol, 149: 158 - 164.

Xu B Y, Su W, Liu J H, et al, 2007. Differentially expressed cDNAs at the early stage of banana ripening identified by suppression subtractive hybridization and cDNA microarray [J]. Planta, 226 (2): 529 - 539.

Yang Z, Tang L, Li M, et al, 2010. Monitoring homologous recombination in rice (*Oryza sativa* L.) [J]. Mutation research Fund Mol Mech Mutagen, 691: 55 - 63.

Yu H, Wang M, Tang D, et al, 2010. OsSPO11-1 isessential for both homologous chromosome pairing and crossover formation in rice [J]. Chromosoma, 119: 625 - 636.

第10章 芭蕉属植物蛋白质组学和代谢组学

Hoang Lan Chi Dinh[①,a,*] Peer M. Schenk[①,b]

摘 要：如今，在“组学”技术“装备”基础上产生了大量数据。蛋白质组学和代谢组学这两种典型的后基因组分析技术已被用于作物发育研究，其中包括香蕉和大蕉在内的芭蕉属植物的当前基因组学项目都辅以高通量代谢组学和蛋白质组学方法。全球范围内关于植物代谢分析的研究都集中在特定条件下植物的小分子，而蛋白质组学则关注于鉴定和定量差异表达的蛋白质。代谢物和蛋白质谱数据的生物信息学分析对较为常用的转录方法进行了补充。本章回顾了芭蕉属植物代谢组学和蛋白质组学研究中不同领域的最新进展。NMR、MS、2-DE、GC-MS、MALDI-TOF等试验和分析工具已被认为是产生芭蕉代谢组学和蛋白质组学数据的必要手段。一般而言，芭蕉属植物的系统生物学的分析结果将整合芭蕉属植物基因组学研究中转录组学、代谢组学、蛋白质组学和生物信息学方法的成果。

关键词：香蕉，“组学”技术，大蕉，系统生物学

10.1 引言

芭蕉属植物包括香蕉和大蕉，是世界第四重要作物。基因组、转录组、蛋白质组、代谢组和相互作用组的总体高通量技术已经发展成为强大的工具，来加深人们对细胞功能和植物系统生物学的理解（Fukushima et al.，2009）。本章回顾在利用组学技术、蛋白质组学和代谢组学对芭蕉属植物基因组学进行的研究上已取得的最新进展。

① School of Biological Sciences，The University of Queensland，St. Lucia，Queensland 4072，Australia.

a. e-mail：dinhhoanglanchi@gmail.com.

b. e-mail：p.schenk@uq.edu.au.

* 通信作者。

蛋白质组学覆盖对各种条件下细胞总体产生的蛋白质的所有研究（Primrose and Twyman，2006）。“蛋白质组”这一概念由威尔金斯和他的同事通过蛋白质分离和鉴定研究基因表达而确立的（下的定义）（Wilkins et al.，1995）。芭蕉属植物蛋白质组学研究基于蛋白质分离、纯化、检测和定量的大规模技术。代谢组学是发现和分析多效性影响下的生物体生物途径内的所有代谢物（Fiehn，2002）。在农业方面，系统生物学的代谢物图谱已被用于优化性状发展（Dixon et al.，2006）。代谢组学定性和定量一个样品内的所有代谢物，而该样品从针对构建所有生化网络的特定生物体中提取（代谢组学是通过定性和定量分析单个生物样本内所有的代谢物来建立生化代谢网络）（Primrose and Twyman，2006）。植物代谢组学是指对所有分子质量小于 1 000 u 的小分子的无偏鉴定和定量。代谢组包括了在基因表达及蛋白质活动作用下进行生物合成的整套蛋白质。这种技术不仅利用典型的代谢物，包括初级代谢物，也涉及如色素、营养素、调味剂和香味的一些有共同特征的次级代谢物分子（Clark and Pazdernik，2009）。

本章概述了芭蕉属植物的蛋白质组学和代谢组学上的一些最新进展，并与其他组学技术一起来补充关于芭蕉属植物基因组学的研究。

10.2 蛋白质组学的研究工具

最新蛋白质组学技术支持蛋白质/肽的分离，以及对某些分子的鉴定和特征描述，该技术包括二维凝胶电泳（2－DE）、十二烷基硫酸钠聚丙烯酰胺凝胶电泳（SDS－PAGE）、质谱（MS）、串联质谱（MS/MS）、毛细管电泳（CE）、表面增强激光解吸/电离（SELDI）、傅里叶变换离子回旋共振（FTICR）和电喷雾电离质谱（ESI－MS/MS）。2－DE 和 MS（2－DE/MS）之间的融合构成植物蛋白质组学研究中最重要的方法（Jorrín et al.，2007；Thiellement et al.，2007）。MS 当前用于识别从 2－DE 分离的大规模蛋白质，其通过等电聚焦（IEF）分离并将蛋白质可视化（O'Farrell，1975），在之后 MS 被用于对分离的蛋白进行识别和特征描述（Fenn et al.，1989）。这种 MS 方法的原理是利用 MS 仪器片段的质量将靶蛋白匹配到数据库中的基因序列上（Roberts，2002）。

至于凝胶技术，2－DE 已成为蛋白质组学中分离蛋白质的经典方法（图 10－1；Thongboonkerd，2007）。2－DE 对蛋白进行操作的两个步骤为蛋白质的分离和可视化。首先，通过第一维度等电点赋予的 p*I* 和 pH 梯度聚丙烯酰胺凝胶，将粗蛋白分离；然后通过以分子质量为基础的 SDS－PAGE，将分离的蛋白质分开；蛋白质分离的最后一步，利用荧光染料来可视化那些分离的蛋白（Patton，2000）。基于凝胶的应用程序产生的数据随后经由 MS 设备进行分析（Weckwerth，2008）。蛋白质鉴定的非凝胶方法包括使用作为鸟枪法研究蛋白质组学的 LC/MS 技术（图 10－1；Thongboonkerd，2007；Weckwerth，2008）。这种技术不仅识别蛋白，还分离蛋白，因而是一种可用于植物蛋白质组学的有前景的工具（Porubleva et al.，2001）。

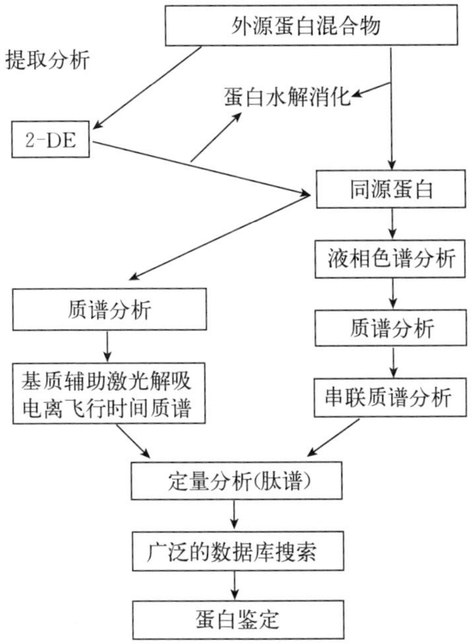

图 10－1　植物蛋白质组学的实验方法

10.3　标记识别和功能分析的蛋白质组学

简而言之，蛋白质组学分析的共同过程包括蛋白质提取，通过 2－DE 进行分离、提取，接着通过 MS 鉴定蛋白质，为下一步生物信息学的蛋白质表达分析提供材料。蛋白质组分析已成为通过建立数量性状位点（QTL）图谱来对数量性状遗传学进行特征描述的有效工具（de Vienne et al.，1999）。目前，2－DE 和 de novo MS 的整合手段是生成遗传标记的最有效工具（Samyn et al.，2007）。因此，遗传作图在芭蕉属植物育种研究中已变得十分流行，许多分子标记都是以这种方式鉴定，并且在过去几年这一方法已得到应用。例如，鉴定出 *Musa balbisiana* 25 个新微卫星标记，这些标记为大蕉和煮食蕉的连锁作图提供了最新信息（Buhariwalla et al.，2005）。另一个实例是使用 *M. acuminata* 基因组的细菌人工染色体（BAC）末端序列中的 325 个简单重复序列（SSR）标记（Cheung and Town，2007）。基于扩增片段长度多态性（AFLP）、同工酶、RFLP（限制性片段长度多态性）和链接到抗叶斑病性状的一个QTL图的SSR数据，研究人员绘制出二倍体香蕉的连锁图（Pillay and Tripathi，2007）。

多种蛋白质组学分析方法已用于芭蕉属植物研究。例如，Carpentier和他的同事（2007）使用NMR和MS来研究在渗透胁迫作用下而累积的芭蕉属植物分生组织蛋白质组。这些技术随后用于分析氨基酸序列数据，并与用序列或结构同源性搜索鉴定跨种一起进行对大型分生组织蛋白质的识别和随后的功能分析。芭蕉属植物分生组织研究的重要性在于产生理想的数据库，供公众访问、查阅所有分生组织蛋白质组信息。最易访问的芭蕉属植物蛋白质组学数据库之一是pdata，网址为：http：//www. pdata. ua. ac. be/musa/。

植物蛋白质组学的另一重要领域是对衍生肽的鉴定。Samyn等（2007）使用多种技术，包括2-DE、MALDI-MS、从头测序和基于同源性的搜索算法鉴定香蕉40个蛋白质和不同的异形体。他也指出了由基因组的持续发展和EST测序导致的数据库中植物蛋白质组的数量不断增多。除了这一研究，使用2-DE释放不同的蛋白质异形体很好地解释了芭蕉属植物基因组中出现的一些遗传差异，包括基因位点上的差异、等位基因变异、亚基相互作用和剪接形式。

对芭蕉属亚种和香蕉栽培种的各种分析（Venkatachalam et al.，2008）显示，芭蕉属植物转录组、蛋白质组和代谢组的分析为作物改良和育种提供合适的策略。芭蕉属植物中蛋白质分析主要的目的是增加蛋白质数据库与功能知识，并辅助标记检索。序列相似性已被用于研究非测序的基因组生物体如香蕉，在这些生物体中数据库是比较和筛选目标蛋白质组的决定性因素。

10.4 代谢组学的研究工具

与蛋白质组学和转录组学预测基因活性相比，代谢组学是基于形态学、生理学和植物的疾病易感性能来描述个体植物在表型水平上的特征（Allwood et al.，2008；Nikolau and Wurtele，2007）。这种非目标剖析技术用于检测和定量多种代谢物，特别是用于识别初级代谢的大中间体（Matthew et al.，2009）。为了确保代谢组学分析的准确性和再现性，可以使用某些技术，如核磁共振（NMR）、气相色谱（GC）、液相色谱（LC），以及质谱法与气相色谱或液相色谱相结合（Nikolau and Wurtele，2007）。其他代谢组学技术包括基质辅助激光解吸/电离（MALDI）（Li et al.，2008），傅里叶变换红外光谱法（Hall et al.，2008），直接注射MS和流动注射MS，基于质量过滤器中的电场分析质量离子的四极 MS，代谢物同一性的串联质谱（MS－MS）（Hanzlik et al.，2009），分析物识别的飞行时间质谱联用（TOF－MS），de novo 方法包括结合 LC、NMR 或 MS 鉴定次生代谢物（Samyn et al.，2007）。Mathew 等（2009）确定了源自植物中重要代谢物的很多信息可以通过 NMR、GC / MS 和薄层色谱（TLC）的平行路径而破译。植物代谢组学的技术程序主要基于质谱法和色谱法之间的融合（图 10－2；Moritz and Johansson，2008；Iijima et al.，2008）。

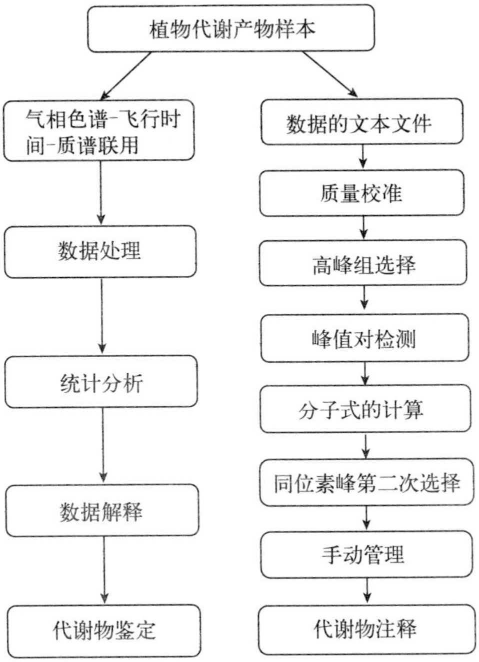

图 10－2　植物代谢组学的试验方法

10.5　代谢组学在表型组学或基因组学辅助育种上的应用

植物蛋白质组学和代谢组学有助于 QTL 作图（Subramaniam，2005）。QTL 作图利用特定的标记或表型变异来澄清位点的数目、表型的影响和互动（Nikolau and Wurtele，2007）。QTL 作图通常使用 DNA 标记测定法评估一个特定的基因座是如何影响与所需表型相关的性状。利用转录组、蛋白质组和代谢组这些植物表达特征，可以显著辅助这一过程。表型、基因型和遗传度的融合度越高，QTL 作图对性状的识别越可靠（Fiehn，2002；Zeng et al.，2008）。

对芭蕉属植物的数量遗传作图目前仍处于起步阶段。然而，研究人员对几个性状如串重、果实品质参数、果实膨大时间和疾病抗性也进行了重要考量（Ortiz，1995）。芭蕉属植物的遗传图谱辅助育种计划始于1990年，是通过对野生二倍体 *M. acuminata* spp. *banksii* ×SF265 进行杂交来进行的。两种 *M. acuminata* 种质杂交而生成的 F_2 代中92个体的90个位点的第一张QTL图谱绘制完成，并于1992年出版（Heslop-Harrison and Schwarzacher，2007）。连锁图和染色体位置目前都融入香蕉遗传图（Roux et al.，2008）。共有120个分子标记（AFLP、RFLP和SSR）用于构建 Calcutta 4×Madang 的遗传图谱。代谢物的自然变化与表型生长密切相关，因此代谢组学中数量遗传学的应用产生了对遗传控制的能力（Keurentje et al.，2008）。育种者使用QTL图创建包含改良多基因性状的如高产及抗生物和非生物胁迫的新型芭蕉属植物。QTL作图的一个实例来自二倍体野生种的芭蕉属植物基因组内的 *R* 基因，这些二倍体野生种包括 *Musa acuminate*（A）、*M. balbisiana*（B）、*M. schizocarpa*（S）、*M. textilis*（T）、*M. velutina* 和 *M. ornate*（Azhar and Heslop-Harrison，2008）。当前，微卫星标记物已被广泛用于香蕉育种计划（Creste et al.，2006）。

研究人员可以容易地对数百个样品进行代谢物分析，而通过NMR可以进行高通量自动分析（Allwoodet al.，2008），此外，NMR还可系统地识别未知化合物（Fiehn，2002）。一般来说，质子核磁共振（H-NMR）用于对除草剂及其生物活性成分进行分类（Schauer and Fernie，2006）。核磁共振还用于识别“体内不同隔室中代谢物的稳态浓度”（Fiehn，2002；Roberts，2002）。至于质谱、吸光度或通过以化学键为基础的技术进行的电磁重定向（EM）辐射，通过快速光谱采集和代谢物的空间测量，提供了全球代谢物指纹图谱（Allwood et al.，2008）。GC-MS的优点是使用化学衍生法提高代谢物的挥发性（Fiehn，2002）。事实上，利用极性非挥发性分析，MS能检测挥发温度高达2 500 ℃的代谢物，包括氨基酸、糖和有机酸等（Allwood et al.，2008）。GC-MS和LC-MS的一个应用是用于识别植物中的盐胁迫（Kim et al.，2007）。科学家将H-NMR、C-NMR、MS和HPLC结合起来，用于研究香蕉植物抗毒素的生物合成途径（Kamo et al.，2000）。高效液相色谱（HPLC）能够识别代谢物包括其组成和量之间的差异。例如，Ascensao和Dubery（2003）采用HPLC鉴定香蕉根内不同的酚类（阿魏酸和芥子酸）和酚醛聚合物。苯丙素生物合成途径反映出植物抗病原诱导子的重要代谢活动。De Ascensao和Dubery（2003）已经开发出了提取和鉴定来自香蕉根部抗 *Fusarium oxysporum* f. sp. *cubense*（*Foc*）诱导子的酚类的方法。

代谢组学持续推动香蕉代谢物发现和利用等方面研究的前进。第一种方法用于作物/食品分析，包括 LC－MS、GC－MS 和核磁共振在内的不同技术对于确定由各种代谢物成分所控制的食品质量和营养成分十分重要（Hall et al.，2008）。第二，代谢组学用于诊断目的，例如，GC－MS 是筛选除莠剂作用方式的最常见技术策略（Schauer and Fernie，2006）。

芭蕉属植物代谢组学在辅助功能基因组学和基因功能注释上有很大潜力。然而，为了代谢组学的进一步应用，有必要控制和标准化这门科学的技术流程和基础设施。代谢组学标准协会（MSI）已经建立了代谢物记载的一些标准（Fiehn et al.，2008）。测量参数包括研究设计、样品制备、数据采集、数据处理、数据分析和数据集成（Fiehn et al.，2008）。

代谢组学在设计更好的作物上起着重要作用，因为与该技术鉴定的关键性状有关的分子签名帮助科学家重新排列代谢相互作用，重新设计/选择拥有期望表型和基因型的作物（Hall et al.，2008）。整个表型和基因型的研究分别对应的是表型组学和基因组学。德国植物研究机构（Institute for Phytosphere Research）主任 Uli Schurr 认为，表型组学是"精准农业和预测育种"中很有潜力的技术（Finkel，2009）。代谢物分析也用于未来植物育种计划的基因组学。例如，对香蕉的苯丙素途径的代谢组学研究和对仙茅科（Haemodoraceae）植物的比较基因组研究揭示了这两个物种之间的密切关系（Kamo et al.，2000）。科学家可以使用来自 Haemodoraceae 模式植物的苯丙素途径信息来修改香蕉果实中影响成熟性状的植物抗毒素的遗传基础。

植物代谢物研究拓宽了对植物蛋白质和遗传学的认识；代谢物谱上可能发生的变化解释了控制代谢途径的相关蛋白质和基因上的变化（Roberts，2002）。国际热带农业研究所（IITA）(Vuylsteke et al.，1997；Crouch et al.，1998）已对芭蕉属物种开展育种计划。NMR 和 MS 可以分析几种代谢产物的状态，如胁迫、营养因子、时间和环境微扰（Dixon et al.，2006）。其他技术如用微阵列分析技术进行代谢物分析，例如，测定转基因（GM）作物的预期和非预期的不良影响（Dixon et al.，2006）。

10.6　结论

蛋白质组学和代谢组学直接测量蛋白质和代谢物的表达，有利于生物过程的确认和建设。这些技术旨在更好地了解芭蕉属植物和细胞功能。

芭蕉属植物基因组中已鉴定出各种蛋白质。例如，编码香蕉成熟性状的基因，包括基因型的 AAA（卡文迪什）、AAB（Rasthali 和 Poovan）、AB（Kanthali）和 ABB（Monthan）（Choudhury et al.，2008）。使用代谢组学和蛋白质组学方法的优点包括选择转移到有更强扩展能力的半自动方式。例如，质谱技术可以高通量地用于蛋白质组学，尽管这需要大量的设备投资。相比传统的以 2-DE 为基础的方法，该技术的优点在于只需要准备更少的蛋白质制剂和消耗品，却有高速率和低数量的反应（Morot-Gaudry et al.，2007）。综合数据库可以破译大量蛋白质图谱。植物王国蕴藏着巨大的化学多样性和动态特性，目前的代谢分析仍然十分有限（Tohge and Fernie，2009）。Sumner（2003）指出基于 MS 的代谢组学面临着动态范围的问题。

与蛋白质组学相比，代谢组学呈现出更高的通量和更经济的方向（Hollywood et al.，2006）。最近的“组学”时代正在越来越多地使用针对代谢物的微阵列、新一代测序和数据库，以及推动植物育种计划的蛋白质组筛选；因而这增强了模式作物如芭蕉属物种的下一世代改良的可能。此外，蛋白质组学和代谢组学的兴起正在吸引大批青年科学家，他们将升级并扩大这些学科的应用领域。蛋白质组学和代谢组学的研究意义是要进一步开拓芭蕉属物种生化途径上的多效性效应。先进的“组学”技术增加了研究芭蕉属植物基因组中许多蛋白质和代谢物的可能性。此外，最近发表的植物代谢组学和蛋白质组学程序可作为基本资源供科学家使用（Nikolau and Wurtele，2007；Thiellement et al.，2007）。芭蕉属植物基因组注释面临的挑战是选择有效集成分析工具用于转录组、蛋白质组和代谢组数据集。用于芭蕉属或植物物种的系统生物学研究的数据库和程序列表总结在本书的芭蕉属植物生物信息学章节。植物蛋白质组学和代谢组学的发展对于通过遗传基础进行系统生物学和植物改良是必要的；但是，这两个学科门类都需要技术管理。代谢组学或 ArMet 架构（http：//www.armet.org/publications/index.html）已被用作数据模型来分析代谢物（Jenkins et al.，2004），而 HUPO（人类蛋白质组组织）的蛋白质组学标准倡议或 PSI 近来用作蛋白质组学研究的评估标准。不同植物物种的基因组序列的许多宝贵资源促进植物蛋白质组学和代谢研究领域的发展，这也有助于芭蕉属植物研究。

由于具有识别强势性状和显性基因的能力，基于“组学”的生物技术能显著提高香蕉和大蕉的遗传开发。转录组学、蛋白质组学、代谢组学和生物信息学的集成提供了全面的芭蕉属植物基因组数据。对蛋白质和代谢物的研究显著促进了世界热带和亚热带地区作物的发展。除 mRNA 之外，蛋白质和代谢物提高了对分子表型包括芭蕉属物种的全面分析和数量分析。蛋白质组学和代谢组学促进芭蕉属植物在分子水平上的发展，间接协助了芭蕉属植物的育种计划，从而提高了这种水果的商业价值。

缩略语

AFLP：扩增片段长度多态性

BAC：细菌人工染色体
CE：毛细管电泳
ESI-MS/MS：电喷雾电离质谱
FTICR：傅里叶变换离子回旋共振
GC：气相色谱
HPLC：高效液相色谱
IEF：等电聚焦
LC：液相色谱
MALDI：基质辅助激光解吸/电离
MS/MS：串联质谱
MS：质谱
NMR：核磁共振
QTL：数量性状位点
RFLP：限制性片段长度多态性
SDS-PAGE：十二烷基硫酸钠聚丙烯酰胺凝胶电泳
SELDI：表面增强激光解吸/电离
SSR：简单重复序列
TOF：飞行时间

致谢

感谢 Kevan Jones 提出的有用建议和澳大利亚研究理事会（DP1094749）的研究经费支持。

参考文献

Allwood J W，Ellis D I，Goodacre R，2008. Metabolomic technologies and their application to the study of plants and plant-host interactions [J]. Physiol Plant，132：117 - 135.

Azhar M，Heslop-Harrison J S，2008. Genomes，diversity and resistance gene analogues in *Musa* species [J]. Cytogenet Genome Res，121：59 - 66.

Buhariwalla H K，Jarret R L，Jayashree B，et al，2005. Isolation and characterization of microsatellite markers from *Musa balbisiana* [J]. Mol Ecol Notes，5：327 - 330.

Carpentier S C，Witters E，Lauken K，et al，2007. Banana（*Musa* spp.）as a model to study the meristem proteome：acclimation to osmotic stress [J]. Proteomics，7：92 - 105.

Chen S，Harmon A C，2006. Advances in plant proteomics [J]. Proteomics，6：5504 - 5516.

Cheung F，Town C D，2007. A BAC end view of the *Musa acuminata* genome [J]. BMC Plant Biol，7：29.

Choudhury S R, Roy S, Saha P P, et al, 2008. Characterization of differential ripening pattern in association with ethylene biosynthesis in the fruits of five naturally occurring banana cultivars and detection of a GCC-box-specific DNA-binding protein [J]. Plant Cell Rep, 27: 1235-1249.

Clark D P, Pazdernik N J, 2009. Biotechnology: applying the genetic revolution [M]. Amsterdam, The Netherlands: Academic Press/Elsevier.

Creste S, Benatti T R, Orsi M R, et al, 2006. Isolation and characterization of microsatellite loci from a commercial cultivar of *Musa acuminata* [J]. Mol Ecol Notes, 6: 303-306.

Crouch J H, Vuylsteke D, Ortiz R, 1998. Perspectives on the application of biotechnology to assist the genetic enhancement of plantain and banana (*Musa* spp.) [J]. Electronic J Biotechnol, 1: 11-22.

De Ascensao A R, Dubery I A, 2003. Soluble and wall-bound phenolics and phenolic polymers in *Musa acuminata* roots exposed to elicitors from *Fusarium oxysporum* f. sp. *cubense* [J]. Phytochemistry, 63: 679-686.

De Vienne D, Leonardi A, Damerval C, et al, 1999. Genetics of proteome variation for QTL characterization: application to drought-stress responses in maize [J]. J Exp Bot, 50: 303-309.

Dixon R A, Gang D R, Charlton A J, et al, 2006. Applications of metabolomics in agriculture [J]. J Agric Food Chem, 54: 8984-8994.

Fenn J B, Mann M, Meng C K, et al, 1989. Electrospray ionization for mass spectrometry of large biomolecules [J]. Science, 246: 64-71.

Fiehn O, 2002. Metabolomics—the link between genotypes and phenotypes [J]. Plant Mol Biol, 48: 155-171.

Fiehn O, Wohlgemuth G, Scholz M, et al, 2008. Quality control for plant metabolomics: reporting MSI-compliant studies [J]. Plant J, 53: 691-704.

Finkel E, 2009. With 'Phenomics' plant scientists hope to shift breeding into overdrive [J]. Science, 325: 380.

Fukushima A, Kusano M, Redestig H, et al, 2009. Integrated omics approaches in plant systems biology [J]. Curr Opin Chem Biol, 13: 532-538.

Hall R D, Brouwerand I D, Fitzgerald M A, 2008. Plant metabolomics and its potential application for human nutrition [J]. Physiol Plant, 132: 162-175.

Hanzlik R P, Fang J, Koen Y M, 2009. Filling and mining the reactive metabolite target protein database [J]. Chem-Biol Interact, 179: 38-44.

Heslop-Harrison J S, Schwarzacher T, 2007. Domestication, genomics and the future for banana [J]. Ann Bot, 100: 1073-1084.

Hollywood K, Brison D R, Goodacre R, 2006. Metabolomics: current technologies and future trends [J]. Proteomics, 6: 4716-4723.

Iijima Y, Nakamura Y, Ogatal Y, et al, 2008. Metabolite annotations based on the integration of mass spectral information [J]. Plant J，54：949 - 962.

Jenkins H，Hardy N，Beckmann M，et al，2004. A proposed framework for the description of plant metabolomics experiments and their results [J]. Nat Biotechnol，22：1601 - 1606.

Jorrín J V，Maldonado A M，Castillejo M A，2007. Plant proteome analysis：a 2006 update [J]. Proteomics，7：2947 - 2962.

Kamo T，Hirai N，Tsuda M，et al，2000. Changes in the content and biosynthesis of phytoalexins in banana fruits [J]. Biosci Biotechnol Biochem，64：2089 - 2098.

Keurentjes J J B，Koornneef M，Vreugdenhil D，2008. Quantitative genetics in the age of omics [J]. Curr Opin Plant Biol，11：123 - 128.

Kim J K，Bamba T，Harada K，et al，2007. Time course metabolic profi ling in Arabidopsis thaliana cell cultures after salt stress [J]. J Exp Bot，58：415 - 424.

Lakshmanan V，Venkataramareddy S R，Neelwarne B，2008. The use of genetic markers for detecting DNA polymorphism，genotype identifi cation and phylogenetic relationships-among banana cultivars [J]. Mol Phylogenet Evol，47：974 - 985.

Li Y，Shrestha B，Vertes A，2008. Atmospheric pressure infrared MALDI imaging mass spectrometry for plant metabolomics [J]. Anal Chem，80：407 - 420.

Matthew T，Zhou W，Rupprecht J，et al，2009. The metabolome of *Chlamydomonas reinhardtii* following induction of anaerobic H_2 production by sulfur depletion [J]. J Biol Chem，284：23415 - 23425.

Moritz T，Johansson A I，2008. Plant metabolomics [M]// Griffiths W J. Metabolomics，metabonomics and metabolite profiling. Cambridge，UK：The Royal Society of Chemistry：254 - 272.

Morot-Gaudry J F，Lea P，Briat J F，2007. Functional plant genomics [M]. Enfield，NH，USA：Science Publishers.

Nikolau B J，Wurtele E S，2007. Concepts in plant metabolomics [M]. The Netherlands：Springer.

O' Farrell P H，1975. High resolution two dimensional electrophoresis of protein [J]. J Biol Chem，250：4007 - 4021.

Ortiz R，1995. *Musa* genetics [M]// Gowen S. Banana and plantains. London，UK：Chapman and Hall：84 - 109.

Patton W F，2000. A thousand points of light：the application of fluorescence detection technologies to two-dimensional gel electrophoresis and proteomics [J]. Electrophoresis，21：1123 - 1144.

Pillay M，Tripathi L，2007. Banana breeding [M]//Kang M S，Priyadarshan P M. Breeding major food staples. Boston，MA，USA：Blackwell Publishing：393 - 428.

Porubleva L，Velden V，Kothari K S，et al，2001. The proteome of maize：use of gene sequence and expressed sequence tag data for identification of proteins with mass fingerprints [J]. Electrophoresis，22：1724 - 1738.

Primrose S B, Twyman R M, 2006. Principles of gene manipulation and genomics [M]. 7th ed. Oxford, UK: Blackwell.

Roberts J K M, 2002. Proteomics and a future generation of plant molecular biologists [J]. Plant Mol Biol, 48: 143 - 154.

Roux N, Baurens F C, Doleze J, et al, 2008. Genomics of banana and plantain (*Musa* spp.), major stable crops in the tropics [M]// Moore P H, Ming R, Delmer D P. Plant genetics and genomics of tropical crop plants, vol 1. New York, USA: Springer: 83 - 111.

Samyn B, Sergeant K, Carpentier S, et al, 2007. Functional proteome analysis of the banana plant (*Musa* spp.) using *de novo* sequence analysis of derivatized peptides [J]. J Proteome Res, 6: 70 - 80.

Schauer N, Fernie A R, 2006. Plant metabolomics: towards biological function and mechanism [J]. Trends Plant Sci, 11: 508 - 516.

Subramaniam S, 2005. Bioinformatics [M] // Dunn M J, Jorde L B, Little P F R, et al. Encyclopedia of genetics, genomics, proteomics and bioinformatics. New York, USA: Wiley: 3117 - 3126.

Sumner L W, Mendes P, Dixon R A, 2003. Plant metabolomics: large-scale phytochemistry in the functional genomics era [J]. Phytochemistry, 62: 817 - 836.

Thiellement H, Zivy M, Damerval C, et al. 2007. Plant proteomics: methods and protocols methods in molecular biology, vol. 355 [M]. Totawa, N J, USA, Human Press, Inc.

Tohge T, Fernie A R, 2009. Web-based resources for mass-spectrometry-based metabolomics: a user's guide [J]. Phytochemistry, 70: 450 - 456.

Thongboonkerd V, 2007. Practical points in urinary proteomics [J]. J Proteome R, 6: 3881 - 3890.

Venkatachalam L, Sreedhar R V, Bhagyalakshmi N, 2008. The use of genetic markers for detecting DNA polymorphism, genotype identifi cation and phylogenetic relationships among banana cultivars [J]. Molecular Phylogenetics and Evolution, 47: 974 - 985.

Vilarinhos A, 2004. Genetic and cytogenetic mapping in bananas: characterization of translocations [J]. InfoMusa, 13: 1.

Vuylsteke D, Ortiz R, Ferris R S B, et al, 1997. Plantain improvement [J]. Plant Breed Rev, 14: 267 - 320.

Weckwerth W, 2008. Integration of metabolomics and proteomics in molecular plant physiology—coping with the complexity by data-dimensionality reduction [J]. Physiol Plant, 132: 176 - 189.

Wilkins M R, Sanchez J C, Gooley A A, et al, 1995. Progress with proteome projects: why all proteins expressed by a genome should be identified and how to do it [J]. Biotechnol Genet Eng Rev, 13: 19 - 50.

Zeng Y, Li J, Wang C, et al, 2008. Genetic mapping of quantitative trait loci [M] // Kole C, Abbott A G. Principles and practices of plant genomics. Ⅰ. Genome Mapping. Northants, UK: Science Publishers: 175 - 204.

第 11 章　生物信息学的作用

Mathieu Rouard①,a,*　Sebastien Christian Carpentier②　Stephanie Bocs③,b
Gaëtan Droc,③,c　Xavier Argout③,d　Nicolas Roux①,e　Manuel Ruiz③,f

摘　要： 生物信息学在当今的植物科学中起到重要作用，这主要得益于高通量测序技术产生的基因组序列呈指数式增长。基因组学的成功也促进了"组学"研究领域的兴起，并产生数据的多样化。在这种情况下，研究人员开发出了各种方法、软件和数据库将生物数据转化成有意义的信息，其中一些信息是科学家每天都需要使用的。相较于模式植物，芭蕉属植物还处在起步阶段，但已开发出许多有用的工具，并为不久后更大型数据集的产生做好了准备。本章将介绍有助于芭蕉属植物研究的资源和工具，也将讨论生物信息学话题，如基因注释、转录组学、蛋白质组学和数据集成。

关键词： 生物信息学，数据库，基因组注释，蛋白质组学，分子标记

11.1　引言

本章是有关芭蕉属植物生物信息学的内容，涵盖全球芭蕉属植物基因组联盟（GMGC）开展的最新活动信息。以下部分描述了遗传、基因组、转录组和蛋白质组学数据如何在各种信息系统中被利用。这些发展旨在促进对重要农

① Bioversity International，Commodities for Livelihood Program，34397 Montpellier，France.
a. e-mail：m. rouard@cgiar. org.
② Department of Biosystems，Division of Crop Biotechnics，Katholic University of Leuven，Leuven，Belgium；e-mail：sebastien. carpentier@biw. kuleuven. be.
③ CIRAD，Biological Systems Department，UMR AGAP-TA 40/03，34398，Montpellier，France.
b. e-mail：stephanie. sidibe-bocs@cirad. fr.
c. e-mail：gaetan. droc@cirad. fr.
d. e-mail：xavier. argout@cirad. fr.
e. e-mail：n. roux@cgiar. org.
f. e-mail：manuel. ruiz@cirad. fr.
*　通信作者。

艺性状的遗传基础的理解，除其他结果之外，这些遗传基础将有助于育种工作。

11.2 基因和基因组数据库

不同于模式植物如拟南芥和水稻的基因组学，芭蕉属植物基因组学还处于起步阶段，尽管大型项目（如全基因组测序）将很快为其产生大量的数据。迄今为止，研究项目已经催生了由 GMGC（http://www.musagenomics.org/）维护的数据库，其中大部分可用信息集中于生物和非生物胁迫或比较基因组学。这些序列是通过筛选细菌人工染色体（BAC）文库获得的。所使用的 BAC 文库是多数栽培种源头的二倍体野生种——提供 A 基因组的 *M. acuminata* 和提供 B 基因组的 *M. balbisiana*，以及源自 *M. acuminata* 的三倍体栽培种 Grande Naine（Vilarinhos et al.，2003；Safar et al.，2004；Ortiz-Vazquez et al.，2005）。

11.2.1 自动注释

研究人员已经对芭蕉属植物基因组部分进行了许多分析，以预测基因和转座因子（TE），并推断基因功能。该预测程序或使用的方法结合在一起构成了促进序列分析的流程。表 11-1 中提供了一个清单，上面列举注解流程所使用的部分软件。研究人员利用 EuGène 软件和水稻特异性参数对基因结构进行预测（Foissac et al.，2008）。该软件结合了 *ab initio* 基因发现程序和外部相似性搜索程序。尽管没有芭蕉属植物的实验数据，但是随着表达序列标签（EST）可用性的不断增强，基因组注释也得到提高。计算分析也有助于识别和注释促成基因组结构和进化的转座因子（TE）。

表 11-1 序列注释流程所用软件

程序	URL
Eugène	http://www.inra.fr/mia/T/EuGene/
Fgenesh	http://linux1.softberry.com/
GeneMark	http://exon.gatech.edu/GeneMark/
SpliceMachine	http://bioinformatics.psb.ugent.be/webtools/splicemachine/
Sim4	http://globin.cse.psu.edu/dist/sim4/
RepeatMasker	http://www.repeatmasker.org/
Repseek	http://wwwabi.snv.jussieu.fr/public/Repseek/
Repet	http://www.inra.fr（在线不可用）

（续）

程序	URL
BLASTX	http://blast.ncbi.nlm.nih.gov/
Exonerate	http://www.ebi.ac.uk/~guy/exonerate/
InterproScan	http://www.ebi.ac.uk/Tools/InterProScan/

注：利用几个基因发现程序来预测基因，如果可能的话，利用 cDNA 比对来对基因加权。使用 Eugène 组合器寻找共有序列。重复因子也被不同的工具，如 Repet 流程检测。利用 TrEMBL 和 Swiss-Prot 蛋白质数据库检索序列。

通过搜索蛋白质数据库，如 UniProtKB（Schneider et al.，2005）和 Interpro（Hunter et al.，2009），研究人员已经对这些基因产物的功能信息进行了预测。UniProtKB/Swiss-Prot 数据库提供了经过审查的条目，因而保证了高质量的注释，作为它的补充部分，UniProtKB/TrEMBL 提供了对于尚未注释序列的见解。

基于对 64 个 BAC（约 7 Mb）的分析，预测出了 1 508 个基因。这个数据集计算出的基本统计数据改进了先前对芭蕉属植物基因组结构的认识（Aert et al.，2004；Cheung and Town，2007；Lescot et al.，2008）。G + C 平均含量约为 40%，而基因密度估计为每 4.7 kb 一个。分析的基因组序列的 27% 由转座因子构成。使用 Repet 流程（Abad et al.，2008）鉴定出 3 140 个 TE。它们主要来自Ⅰ类（LTR：长末端重复序列，LARD：大反转录转座子衍生物，LINE：长分散核因子，SINE：短分散核因子）（Wicker et al.，2007）。这些 TE 中，39%没有很好地进行特征描述。这些结果凸显了手动注释对于充分利用数据库内容的重要性。

11.2.2　数据存储和浏览器

先前描述的自动注释被存储在由通用模式生物系统数据库项目（GMOD）提供的 Chado 模式下的 PostgreSQL 数据库中。然后，建立了基于 Web 的基因组浏览器 GBrowse（Stein et al.，2002），来呈现由自动分析生成的参考序列的基因组特征（图 11－1）。基因组浏览器使用户能够访问基因组区域，此外还能显示不同的信息轨迹，包括基因和重复因子的自动预测和手动编制。用户可以搜索特定的基因产物，然后下载序列。还编制了具有显著相似性的表达序列（芭蕉属植物 EST）和蛋白质序列（UniprotKB/Swiss-prot 和 UniprotKB/TrEMBL）。GMGC 网站运行最新版本的 GBrowse，其中包括新的人性化的功能。例如，可以使用拖放功能来移动轨迹并以便捷的方式进行组织。GBrowse 提供了许多选项来自定义视图并方便浏览基因组。通过 GMGC 网站可以访问基因数据库。

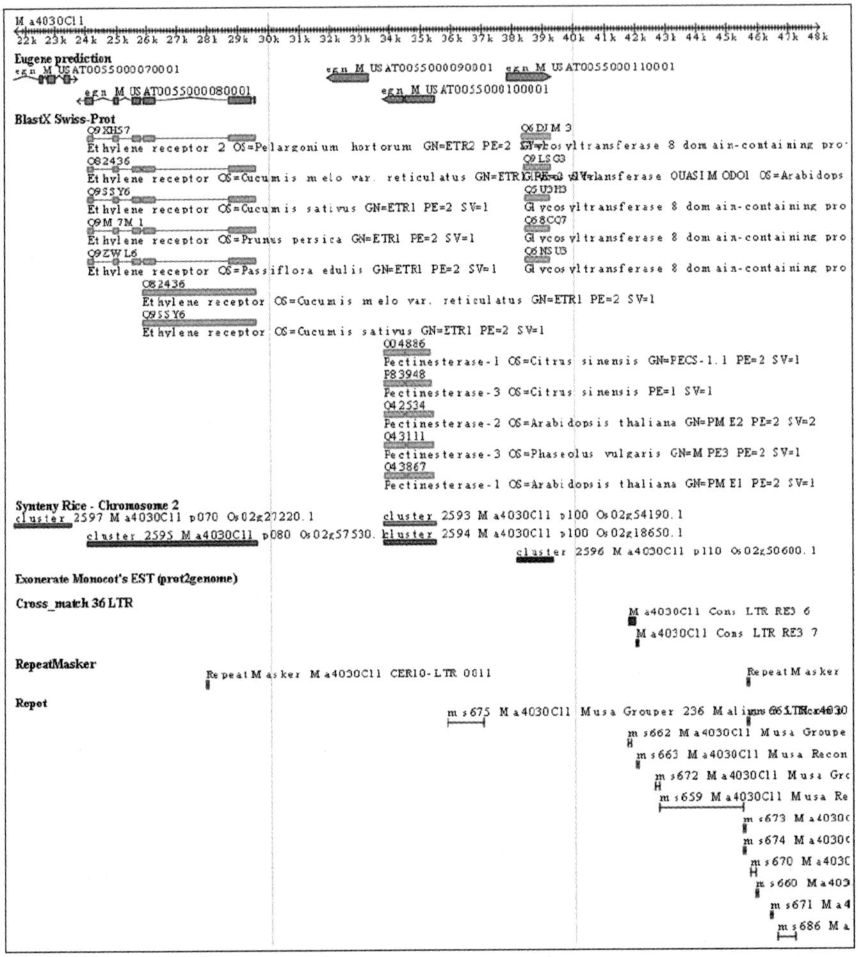

图 11-1 显示芭蕉属植物 BAC 序列的 GNPAnnot 基因组浏览器的屏幕截图

[根据分析的类型（如 Eugene、Repet、BLASTX）将基因组特性分派到不同的轨道。点击其中一个功能，会弹出一个窗口，提供详细报告的链接。与 Apollo 和 Artemis 软件的链接可以在专门用于手动管理的轨道上使用]

11.2.3 手动注释

手动检查能提高预测基因结构和蛋白质功能的准确性。验证蛋白质编码基因结构的最可靠证据是，当位点序列与其同源全长的 cDNA 一致（Aubourg et al.，2006），在功能上二者是同源的关系时。例如，GreenPhyl（Conte et al.，2008a，b）提出了基于系统发育学方法对直系同源基因的预测，该方法允许将注释从研究充分的基因转移到未鉴定的基因。这些程序对于芭蕉属植物基因组研究十分重要，因为基因组测序正在进行，几个项目已经产生了 454/Roche 和 Solexa/Illumina 数据。因此，基因数据库也被结合到用于序列注释的主流工具（图 11－2）Artemis（Carver et al.，2008）和 Apollo（Lewis et al.，2002）中。这两种工具都可用，不过通常情况下，用户群一旦熟悉其中一个，就很难说服他们采用另一个。到目前为止，注释器加载了存储在计算机内的平面文件，很少共享和同步数据。通过点击弹出窗口中的链接，注释器可以从基因组浏览器中搜索目标基因组特征，并发动注释编辑器。例如，这些序列注释工具可用于修改，例如边界基因和转座因子。上游和下游的非翻译区（例如，启动子、调控因子）也可以被注释。注释直接被读取并保存在远程服务器上。这些基础设施将有力地推动芭蕉属植物基因组学合作倡议。

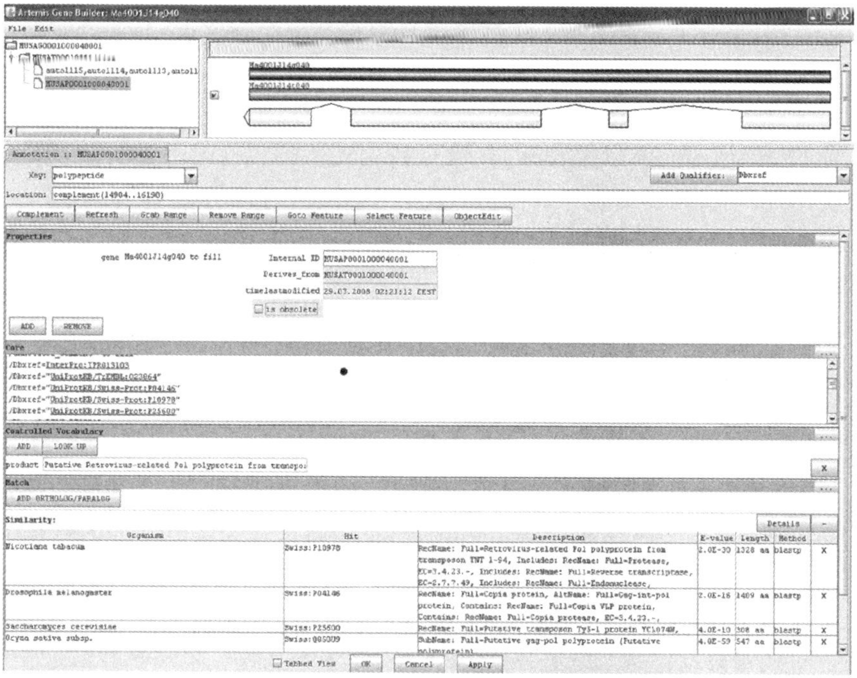

图 11－2　Artemis 基因生成器

（在基因组浏览器中选择目标基因后，用户可以可视化，编辑和验证基因。还可以编辑基因模型的坐标，进行功能注释。Ontological terms 也可以与注释相关联）

11.3　基因表达数据库

EST 数据库是转录组学研究的起点，因为大量的 EST 集一旦聚集，代表基因组中基因编码组件的潜在重要比例。此外，它们在鉴定基因区域和改善基因预测上有很大价值。因此，与许多其生物体一样，开发这些资源一直是GMGC对芭蕉属植物基因组进行表征的策略之一（Roux et al.，2008）。公共的EST序列可在TIGR Plant TranscriptAssemblies上获得（Childs et al.，2007），而在ESTtik上可以获得表达序列标签处理和调查工具包（Argout et al.，2008），更多细节可访问GMGC网站。芭蕉属植物基因组联盟特别采用后者来提供EST的不同访问级别，这取决于EST是否在公共领域，是否只供GMGC成员或已经签署材料转让协定的成员访问。而对于GMGC成员，鼓励他们尽可能地将数据发布在公共领域。

11.3.1 表达序列标签（EST）数据库

ESTtik是一种信息系统，由数据处理流程和一个探索EST集合的网站构成（http://esttik.cirad.fr）。考虑到EST中发生序列错误和冗余的概率高，因此需要一个可以解决这些问题的流程来处理序列。图11-3中显示了碱基识别，序列清除，组装和聚类等步骤。利用BLAST注释所获得的重叠群（Altschul et al.，1990），Gene Ontology terms被自动分配到转录物（Ashburner et al.，2000）。这些来自多种种质和A、B基因组的序列虽然数量仍然不多，但许多团体为之做出了贡献。目前，芭蕉属植物EST数据库包含约100 000个序列。EST的长度显著短于植物基因的平均长度（Rudd，2003），平均为609 bp。数据集被划分为几个项目，以便于鉴定具有不同表达水平的基因（例如，不受胁迫和胁迫下的叶）。研究人员已经开始使用EST序列开发SSR标记，现在已有超过2 000个SSR被发现。

通过GMGC网站访问的ESTtik数据库提供了搜索个体序列的方法，其方法是以序列各自的cDNA库和它们的初步功能注释为基础，因为数据库储存着由NCBI维护的非冗余肽序列公共数据库（NR）中的10个最佳BLAST位点数。将翻译过的EST与蛋白质序列（BLASTX）进行比对，因为已知蛋白质比对能更好地识别亲缘关系较远的蛋白质编码基因。共同的祖先通常会导致物种之间共享高度保守的基序。结构的保存可能意味着功能的保持。

例如，在ESTtik中输入关键词“ethylene”，即调节果实成熟的激素，将获得30多个被注释为推测的乙烯反应因子或者转录因子的序列。结果中可以过滤出含有SSR的序列。随后可以下载聚类数据或原始数据来进行验证基因功能的互补实验。然而，由于大型公共数据库中包含大量的序列，它们都没有被很好地描述，因此建议仔细查看类似的EST序列。该网站含有一个BLAST服务器，使得用户可以基于自身序列的相似性在该数据库搜索。

对所有序列进行的暂时组装（Town et al.，未见刊）产生了12 385个重

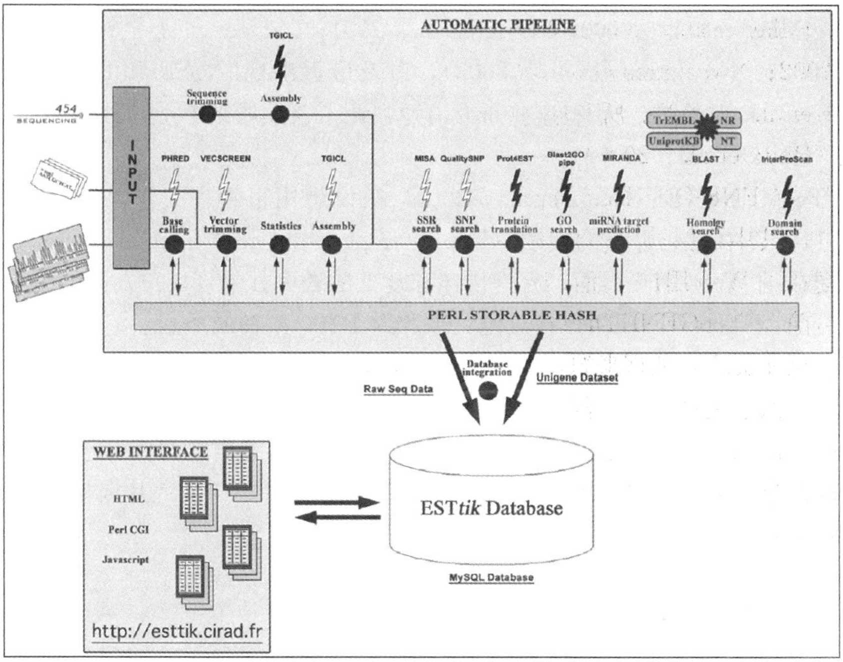

图 11-3 ESTtik 流程

（执行一系列程序，以评估色谱图的质量和核苷酸，然后进行编辑，并将输入的 cDNA 序列组装成一个非冗余数据集；这一 Unigene 集随后用于开发 SSR 和 SNP，而 Gene Ontology terms 和 BLAST 注释被分配到集群中）

叠群和总共 26 918 个非重复序列基因中有 14 533 个单元素集合。预计整个集合很快将出现在公共领域，而这也将成为发现新基因的珍贵资源。

11.4 分子标记和遗传图谱数据库

在比利时的生物多样性国际转运中心（ITC）保存着代表作物多样性的超过 1 200 多种的芭蕉属品种。大多数保存在种质库的种质属于公共领域，可免费地用于研究和育种。因此，使用分子标记的遗传学研究对鉴定保存材料的特征和检查其完整性非常有帮助。此外，评估收集的多样性是促成合理化的先决条件。

11.4.1 遗传多样性

自 20 世纪 80 年代以来，已经使用分子标记来研究香蕉野生和栽培种质，评估其遗传多样性。现已产生了各种各样的标记，包括 RAPD、同工酶、

RFLP、AFLP（Horry and Jay，1988；Jarret et al.，1992；Lebot et al.，1993；Pillay et al.，2000；Wong et al.，2001；Carreel et al.，2002；Ude et al.，2002；Nwakanma et al.，2003），以及最近出现的 SSR 和 DArT（Hippolyte et al.，2010）。所有这些研究清楚强调了芭蕉属组下不同种之间的遗传分化（Roux et al.，2008）。

TropGENE-DB（Ruiz et al.，2004）被设想用来管理关于热带作物的遗传信息，其中包括香蕉的模块（http：//tropgenedb. cirad. fr/en/banana. html）。已经创造了 Web 用户界面，允许快速和复杂的查询并呈现用户友好的结果。

目前，TropGENE-DB 香蕉模块包含约 1 100 项种质条目。它们与详细的表型信息相关联：地理来源，收集描述，农业形态学数据，疾病、害虫和非生物反应，亲子关系，育种数据等。相关的分子和遗传数据包括各种标记（例如 RFLP、AFLP 、微卫星、同工酶）的基因型、遗传图谱、与探针连锁的标记本身的信息、引物、序列数据等。也可获得凝胶图片、果实照片或植物疾病反应这些信息。

11.4.2 遗传作图

芭蕉属植物遗传图谱绘制的第一次尝试（Fauré et al.，1993；Noyeret al.，1997）现保存于 TropGENE-DB。图谱对比浏览器 CMap，发展为 GMOD 项目的一部分（http：//gmod. org/wiki/Cmap），现已被纳入 TropGENE-DB，支持这些连锁图谱数据的比较分析（表 11－2，图 11－4）。CMap 允许用户在一个给定的比较中选择任意图谱作为参考图谱。然后可将对比图谱添加到参考图谱的左边或右边，而该对比图谱来自数据库所有图谱中与参考图谱的对应集最小的图谱。对比图谱时，对应特征之间绘制出表明特征之间关系的线。事实上，CMap 能突出非同源和同源重复、同线重复、倒位导致的冲突和易位的证据。

表 11－2　TropGENE-DB 香蕉模块中标记类型的数量

标记类型	标记数
AFLP	358
同工酶	8
SSR	126
RAPD	8
RFLP	151

注：该数据库主要包含 CIRAD 产生的标记，不过也与国际生物多样性组织签署了谅解备忘录（MOU），任何愿意分享数据的合作伙伴也可将数据纳入数据库中。

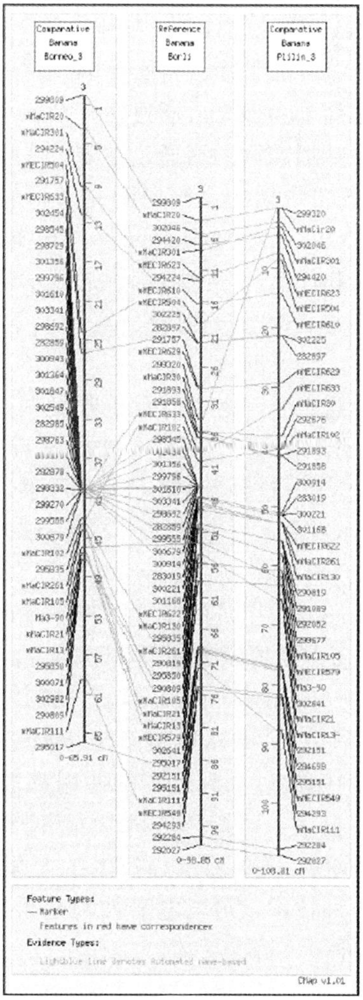

图 11 - 4　在 CMap 中两个香蕉连锁群之间对应关系

11.5　蛋白质组学工具和蛋白质数据库

通过多种技术可以进行基因表达分析和理解基因功能。基于 RNA 的系统生物学方法已经在很大程度上应用于经典的模式生物。这些所谓的转录组学方法是非常强大和高度自动化的，同时可以对数百个基因进行大规模筛选。然而，这些方法的成功很大程度上依赖于基因组学的进展。成功的技术，如 cDNA 微阵列、cDNA 扩增片段长度多态性（AFLP）和基因表达系列分析（SAGE）实际上仅限于已经在很大程度上表征的模式生物或物种。由于缺少基因组信息或由于与相关模式生物的序列差异，那些基于转录物的技术无法用于非模式生物，如香蕉（Carpentier et al.，2008b）。两种物种之间的基因序列很少相同，而直系同源基因通常带有核苷酸替换。检查基因表达的另一种方法是研究其最终产物蛋白质。蛋白质序列更为保守，通过对比众所周知的直系同源蛋白质可以非常有效地进行非模式基因产物的高通量鉴定。“蛋白质组学是通过对特定的生物/生理环境中发现的蛋白质的大规模研究来了解基因功能，并对活细胞的分子过程进行表征”（Liska and Shevchenko，2003a）。顾名思义，蛋白质组学与基因组学、转录组学和代谢组学这些同源的技术一脉相连。蛋白质组是在特定条件下，在特定时间点，存在于细胞、组织或生物体中的完整的一组蛋白质。

为了启动香蕉的蛋白质组学，必须着重于开发 Web 实验室技术（Carpentier et al.，2005）和使用不同的软件工具进行统计分析（Carpentier et al.，2008a），以及进行蛋白质鉴定。这种鉴定可以通过两种不同的方法来实现：①高通量依赖于数据库的检索（跨物种并基于 EST）（Carpentier et al.，2007）；②独立于数据库的从头测序结合容错检索（Samyn et al.，2007）。

本章将着重介绍蛋白质鉴定工具的实施。

20 世纪 80 年代末，蛋白质鉴定方法发生了转变。在那以前，Edman 测序是蛋白质鉴定的主要方法（Edman and Begg，1967）。为了确定肽的序列，标记 N 末端氨基酸，并以循环的方式从肽裂解出该氨基酸直到获得完整的肽序列。虽然 Edman 的从头测序从过去到现在的使用都取得相当大的成功，但是该方法仍相对较慢，不够灵敏。对矩阵辅助激光解吸电离（MALDI）和电喷雾离子化（ESI）质谱（MS）这些软电离技术的使用上，随后提供了更快更灵敏的工具。第一次，一种技术，可以电离大的分子（如肽），而这些分子随后可以通过质谱（MS）进行分析。

11.6 依赖于数据库的检索

对香蕉基因组表征的描述仍相对落后。举例说明，利用带有 organism=*Musa* * 限制的序列检索系统对 UniprotKB（Swissprot 和 TrEMBL）数据库（2009 年 3 月）进行查询，仅返回 1 302 项条目。与之形成对比的是，利用带有 organism=*Arabidopsis* * 这一限制的序列检索系统来检索，返回的条目有 88 312 项。每当一个基因组的表型描述很少时，跨物种鉴定则成为鉴定蛋白质的唯一选择（Wilkins and Williams，1997；Lester and Hubbard，2002；Mathesius et al.，2002；Liska and Shevchenko，2003a；Liska and Shevchenko，2003b；Witters et al.，2003；Carpentier et al.，2008c）。这种方法通过

比较目标蛋白和得到充分鉴定的物种的直系同源蛋白质来鉴定蛋白质。若将多个肽的质量信息用于比较亲缘物种的直系同源蛋白质，则可以成功进行跨物种鉴定。

肽质量指纹图谱（PMF）可用于蛋白质鉴定（Wilkins and Williams，1997）。肽质量指纹分析背后的概念十分简单，由几个组在大约相同的时间独立地实施（Henzel et al.，1993；James et al.，1993；Mann et al.，1993；Pappin et al.，1993；Yates et al.，1993）。通过PMF，将各种蛋白质消化成较小的肽（例如，通过胰蛋白酶），随后测量这些蛋白质水解肽的质量。在数据库检索中，根据酶或底物的特异性，将每个蛋白质序列进行计算机模拟消化。将每个单独数据库条目的计算机模拟消化肽与测量后的质谱进行比较。来自蛋白质消化物的所有肽中仅有一个子集的肽需要配对以获得一次有效命中。由于翻译后修饰（PTM）而导致氨基酸置换或质量移位的肽不在所选窗口或对应的所选的PTM，这种肽不被识别，并且不会有助于鉴定。一些软件工具可用于PMF（Fenyo，2000）。用于肽作图的最简单和最明显的计分方法是计算对应于数据库中每个蛋白质理论肽质量的测量肽质量的数目。然而，这些方法用于更大的蛋白质时会出现偏离。更精细的方法考虑了同样的基本原则，也试图更好地利用质谱信息来弥补蛋白质大小的示例影响。MOWSE（分子质量搜索）得分（Pappin et al.，1993）考虑了数据库中肽的相对丰度。就这一点而论，非冗余肽将有助于获得更高的分数。

然而，当生物体系统发育的亲缘关系越远时，直系同源基因保留一致性的比例更低。因此，发现显著和保守肽的概率降低，PMF无法鉴定或导致假阳性位点（Mathesius et al.，2002）。所以额外的信息至关重要，需要在MS/MS模式下测量衍生的碎片的离子质量。现已开发了几个精细的算法，如Mascot（Perkins et al.，1999）和SEQUEST（Eng et al.，1994），用于从肽MS/MS数据鉴定蛋白质。通过关联未解译的MS/MS光谱与从可用数据库中包含的相同质量的肽衍生的理论（预测）产物的离子光谱来鉴定肽。与可用序列相比，更具体的MS/MS分析为具有较低序列一致性的蛋白质提供了可靠的跨物种鉴定。只有少数的直系同源肽可能可以进行可靠的鉴定。

Mascot支持PMF和MS/MS离子搜索（Perkins et al.，1999）。相同的仪器获得的肽质量指纹和MS/MS谱，以及相结合的解译显著提高了识别率。基于Mascot概率的计分的主要优势在于可以估算假阳性位点的概率。概率（P）通过这种方式来计算，而这种方法中观察到实验数据集和每个序列数据库条目之间的匹配是随机事件。之后这一概率被报告为基于MOWSE计分的概率：$-10\lg P$。然而，报告大数据集时，不能低估多路测试的问题，Mascot计算的概率的正确性应该通过重复检索随机的诱饵数据库来进行验证。在相同的 Mascot 搜索中，采用自动化方法，利用 PMF 和系列信息将未解译的光谱与来自不同数据库的理论产物光谱相关联。许多不同的蛋白质序列数据库都可用，并能产生更佳匹配的其他信息。多个数据库的查询对鉴定进行了良好验证，在一些情况下，也可使假阴性和显著阳性位点之间产生差异。然而，减少假阳性分布仍然是自动化蛋白质鉴定中的一个主要问题。虽然香蕉基因组的特征描述还不充分，但是已有几个 EST 数据库可用。Mascot 在所有 6 个阅读框中翻译核酸数据库，并在不寻找起始密码子的情况下，预测理论碎片离子的一切可能，并在找到终止密码子后重新启动。EST 数据库对于确定杂交种的阳性结果非常有用，并且在对直系同源蛋白质搜索不成功的情况下，可以进行阳性鉴定。

11.7 数据库独立检索

当直系同源基因保持低比例的同一性时，发现显著并保守的肽的概率下降，并且 PMF-MS/MS 方法可能会失败。在固定的肽质量和肽片段的质量公差窗口中，所有算法需要所分析的肽与数据库条目无容错匹配。如果将这些公差窗口设置的太窄，可能会导致假阴性结果，而设置的过于宽松可能会导致假阳性结果。可以直接从光谱中提取出蛋白质识别信息而不依赖现有的数据库的方法将是非常有价值的。现今的情况是，获取的每个个体香蕉蛋白质的从头序列，之后能用于基于同源性的搜索算法，以进行鉴定。来自每个点的从头衍生的（部分）序列信息被组合在一个搜索查询中并使用搜索算法对其进行分析，这些方法包括 MS-BLAST（Shevchenko and Shevchenko，2001）、FASTS（Mackey et al.，2002）和 MS-Homology（Clauser et al.，1999）。FASTS 搜索数据库使用未知顺序的肽序列来评估所有可能的肽序列。由于 FASTS 所使用的查询肽的真正顺序未知，因此 FASTS 的要求只是这些肽不重叠。该算法基于启发式 FASTA 比较策略，并使用对准概率而不是相似性得分作为对比的标准（Mackey et al.，2002）。MS-BLAST 使用的是冗余的、简并的、部分失准的肽序列数据。MS-BLAST 不允许肽个体中有间隙，却允许肽之间出现间隙，而且可以是任意长度。因此，通过 MS/MS 解译获得的所有肽序列是组成顺序随意排列的单一搜索字符串（Shevchenko and Shevchenko，2001）。MS-Homology（Protein Prospector 4.0.5）是一个针对选定的数据库将一些从头衍生的肽序列进行比较的程序。在获得每个序列允许最大数目的氨基酸替代的指示后，输入这些肽序列。另外，也可以输入序列的一部分作为群集，以及容许因子（Clauser et al.，1999）。数据库搜索会寻找含有与列出序列相同或同源的肽的蛋白质。所有搜索算法得出的结果取决于进行测序的肽的数目和输入的序列信息的准确度，以及数据库的完整性和输入肽的物种序列的变异性。若研究人员试图通过序列相似性搜索来鉴定蛋白质，从酶解的蛋白质中鉴定出的肽的数量决定了这一鉴定是否成功。据统计，随着对更多的肽进行分析和匹配，鉴别出的蛋白质与数据库序列的相似性更小，同一性约为 50%（Mackey et al.，2002）。

非模式生物体的优化工作流程包括：①结合不同互补蛋白质分级分离、分离和定量技术，最大限度地提高分辨率和尽可能覆盖蛋白质组；②使用不同的互补 MS 技术和容错数据库检索。通过结合依赖于和独立于数据库的手段而进行的跨物种鉴定有望促进全球生物多样性组织进行的国际芭蕉属种质资源收集，目前收集的种质资源在联合国粮农组织（FAO）赞助下，存放在热带作物改良实验室（比利时的天主教鲁汶大学）。为了推动未来对香蕉组织的蛋白质组分析，并方便数据访问，研究人员利用二维凝胶电泳（2－DE）构建了个在线数据库（http://www.pdata.ua.ac.be/*Musa*/）。

11.7.1　不同数据的集成

科学家面临的一个关键问题是多样并分散的生物数据的整合和使用多种分析工具对其进行分析（Stein，2003）。全球挑战计划（GCP）生物信息学平台旨在通过提供信息平台来缓解此问题，该平台允许通过商定的域模型和可相互操作的应用程序工作台进行数据集成（Bruskiewich et al.，2006；Bruskiewich et al.，2008；Wanchana et al.，2008）。在这方面，已经开发了 GenDiversity 遗传多样性的可视化和分析工作台。由于篇幅所限，本节的内容主要对 GenDiversity 的特点进行描述。

11.7.2　GenDiversity：遗传多样性分析工作台

现已存在涉及数千种质基因型的非常大型的数据集。遗传学家的当前目标是通过对这些数据进行深入分析来解答关于驯化、物种遗传组织和不同作物模型之间比较的广泛问题。此外，对基因型、通行基因和表型等多级数据进行整合将有助于鉴定涉及复杂耐受基因的新颖多样的变体。

GenDiversity，是一个查询和分析的 web 应用，用于：①从不同的数据库中查询分散的基因分型数据；②组合两个或多个不同的基因分型研究；③将输出转换成不同的文件格式以进行多样性分析：DARwin（Perrier，2006），Convert（Glaubitz，2004），Structure（Falush et al.，2007），Haploview（Barrett et al.，2005）；④通过通行数据进行过滤；⑤计算等位基因频率；⑥计算连锁不平衡；⑦校验重复。SSR 和 SNP 基因分型研究目前已经可用。

这些数据来自 TropGENE-DB 数据库（Ruiz et al.，2004）和 GCP 中央储存库，其中存储由 GCP 项目生成的所有数据，网址为：http://gcpcr.grinfo.net/。经由

互联网可以访问 GenDiversity（http://gendiversity. cirad. fr/Home）。

（1）基于种质名/标记名的过滤。对于每个基因分型研究，用户可以访问分析的种质和标记的名单。用户可在列表中选择他们想要提取的组（种质×标记）。他们也可以输入一列种质和/或一列标记，用分号隔开（图 4－1）。用户可以定义所研究的物种的倍性。倍性被默认设置为二倍体。如果 GenDiversity 检测到真正的基因分型数据和倍性之间不一致，将会发出警告。用户可以定义允许丢失的数据比率的阈值。当所有种质的丢失数据率高于用户指定的比率，则这些种质都将在结果矩阵中进行预先检验。而预先检验的数据可以轻松删除。默认情况下，GenDiversity 会预先检查丢失了至少一个数据的所有种质。

（2）基于通行数据的过滤。用户可以过滤基因分型数据，方法是通过种质名称、标记名称和/或通行信息：种质 ID、收集、品种、谱系、收集地、描述符等（图 11－5）。

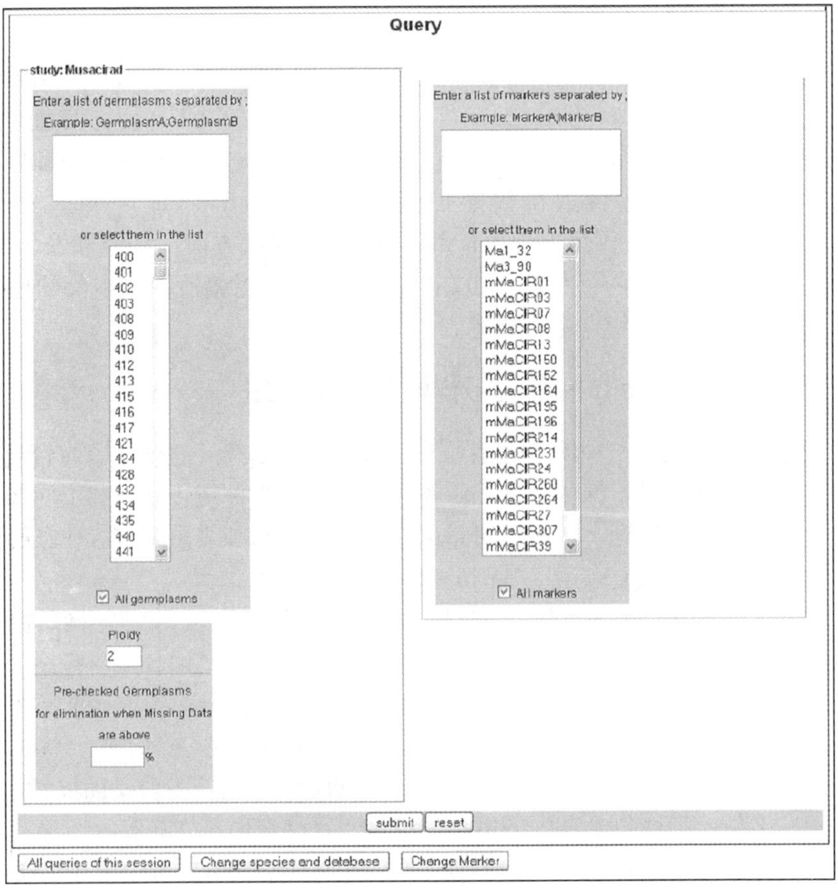

图 11－5　芭蕉属植物基因分型研究的种质和标记名单

在点击“通行和表型分型数据的过滤”后，GenDiversity 指示出哪些通行研究是可用的。用户只能选择一项研究。基于通行数据的过滤只可用于一个研究。选择通行研究后，会显示出一列通行属性。用户可以选择要用于过滤数据的属性。只有具有数值的属性可用复选框选择。对于每个选定的属性，会出现显示不同值的列表，而括号内会标明命中数。之后，用户可以通过选择数值进行过滤（图 11－6）。

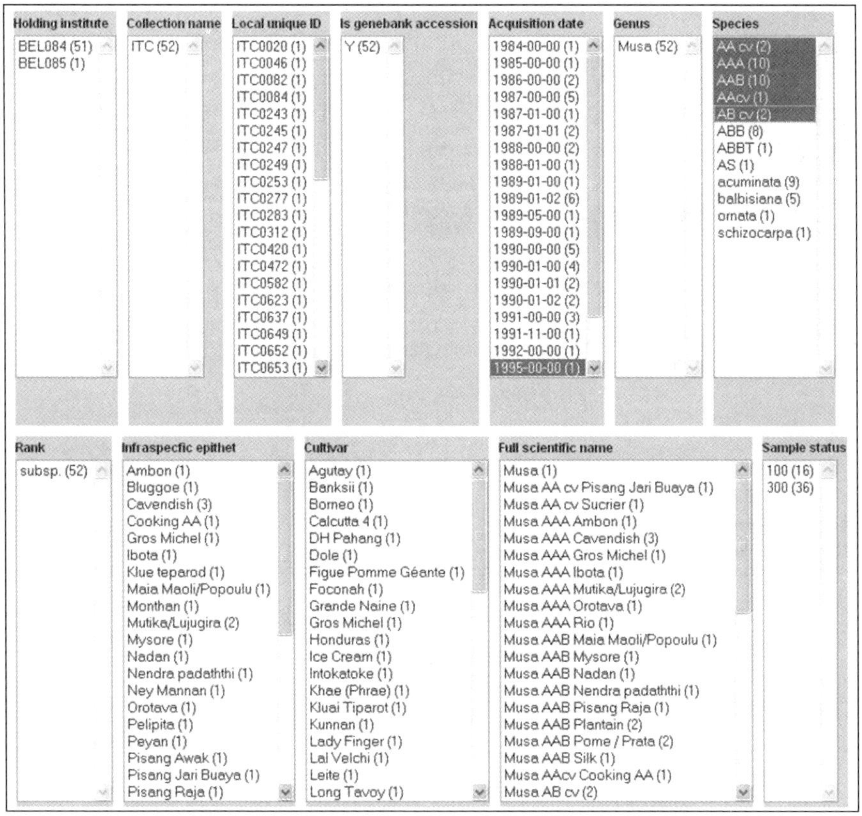

图 11－6　用户可以选择用于过滤数据的通行属性
（对于每一个属性，会出现显示不同值的列表，而括号内会标明命中数）

（3）结果显示。在结果显示中，用户可以链接到矩阵、统计数字、等位基因频率、连锁不平衡值及单态位点指示（图 11－7）。丢失的数据会被突出显示。几个功能可供使用：标记的数据缺失率，种质的数据缺失率，根据数据缺失率默认选中复选框删除种质，复选框全选或取消全选所有标记，复选框全选或取消全选所有种质和所有缺失数据的比率。

点击“删除选中的种质和标记”按钮，用户就可以删除选定的种质和标记。其结果是产生去掉之前选定的标记和种质的一个新矩阵。如果用户对新矩阵满意，可以将其以多种格式导出：DarWin、Structure、Convert 和 Haploview（适于 SNP）。

Germplasm 种质	mMaCIR01		mMaCIR03		mMaCIR07		mMaCIR08		mMaCIR13		mMaCIR150	
400	-99	-99	122	124	-99	-99	261	267	286	286	257	261
401	254	264	122	124	-99	-99	261	267	286	296	257	257
402	258	266	122	127	158	160	261	261	286	286	257	257
403	250	254	120	124	170	170	261	264	286	298	257	259
408	254	258	122	124	158	170	261	267	286	286	257	261
409	254	258	122	124	158	170	261	267	286	286	257	261
410	258	266	122	127	158	160	261	261	286	286	257	257
412	-99	-99	122	122	-99	-99	261	261	286	286	257	257
413	256	298	122	127	158	170	261	261	270	270	257	264
415	258	298	120	122	158	172	261	261	286	286	257	257
416	292	304	121	127	160	166	270	270	292	292	261	261
Missing Data 丢失数据	18.2%	18.2%	0.0%	0.0%	27.3%	27.3%	0.0%	0.0%	0.0%	0.0%	0.0%	0.0%
Monomorphic 单态	No		No		No		No		No		No	
☐	☐		☐		☐		☐		☐		☐	
	mMaCIR01		mMaCIR03		mMaCIR07		mMaCIR08		mMaCIR13		mMaCIR150	

Alleles frequencies are here

Linkage disequilibrium are here

图 11-7　基因分型研究的矩阵，含有对应标记名的种质名和单元格中的等位基因值
（丢失数据用黑色标注）

（4）不同结果矩阵的融合。如果同一页上有几个矩阵来自不同的基因分型研究，用户可以通过以下 4 个选项将它们合并：①整合融合结果而不反映其共同点。这将在没有信息的交叉点处产生数据丢失（例如某一种质只出现在一项研究中，而某一标记只出现在另一项研究中）。②只使用常见种质的融合。对于没有这两项研究的种质会产生信息的丢失。③只使用常见标记的融合。④使用常见种质和标记的融合。这对于检查不同研究之间的数据一致性十分重要。

对于同一对种质或标记，这些融合的结果可能突出不同的研究之间的不一致性。

缩略语

AFLP：扩增片段长度多态性

BAC：细菌人工染色体

BLAST：基本局部比对搜索工具
EST：表达序列标签
GCP：全球挑战计划
GMGC：全球芭蕉属植物基因组联盟
GMOD：通用模式生物系统数据库项目
LARD：大反转录转座子衍生物
LINE：长分散核因子
LTR：长末端重复序列
MOWSE：分子质量搜索
MS：质谱
PMF：肽质量指纹图谱
PTM：翻译后修饰
RAPD：随机扩增多态性 DNA
RFLP：限制性片段长度多态性
SAGE：基因表达系列分析
SINE：短分散核因子
SNP：单核苷酸位置
SSR：简单重复序列
TE：转座因子

致谢

基因注释平台是法国国家研究机构（ANR）支持的 GNPAnnot 项目（http://www.gnpannot.org）的一部分，并涉及以下组织：CIRAD、INRA 和国际生物多样性组织。

GenDiversity 由全球挑战计划（GCP）支持。我们还要感谢 Anne Vezina 对论文编辑的建议。

参考文献

Abad P, Gouzy J, Aury J-M, et al, 2008. Genome sequence of the metazoan plant-parasitic nematode *Meloidogyne incognita* [J]. Nat Biotech, 26: 909.

Aert R, Sági L, Volckaert G, 2004. Gene content and density in banana (*Musa acuminata*) as revealed by genomic sequencing of BAC clones [J]. Theoretical and Applied Genetics, 109: 129-139.

Altschul S F, Gish W, Miller W, et al, 1990. Basic local alignment search tool [J]. Journal of Molecular Biology, 215: 403 - 410.

Argout X, Fouet O, Wincker P, et al, 2008. Towards the understanding of the cocoa transcriptome: production and analysis of an exhaustive dataset of ESTs of *Theobroma cacao* L. generated from various tissues and under various conditions [J]. BMC Genomics, 9: 512.

Ashburner M, Ball C A, Blake J A, et al, 2000. Gene ontology: tool for the unification of biology [J]. Nature Genetics, 25: 25 - 29.

Aubourg S, Delseny M, Lecharny A, 2006. L'organisation des génomes végétaux révélée par leur annotation [J]. Biofuture, 265: 33 - 37.

Barrett J C, Fry B, Maller J, et al, 2005. Haploview: analysis and visualization of LD and haplotype maps [J]. Bioinformatics, 21: 263 - 265.

Bruskiewich R, Davenport G, Hazekamp T, et al, 2006. Generation Challenge Programme (GCP): standards for crop data [J]. Omics: a Journal of Integrative Biology, 10: 215 - 219.

Bruskiewich R, Senger M, Davenport G, et al, 2008. The generation challenge programme platform: semantic standards and workbench for crop science [J]. International Journal of Plant Genomics.

Carreel F, Gonzalez de León D, Lagoda P, et al, 2002. Ascertaining maternal and paternal lineage within *Musa* by chloroplast and mitochondrial DNA RFLP analysis [J]. Genome, 45: 679 - 692.

Carpentier S C, Witters E, Laukens K, et al, 2005. Preparation of protein extracts from recalcitrant plant tissues: an evaluation of different methods for two-dimensional gel electrophoresis analysis [J]. Proteomics, 5: 2497 - 2507.

Carpentier S C, Witters E, Laukens K, et al, 2007. Banana (*Musa* spp.) as a model to study the meristem proteome: acclimation to osmotic stress [J]. Proteomics, 7: 92 - 105.

Carpentier S C, Coemans B, Podevin N, et al, 2008a. Functional genomics in a non-model crop: transcriptomics or proteomics [J]? Physiologia Plantarum, 133: 117 - 130.

Carpentier S C, Panis B, Swennen R, et al, 2008b. Finding the significant markers: statistical analysis of proteomic data [J]. Methods in Molecular Biology (Clifton, NJ), 428: 327 - 347.

Carpentier S C, Panis B, Vertommen A, et al, 2008c. Proteome analysis of non-model plants: a challenging but powerful approach [J]. Mass Spectrometry Reviews, 27: 354 - 377.

Carver T, Berriman M, Tivey A, et al, 2008. Artemis and ACT: viewing, annotating and comparing sequences stored in a relational database [J]. Bioinformatics. Oxford, England, 24: 2672 - 2676.

Cheung F, Town C D, 2007. A BAC end view of the *Musa acuminata* genome [J]. BMC Plant Biology, 7: 29.

Childs K L, Hamilton J P, Zhu W, et al, 2007. The TIGR plant transcript assemblies database [J]. Nucleic Acids Research, 35: D846 - 851.

Clauser K R, Baker P, Burlingame A L, 1999. Role of accurate mass measurement (+/— 10 ppm) in protein identification strategies employing MS or MS/MS and database searching [J]. Analytical Chemistry, 71: 2871 - 2882.

Conte M G, Gaillard S, Droc G, et al, 2008a. Phylogenomics of plant genomes: a methodology for genome-wide searches for orthologs in plants [J]. BMC Genomics, 9: 183.

Conte M G, Gaillard S, Lanau N, et al, 2008b. GreenPhylDB: a database for plant comparative genomics [J]. Nucleic Acids Research, 36: D991 - 998.

Edman P, Begg G, 1967. A protein sequenator [J]. European Journal of Biochemistry/FEBS, 1: 80 - 91.

Eng J K, McCormack A L, Yates Iii J R, 1994. An approach to correlate tandem mass spectral data of peptides with amino acid sequences in a protein database [J]. Journal of the American Society for Mass Spectrometry, 5: 976 - 989.

Falush D, Stephens M, Pritchard J K, 2003. Inference of population structure using multilocus genotype data: linked loci and correlated allele frequencies [J]. Genetics, 164: 1567 - 1587.

Falush D, Stephens M, Pritchard J K, 2007. Inference of population structure using multilocus genotype data: dominant markers and null alleles [J]. Molecular Ecology Notes, 7: 574 - 578.

Fauré S, Noyer J L, Horry J P, et al, 1993. A molecular marker-based linkage map of diploid bananas (*Musa acuminata*) [J]. Theoretical and Applied Genetics, 87: 517 - 526.

Fenyo D, 2000. Identifying the proteome: software tools [J]. Current Opinion in Biotechnology, 11: 391 - 395.

Foissac S, Gouzy J, Rombauts S, et al, 2008. Department of plant B, genetics genome annotation in plants and fungi: EuGene as a model platform [Z].

Glaubitz J C, 2004. convert: A user-friendly program to reformat diploid genotypic data for commonly used population genetic software packages [J]. Mol Ecol Notes, 4: 309 - 310.

Henzel W J, Billeci T M, Stults J T, et al, 1993. Identifying proteins from two-dimensional gels by molecular mass searching of peptide fragments in protein sequence databases [J]. Proc Nat Acad Sci USA, 90: 5011 - 5015.

Hippolyte I, Bakry F, Seguin M, et al, 2010. A saturated SSR/DArT linkage map of *Musa acuminata* addressing genome rearrangements among bananas [J]. BMC Plant Biology, 10: 65.

Horry J, Jay M, 1988. Distribution of anthocyanins in wild and cultivated banana varieties [J]. Phytochemistry, 27: 2667 - 2672.

Hunter S, Apweiler R, Attwood T K, et al, 2009. InterPro: the integrative protein signature database [J]. Nucleic Acids Research, 37: D211 - 215.

James P, Quadroni M, Carafoli E, et al, 1993. Protein identification by mass profile fingerprinting [J]. Biochemical and Biophysical Research Communications, 195: 58 - 64.

Jarret R, Gawel N, Whittemore A, et al, 1992. RFLP-based phylogeny of *Musa* species in Papua New Guinea [J]. Theor Appl Genet, 84: 579-584.

Lebot V, Aradhya K M, Manshardt R, et al, 1993. Genetic relationships among cultivated bananas and plantains from Asia and the Pacific [J]. Euphytica, 67: 163-175.

Lescot M, Piffanelli P, Ciampi A, et al, 2008. Insights into the *Musa* genome: syntenic relationships to rice and between *Musa* species [J]. BMC Genomics, 9: 58.

Lester P J, Hubbard S J, 2002. Comparative bioinformatic analysis of complete proteomes and protein parameters for cross-species identification in proteomics [J]. Proteomics, 2: 1392-1405.

Lewis S E, Searle S M J, Harris N, et al, 2002. Apollo: a sequence annotation editor [J]. Genome Biology, 3: RESEARCH0082.

Liska A J, Shevchenko A, 2003a. Combining mass spectrometry with database interrogation strategies in proteomics [J]. Trends in Analytical Chemistry, 22: 291-298.

Liska A J, Shevchenko A, 2003b. Expanding the organismal scope of proteomics: Cross-species protein identification by mass spectrometry and its implications [J]. PROTEOMICS, 3: 19-28.

Mackey A J, Haystead T A J, Pearson W R, 2002. Getting more from less: algorithms for rapid protein identification with multiple short peptide sequences [J]. Molecular & Cell Proteomics, 1: 139-147.

Mann M, Hojrup P, Roepstorff P, 1993. Use of mass spectrometric molecular weight information to identify proteins in sequence databases [J]. Biological Mass Spectrometry, 22: 338-345.

Mathesius U, Imin N, Chen H, et al, 2002. Evaluation of proteome reference maps for cross-species identification of proteins by peptide mass fingerprinting [J]. Proteomics, 2: 1288-1303.

Noyer J L, Dambier D, Lanaud C, et al, 1997. The saturated map of diploid banana *Musa acuminata* [C]. Plant and Animal Genomes V Conference, January 12-16, San Diego, CA, USA.

Nwakanma D C, Pillay M, Okoli B E, et al, 2003. Sectional relationships in the genus *Musa* L. inferred from the PCR-RFLP of organelle DNA sequences [J]. Theor Appl Genet, 107: 850-856.

Ortiz-Vázquez E, Kaemmer D, Zhang H B, et al, 2005. Construction and characterization of a plant transformation-competent BIBAC library of the black Sigatoka-resistant banana *Musa acuminata* cv. Tuu Gia (AA) [J]. Theor Appl Genet, 110: 706-713.

Pappin D J, Hojrup P, Bleasby A J, 1993. Rapid identification of proteins by peptide-mass fingerprinting [J]. Current Biology, 3: 327-332.

Perkins D N, Pappin D J, Creasy D M, et al, 1999. Probability-based protein identification by searching sequence databases using mass spectrometry data [J]. Electrophoresis, 20:

3551 - 3567.

Perrier X, Jacquemoud-Collet J P, 2006. DARwin software [Z].

Pillay M, Nwakanma D C, Tenkouano A, 2000. Identification of RAPD markers linked to A and B genome sequences in *Musa* L [J]. Genome, 43: 763 - 767.

Roux N, Baurens F-C, Doležel J, et al, 2008. Genomics of banana and plantain (*Musa* spp.), major staple crops in the tropics [M]//Moore P H, Ming R. Genomics of tropical crop plants. New York: Springer New York: 83 - 111.

Rudd S, 2003. Expressed sequence tags: alternative or complement to whole genome sequences [J]? Trends in Plant Science, 8: 321 - 329.

Ruiz M, Rouard M, Raboin L M, et al, 2004. TropGENE-DB, a multi-tropical crop information system [J]. Nucleic Acids Research, 32: D364 - 367.

Šafář J, Noa-Carrazana J C, Vrána J, et al, 2004. Creation of a BAC resource to study the structure and evolution of the banana (*Musa balbisiana*) genome [J]. Genome, 47: 1182 - 1191.

Sàgi L, Volckaert G, Aert R, 2004. Gene content and density in banana (*Musa acuminata*) as revealed by genomic sequencing of BAC clones [J]. Theor Appl Genet, 109: 129 - 139.

Samyn B, Sergeant K, Carpentier S, et al, 2007. Functional proteome analysis of the banana plant (*Musa* spp.) using de novo sequence analysis of derivatized peptides [J]. J Proteome Research, 6: 70 - 80.

Schneider M, Bairoch A, Wu C H, et al, 2005. Plant protein annotation in the UniProt Knowledgebase [J]. Plant Physiology, 138: 59 - 66.

Shevchenko A, Shevchenko A, 2001. Evaluation of the effi ciency of in-gel digestion of proteins by peptide isotopic labeling and MALDI mass spectrometry [J]. Analytical Biochemistry, 296: 279 - 283.

Stein L D, 2003. Integrating biological databases [J]. Nat Rev Genet, 4: 337 - 345.

Stein L D, Mungall C, Shu S, et al, 2002. The generic genome browser: a building block for a model organism system database [J]. Genome Research, 12: 1599 - 1610.

Ude G, Pillay M, Nwakanma D, et al, 2002. Analysis of genetic diversity and sectional relationships in *Musa* using AFLP markers [J]. Theoretical and Applied Genetics, 104: 1239 - 1245.

Vilarinhos A D, Piffanelli P, Lagoda P, et al, 2003. Construction and characterization of a bacterial artifi cial chromosome library of banana (*Musa acuminata* Colla) [J]. Theor Appl Genets, 106: 1102 - 1106.

Wanchana S, Thongjuea S, Ulat V J, et al, 2008. The generation challenge programme comparative plant stress-responsive gene catalogue [J]. Nucleic Acids Research, 36: D943 - 946.

Wicker T, Sabot F, Hua-Van A, et al, 2007. A unified classification system for eukaryotic transposable elements [J]. Nat Rev Genet, 8: 973 - 982.

Wilkins M R, Williams K L, 1997. Cross-species protein identifi cation using amino acid composition, peptide mass fingerprinting, isoelectric point and molecular mass: a theoretical evaluation [J]. J Theor Biol, 186: 7-15.

Witters E, Laukens K, Deckers P, et al, 2003. Fast liquid chromatography coupled to electrospray tandem mass spectrometry peptide sequencing for cross-species protein identification [J]. Rapid Communications in Mass Spectrometry: RCM, 17: 2188-2194.

Wong C, Kiew R, Loh J P, et al, 2001. Genetic diversity of the wild banana *Musa acuminata* Colla in Malaysia as evidenced by AFLP [J]. Ann Bot, 88: 1017-1025.

Yates J R, Speicher S, Griffin P R, et al, 1993. Peptide mass maps: a highly informative approach to protein identification [J]. Analytical Biochemistry, 214: 397-408.

第 12 章 香蕉黑叶斑病致病因子——斐济球腔菌的基因组学

Cláudia Fortes Ferreira[①,a,*] Hermínio Souza Rocha[①,b]
Ricardo Franco Cunha Moreira[②] Lahyre Izaete Silveira Gomes[③]

摘 要： 黑叶斑病，也被称为黑条叶斑病（BLSD），是由斐济球腔菌（*Mycosphaerella fijiensis*）所引起。这种经空气传播的真菌叶斑病被认为是对香蕉危害最大的疾病之一。这一疾病十分凶险，可导致减产 20%～90%，它在全球范围内正在替代香蕉生球腔菌（*Mycosphaerella musicola*）——引发黄叶斑病的病原体。建立在广泛合作基础之上的对这一病菌的基因组测序将为香蕉叶斑病的全球疾病控制战略带来新思路。虽然斐济球腔菌的基因组序列初稿已于 4 年前发布，但是这几年科学家仍旧开展了很多工作，未来也将付出更多努力。本章将重点谈论基因组学的主要议题和其他用于促进斐济球腔菌研究的工具；特别要提及生物技术和 WGS（全基因组测序）在最近几年取得的重要进步。

关键词： 香蕉黑叶斑病，遗传学，基因组学，疾病控制策略

12.1 引言

香蕉黑叶斑病复合体涉及在香蕉中发现的 3 种真菌：斐济球腔菌、香蕉生

① Molecular Biology Department，Rua Embrapa，C. P. 007，44380-000，Cruz das Almas，Bahia，Brazil.

② Federal University of Bahia Recôncavo，Center for Agricultural Sciences，Environmental and Biological，Rui Barbosa，710-University Campus，CEP 44380-000，Cruz das Almas，Bahia，Brazil；e-mail：ricardofcm@ufrb. edu. br.

③ Federal University of Viçosa-UFV，Department of Phytopathology，Rua PH Rolfs 36570-000 Viçosa-MG，Brazil；e-mail：lahyreizaete@yahoo. com. br.

a. e-mail：claudiaf@cnpmf. embrapa. br.

b. e-mail：herminio@cnpmf. embrapa. br.

* 通信作者。

球腔菌和芭蕉球腔菌（*M. eumusae*）(Jones，2003)。黑叶斑病（图 12－1），又称黑条叶斑病（BLSD），由斐济球腔菌（无性型斐济假尾孢菌属）所引起；它是单倍体的、有性的、异宗交配系统的半活体寄生丝状子囊菌，被认为是对香蕉最具毁灭性的疾病之一（Stover and Simmonds，1987；Mourichon and Fullerton，1990；Mobambo et al.，1993；Cordeiro et al.，2004；Conde-Ferraez et al.，2007）。这种经空气传播的真菌叶斑病降低叶面光合作用，减少果实大小并诱导易感香蕉品种的早熟。斐济球腔菌是相对较新的病原体，并仍在迅猛地对香蕉种植的新区域进行定殖。这一病原体来势汹汹，可导致作物减产 20％～90％（Stover and Simmonds，1987；Mobambo et al.，1993），它在全球范围内正在替代引发黄叶斑病的香蕉生球腔菌病原体。控制 BLSD 的主要方法仍然依赖于杀菌剂的过量施用（在一些中美洲国家每年使用量高达 70 倍），耗费了近 27％的总生产成本（Mobambo et al.，1993），并且导致严重的健康问题和环境问题，因此，这种疾病被认为是在香蕉疾病中全球代价最高的疾病之一（Carlier et al.，2000）。

图 12－1　黑条叶斑病在香蕉品种 BRS Maravilha 上的症状
（图片来自 Edson Perito Amorim）

本章将重点谈论基因组学的主要议题和其他用于促进斐济球腔菌研究的工具，包括先进的生物技术手段和用来研究这一病原体的全基因组测序（WGS）战略。

12.2　用于斐济球腔菌研究的分子标记和其他生物技术工具的发展

过去 20 年，DNA 标记技术变革了植物病原体基因组分析的策略，并且广泛应用于分子植物病理学的多个领域。分子标记的出现为更快、更准确地鉴定植物病原体和它的早期发现提供了可能，并且已广泛应用于病原真菌的研究（Michelmore and Hulbert，1987；Meng et al.，1999；Armengol et al.，2010；Bhat and Browne，2010；Daval et al.，2010；Dubey et al.，2010；Gladieux et al.，2010；Njambere et al.，2010）。下文将阐述检测和识别分子技术的应用，遗传多样性研究及斐济球腔菌研究。

引发香蕉叶斑病的真菌在形态学上十分相似，这使得单独的形态学评估十分费力、模糊不清，有时还不可靠，而这阻碍了对斐济球腔菌进行适当控制的预防性管理策略的制订（Arzanlou et al.，2008）。

涉及利用分子标记来检测和鉴定斐济球腔菌的分子研究起始于 20 世纪 90 年代初期。Johanson 和 Jeger（1993）使用聚合酶链式反应（PCR）方法来迅速、准确地区分香蕉生球腔菌和斐济球腔菌，该方法利用的是来自一个可变区的引物，该可变区经鉴定存在于这两种物种的核糖体 DNA（rDNA）的内部转录间隔（ITS）1 区的核苷酸序列中。要达到同样的目的也可使用基于 PCR 技术的随机扩增多态性 DNA（RAPD）（Johanson et al.，1994）。

Arzanlou 等（2007）开发出了快速、可靠的、物种特异性的、基于分子诊断的工具来检测和定量球腔菌属（*Mycosphaerella*）的 3 个种，该工具使用了在肌动蛋白基因基础上发展出的 PCR 引物；可以检测低至 100 pg/μL 的斐济球腔菌 DNA。此外，研究人员开发出基于 β-微管蛋白基因的 TaqMan 实时 PCR 测验法，能检测低至每个 *Mycosphaerella* sp. 1 pg/μL 的 DNA。这些测验法加强对香蕉感染地区病原体的检测，并且可以迅速、快捷地用于隔离策略的订制，以达到控制目的。由于香蕉叶斑病复合体十分复杂，因此需要这些新型测验法，以便全球更多的实验室可以快速地诊断，特别是使用传统 PCR 方法快速地诊断。

交配型基因也呈现出有趣的新结果（Conde-Ferràez，2007；Arzanlou et al.，2010；Conde-Ferràez et al.，2010）。Conde-Ferràez 等（2007）分离了斐济球腔菌的交配型位点（*mat1 -1* 和 *mat1 -2*）并描述了其表征。这类的特性描述需要了解生殖和重组过程，流行病学及这种病菌与其他物种的相互作用。此外，该结果证实了这两种等位基因在内倒位区的高度同一性，暗示了独特倒位事件的出现，而这些独特的倒位事件可能对斐济球腔菌的进化产生关键作用（Conde-Ferràez et al.，2007）。

近日，Conde-Ferràez 等（2010）开发了多重 PCR 方法来鉴定墨西哥斐济球腔菌交配型的分布和频率，并且报告了不同大小片段的 *mat1 -1* 和 *mat1 -2* 基因。研究人员将这些相同引物用于扩增香蕉生球腔菌的等位基因的片段，并对其进行测序，发现这些片段显示出与斐济球腔菌的片段高度相似，这意味着该方法适用于这两个物种。

研究人员通过 PCR 和染色体步移法对球腔菌属真菌的交配型位点进一步分析（Arzanlou et al.，2010）。研究人员还将序列与之前公布的斐济球腔菌及其他球腔菌属的交配型位点进行了比较。科学家报道了两个球腔菌属真菌特异的新基因（*MATORF1* 和 *MATORF2*）。这些基因编码的蛋白质显示出种间同源性高于种内同源性（Arzanlou et al.，2010）。这些研究结果对更好地了解这些病原体的进化路径和配对过程下潜藏的动力学和机制，以及对种群结构的分析，共同进化和病原体分化的研究具有至关重要的意义。

了解发病周期和毒力因子（ariaud et al.，2009）对制订有效的作物保护策略至关重要，这些策略包括通过传统植物育种、基因工程或使用生物控制策略来开发抗性基因型作物（González-Fernández et al.，2010）。

众所周知的是，环境变量，如气温、大气压、光、湿度，会影响真菌的空气生物学和疾病的流行病学。了解病情进程、扩散模式和对病原体种群结构进行分析是实施适当的疾病控制策略的关键（综述见 Chuchill，2010），特别是对于地处热带地区的国家。此外，考虑到斐济球腔菌的异宗结合性质，在不断变化的环境下，甚至是在使用了杀真菌剂的环境下，重组事件会增加适应性更强的分化的发展机会，而这对于种群结构和遗传多样性的研究至关重要。

另一方面，随着分子生物学技术的发展，许多复杂问题迎刃而解，例如菌种的来源，其种群结构的变化和该疾病的种群动态。有学者也研究了获取斐济球腔菌毒力信息的方法（Donzelli and Chuchill，2007）。分子生物学的重要应用包括种群遗传多样性分析，遗传变异特征描述，植物病原体类群的分类和系统发育，基因作图和标记，基因组指纹识别和基因组进化。

研究人员利用不同的分子标记在世界各地展开了对斐济球腔菌的多项遗传研究。限制性片段长度多态性（RFLP）被用于东南亚、非洲、拉丁美洲和太平洋群岛的斐济球腔菌分离种群的全球种群研究（Carlier et al.，1996）。这项研究表明，这些种群中发生随机有性繁殖，而东南亚是这种真菌的起源中心，并在距今较近的时候从那里传播到世界其他地区。

Zapater 等（2004）开发出 PCR-RFLP 标记，应用于研究斐济球腔菌的种群结构。这些标记用于源自不同地方的分离种群，经证实，这些标记很容易转移到香蕉种植国家的实验室，因为所采用的 2.0％琼脂糖凝胶只需要基本的实验室设备。更重要的是，传统 PCR 方法的快速发展，使得这一方法易于转移到发展中国家的实验室中，而在这些发展中国家的香蕉种植园正遭受这些疾病的肆虐。

近年来研究人员越来越多地使用微卫星分子标记，因其高多态性，以及特异性位点。现已开发出斐济球腔菌的许多微卫星标记，并用于种群研究

（Yang and Zhong，2008；Zapater et al.，2008）。

研究人员使用二核苷酸和三核苷酸微卫星标记对尼日利亚斐济球腔菌的分离种群的遗传变异性进行了研究（Muller et al.，1997）。这些标记显示出病变植株、同一植物上的病灶之间、不同栽培品种之间、地理位置之间的遗传变异。

1999 年，Neu 等通过 11 个微卫星标记对尼日利亚和墨西哥分离种群的特征进行了描述，并证实了 Carlier 等（1996）发现的结果，即两个地方分离种群之间表现出区域内的低多样性和很大的地域差异。

Rivas 等（2004）利用 PCR-RFLP 和微卫星标记来分析最近遭受该病原体侵袭的地区的种群结构，从而推断出斐济球腔菌在大陆范围的基因流动和传播模式。通过分析从拉丁美洲、加勒比海地区、非洲的 13 个国家的香蕉种植园中采集的分离种群，作者报道了伴随着近期斐济球腔菌在拉丁美洲-加勒比海地区入侵的始祖效应，这暗示了本病在大陆范围的传播是随机的，而这既可能是子囊孢子的有限传播所致，也可能是受感染的植物材料的易位所致。这个信息很有价值，而这些技术可以应用于在全球范围开展的流行病学调查和疾病监测计划。

在开发的斐济球腔菌的新标记中，一些是基于反转录转座子 IRAP（反转录转座子间扩增多态性）和 REMAP（反转录转座子-微卫星扩增多态性），它们首次应用于描述巴西的斐济球腔菌种群的表征。这些标记不但高多态性，而且低成本。研究人员利用不同分子标记和数据对巴西的斐济球腔菌种群进行了特征描述，数据显示存在 3 个有结构的种群，尽管该病原体在不久前才传入该国（Silva，GF，个人交流）。

Fahleson 等（2009）最近使用 PCR-RFLP 和微卫星标记对从乌干达采集的分离种群进行特征分析，结果显示维多利亚湖流域的斐济球腔菌种群构成了一个同源种群，而这可能是由最近的一个创始人事件的人口减少所造成的。

科学家利用 19 个微卫星标记对目前定殖在哥斯达黎加和喀麦隆的斐济球腔菌种群结构进行了分析。经报道，两国的种群之间含有低到中度的遗传分化，而喀麦隆的种群因距离而分离，这表明了在香蕉生产区，种群因孢子的逐渐扩散而不断扩展分布区（Halkett et al.，2010）。这份报告开辟了新的视角，丰富了关于定殖路径和传播过程的数据。

单核苷酸多态性（SNP）被应用到 rDNA 来确定从尼日利亚采集的不同地理位置的斐济球腔菌种群的遗传结构，并鉴定出斐济球腔菌的 14 个 SNP 单倍型。这项工作还表明斐济球腔菌已基本取代了香蕉生球腔菌，而种群中芭蕉球腔菌的表现度更小（Zandjanakou-Tachin et al.，2009）。

除了分子标记，其他工具，如 BAC（细菌人工染色体）文库被视为真正的遗传信息库，用于物理作图、图位克隆和基因组测序，而大型插入 BAC 文库的方法被视为现代基因组研究的重要工具（Canto-Canché et al.，2007）。这些文库对于与代谢途径相关的基因簇的分离和表征描述，以及对于同线性分析和基因拷贝数都十分关键。斐济球腔菌的一个 BAC 文库已建成，由 1 920 个克隆体构成，代表基因组当量的 4.9 倍。该策略利用非原生质体技术消除了以前使用原生质体技术产生的一些局限性，如费时费力，且一定程度上单凭经验。根据作者表述，这是首次使用非原生质体技术构建的 BAC 文库，给基因组学研究，例如图位克隆技术和其他研究提供了一个经济、便捷的方法（Canto-Canchè et al.，2007）。

研究人员在不同培养环境下构建了两个斐济球腔菌分离种群（MF5-Hilo 和 MF11-Hilo）的 3 个 cDNA 文库。之后生成了总共 3 771 个 EST，集合成了有 1 945 个单一序列的一个序列集。科学家对与病原性有关的 50 个推定基因进行鉴定（Cho et al.，2008），并将 EST 与来自 18 种真菌的测序过的基因组的基因数据进行了比较。有关斐济球腔菌的基因发掘的研究十分缺乏，因而这项工作将为阐明与这种病原体的病原性、抗药性、生命周期有关的基因铺平道路。

遗传连锁图谱这一重要的工具可用于进化研究和对目的基因，如负责植物抗性的基因进行定位克隆（Drader and Kleinhofs，2010；Kang and Rouf Mian，2010；Ubayasena et al.，2010）。科学家利用分子标记已构建出许多致病真菌物种的遗传连锁图谱（Kema et al.，2002；Zhong et al.，2002；Kuhn et al.，2006；Xu et al.，2009）。

斐济球腔菌的首张遗传连锁图谱于 2008 年公布。研究人员利用 298 个 AFLP 标记、16 个 SSR 标记和 MAT 位点对 CIRAD86（喀麦隆）和 CIRAD139A（哥伦比亚）分离种群之间的杂交种进行了分析。该图谱横跨了超过 23 个连锁群，覆盖了 1 878 cM（Manzo-Sánchez et al.，2008）。使用遗传图谱数据估算出单倍体基因组的总大小为 4 298.2 cM。

其他方法和标记也已用于研究斐济球腔菌的许多方面，如电泳核型分布（Rodriguez-Garcia et al.，2006），多基因的核苷酸序列数据集和形态学（Arzanlou et al.，2008）。

12.3 斐济球腔菌的基因组测序

目前，测序技术的进步已产生大量关于致病真菌基因的信息，这些基因对植物感染和定殖十分重要。更好地了解真菌与植物的相互作用是开发有效控制策略的关键。由于对真菌发病机制下潜藏的分子机制的了解仍然有限，所以比较和功能基因组学研究为理解宿主—病原体的相互作用提供了新视角（Xu et al.，2006）。

成立于2003年的国际球腔菌基因组学协会（IMGC；http://imgc.inibap.org/）不断努力，将禾生球腔菌（*M. graminicola*）（Goodwin et al.，2004）作为模型，获得了更多关于斐济球腔菌的遗传和基因组研究信息。以该协会为首启动了一个国际项目，旨在对斐济球腔菌毒株、CIRAD86和40 000个EST的基因组进行测序（Kema，2009）。

能源联合基因组研究所（JGI）已于2007年8月公布了斐济球腔菌基因组的图谱初稿，并于2010年5月公布了二稿。

1.0版的全基因组鸟枪法读取产生的覆盖范围为约7.11倍，602 555个读取集合成395个基因组支架，共计73.4 Mb。在公布了74.1 Mb的第二版基因组后，总共13 903个基因的结构和功能被注释（http://genome.jgi-psf.org/Mycfi 2/Mycfi 2.info.html）。如果要更彻底地了解斐济球腔菌测序项目的整个概念，请参阅Kema（2009）。要想了解更多有关巴西研究人员在MyGene项目的卓越贡献（该项目包括禾生球腔菌和斐济球腔菌的基因组的比较研究），以及了解对斐济球腔菌致病基因的注释的更多信息，请参见Martins等（2008）。

自从2010年5月2.0版本发布以来，有JGI公开的斐济球腔菌基因组序列信息已经呈现显著的效果，预计更多信息将很快面世。

斐济球腔菌基因组图谱的初稿被用来开发14个具有高度多态性的微卫星引物，这些引物中平均每个位点有6.43个等位基因（Yang and Zhong，2008）。

最近，研究人员开发出来自斐济球腔菌基因组数据库的VNTR引物。5个VNTR标记显示多个等位基因在来自世界不同地区（印度尼西亚、坦桑尼亚、布隆迪、菲律宾、加蓬、哥伦比亚、喀麦隆和新喀里多尼亚）的分离种群的参考集和哥斯达黎加香蕉种植园的一个种群中进行了优化。参考集的PIC值介于0.641 4～0.754 4，而种群集的PIC值介于0.040 0～0.737 3，而80%的PIC值高于0.60（Garcia et al.，2010）。这些数值表明这些标记的信息量大，可用于抗药性管理，并用于制订育种策略来控制香蕉黑叶斑病。这是关于直接从斐济球腔菌基因组序列开发出的VNTR－微卫星的第一个报告（V1.0；Garcia et al.，2010）。

研究人员利用87个微卫星标记、3个VNTR标记、交配型位点和覆盖1 417 cM的235个DArT标记来构建源自CIRAD86和CIRAD139A分离种群之间杂交体的个体的遗传连锁图谱，其中包含19个连锁群（LOD>10），而该图谱的构建已到了最后阶段，不久后应该就会公布（Ferreira et al.，2009）。

研究植物和病原体的相互作用及了解抗药性的分子机制（Ma and Michailides，2005）对于育种策略和其他控制手段都至关重要。科学家利用抑制性消减杂交（SSH）cDNA 文库对源自斐济球腔菌和香蕉交互作用的 EST 进行了分析，鉴定出在宿主和病原体感染后期诱导的转录物（Portal et al.，2011）。Portal 等（2011）鉴定出与植物和真菌的相互作用有关的特异性基因，而定量 PCR 试验揭示了编码病程相关（PR）蛋白的抗真菌基因，还揭示了 GDSL（即 Gly-Asp-Ser-Leu，甘氨酸—天冬氨酸—丝氨酸—异亮氨酸）类脂肪酶仅能被短暂诱导，接种后天数只有 30 d，这表明真菌可能会积极抑制植物防御。这些数据也说明了可以将新信息汇入基因组库，增加宿主和病原体相互作用的数据。

12.4 未来展望

科学家对真菌病原体及它们与植物相互作用的研究使用了多种方法，从传统遗传学、细胞生物学、生物化学到现代、全面并且高通量的组学技术，再配以合适的生物信息学工具（González-Fernández et al.，2010）。

近年来，植物病原真菌的研究已取得迅速发展，特别是涉及发病机制和宿主回避性的复杂机制的研究，而这主要归功于这些真菌的基因组序列和用于功能基因组学分析的资源的可用性（Ma et al.，2005；Stergiopoulos et al.，2010；Portal et al.，2011）。另外，也归功于转录组学、蛋白质组学、代谢组学等研究的发展，这些研究与定向诱变或转基因研究相结合解决了分子宿主—病原体串扰的问题。

如上所述，近年来揭示斐济球腔菌基因组的研究已取得了巨大进展，这主要归功于生物技术工具的快速革新。科学家已开发出用于研究宿主—病原体相互作用的斐济球腔菌的转化系统，也对在植物感染的多个阶段表达绿色荧光蛋白（GFP）的转化体的生长进行了表征描述（Balint-Kurti et al.，2001）。这份报告对于了解宿主—病原体感染模式中潜藏的机制是十分重要的。

众所周知，植物性病原真菌暴露在宿主植物或杀真菌剂分泌的活性氧簇（ROS）中。Beltran-Garcia 等（2009）分析了斐济球腔菌对过氧化氢（H_2O_2）和百草枯的氧化应激反应。对 H_2O_2、百草枯和氢醌的氧化应激反应具有过氧化氢酶和超氧化物歧化酶（SOD）活性特征。结果表明，过氧化氢酶不但对于 H_2O_2 快速分解十分重要，并且直接关系到细胞活性，而通过氧化还原或百草枯、对苯二酚形成的超氧阴离子对斐济球腔菌细胞活性的影响比 H_2O_2 还大（Gàrcia-Beltràn et al.，2009）。关于这类研究的文献很少，但是对理解病原体暴露在杀菌剂中的过程中涉及的防御机制十分重要，病原体通过分泌自身毒素来促进它们在植物宿主组织中的渗透及成功定殖（Gàrcia-Beltràn et al.，

2009)。科学家还研究了香蕉植物抵抗斐济球腔菌的防御机制。Cavalcante 等（2011）评估了过氧化物酶和 H_2O_2 的作用，以及与抗性基因型 Calcutta 4 高度灵敏反应的关联。

最近，研究人员从健康香蕉（cv. Grande Naine）叶生的水浸提液中检测出抗斐济球腔菌活性，这暗示该植物产生了植物保护剂（Cruz-Cruz et al.，2010）。

毫无疑问的是，植物—病原体的相互作用还有很多疑问需要解答，而与芭蕉属植物对黑叶斑病耐受有关的甲基化过程（Gimenèz et al.，2006），分泌组分析（Chuc-Uc et al.，2011）和基本生物学工具的发展能帮助阐明构成香蕉黑叶斑病复合体的 3 种主要物种的其他未知方面（Diaz-Trujillo et al.，2009）。

随着最近植物病原体的基因组序列数据库的扩展，位点对表型或基因对表型的反向遗传学工具，如剔除、RNAi、ATMT、REMI 和 TILLING 已经成为越来越有吸引力的方法，用以阐明其宿主—病原体相互作用（亲和或不亲和）的分子基础，植物病原菌的发展，毒力及致病性（Bhadauria et al.，2009）。

这些反向遗传学工具可以有效地解码基因组信息，将其转化为生物信息（Weld et al.，2006；Nakayashiki and Nguyen，2008）。在后基因组时代，通过同源重组进行的基因靶向（剔除）已成为确定基因功能的最有影响力的反向遗传学工具。

显然，对斐济球腔菌病原体的研究已经取得了很大进展，而且在某种程度上，大部分的进步源自大家的共同努力，这得益于成功的伙伴关系（Kema，2009）。随着斐济球腔菌测序基因组第二个版本的发布，未来几年有望在育种和对这种破坏性疾病的控制策略上取得更多进展。基因组序列是理解这种病原体的第一步，而这一数据生成的新信息只算是触及了香蕉叶斑病复合体的冰山一角。

基因组测序流程提供的信息将为扩大对涉及病原体定殖基因和升级控制策略的研究设置标准，因为对香蕉黑叶斑病的控制面临着许多挑战（Churchill，2010）。鉴于基因组已经得以测序的其他物种的研究经验，下一步肯定会涉及蛋白质组学研究，因为蛋白质组学研究将会提炼所产生的所有信息，并为育种者提供更多有用信息，推进世界各地的香蕉育种计划。

缩略语

ATMT：根癌农杆菌介导转化

BAC：细菌人工染色体

BLSD：黑条叶斑病

DArT：多样性阵列技术
EST：表达序列标签
GFP：绿色荧光蛋白
PCR：聚合酶链式反应
RAPD：随机扩增多态性 DNA
REMI：限制性内切酶介导的整合
RFLP：限制性片段长度多态性
RNAi：干扰
ROS：活性氧簇
SNP：单核苷酸多态性
SSH：抑制性消减杂交
TILLING：定向诱导基因组局部突变技术
VNTR：可变数目串联重复
WGS：全基因组测序

参考文献

Armengol J，Vicent A，León M，et al，2010. Analysis ofpopulation structure of *Rosellinia necatrix* on *Cyperus esculentus* by mycelial compatibility and inter-simple sequence repeats (ISSR)［J］. Plant Pathol，59：179－185.

Arzanlou M，Abeln E C A，Kema G H J，et al，2007. Molecular diagnostics for the sigatoka diease complex of banana［J］. Phytopathol，97：1112－1118.

Arzanlou M，Groenewald J Z，Fullerton R A，et al，2008. Multiple genealogies and phenotypic characters differentiate several novel species of *Mycosphaerella* and related anamorphs on banana［J］. Persoonia，20：19－37.

Arzanlou M，Crous P W，Zwiers L H，2010. Evolutionary dynamics of mating-type loci of *Mycosphaerella* spp. Occurring on banana［J］. Eukar Cell，9 (1)：164－172.

Balint-Kurti P J，May G D，Churchill A C L，2001. Development of a transformation system for *Mycosphaerella* pathogens of banana：a tool for the study of host/pathogen interactions［J］. FEMS Microbiol Lett，195：9－15.

Bhadauria V，Banniza S，Wei Y，et al，2009. Reverse genetics for functional genomics of phytopathogenic fungi and Oomycetes［J］. Comp Funct Genom Article ID 380719，11.

Bhat R G，Browne G T，2010. Specific detection of *Phythophthora cactorum* in diseased strawberry plants using nested polymerase chain reaction［J］. Plant Pathol，59：121－129.

Beltran-Garcia M J，Manzo-Sanches G，Guzmán-Gonzalez S，et al，2009. Oxidative stress response of *Mycosphaerella fijiensis*，the causal agent of black leaf streak disease in banana plants，to hydrogen peroxide and paraquat［J］. Can J Microb，55：887－894.

Carlier J, Lebrun M H, Zapater M F, et al, 1996. Genetic structure of the global population of banana black leaf streak fungus, *Mycosphaerella fijiensis* [J]. Mol Ecol, 5: 499-510.

Carlier J E, Fouré F, Gauhl D R, et al, 2000. Black leaf streak [M]//DR Jones. Diseases of banana, abacá and enset. Wallingford, UK: CABI Publishing: 37-79.

Canto-Canché B, Guillén-Maldonado D K, Peraza-Echeverría L, et al, 2007. Construction and characterization of a bacterial artificial chromosome library of the causal agent of black Sigatoka fungal leaf spot disease of banana and plantain, *Mycosphaerella fijiensis* [J]. Mol Biotechnol, 36: 64-70.

Cavalcante, M J B, Escoute J, Madeira J P, et al, 2011. Reactive oxygen species and cellular interactions between *Mycosphaerella fijiensis* and banana [J]. Tropical Plant Biology, 4: 134-143.

Cho Y, Hou S, Zhong S, 2008. Analysis of expressed sequence tags from the fungal banana pathogen *Mycosphaerella fijiensis* [J]. The Open Mycol J, 2: 61-73.

Chuc-Uc J, Brito-Argáez L, Canto-Canché B, et al, 2011. The in vitro secretome of *Mycosphaerella fijiensis* induces cell death in banana leaves [J]. Plant Physiol Biochem, DOI: 10.1016/j.plaphy.2011.02.006.

Churchill A C L, 2010. *Mycosphaerella fijiensis*, the black leaf streak pathogen of banana: progress towards understanding pathogen biology and detection, disease development, and the challenges of control [J]. Mol Plant Pathol, 1: 22.

Conde-Ferràez L, Waalwijk C, Canto-Canchè B B, et al, 2007. Isolation and characterization of the mating type locus of *Mycosphaerella fijiensis*, the causal agent of black leaf streak disease of banana [J]. Mol Plant Pathol, 8: 111-120.

Conde-Ferráez L, Grijalva-Arango R, Canto-Canché B B, et al, 2010. The development of mating type-specific primers for *Mycosphaerella fijiensis*, the causal agent of black sigatoka of banana, and analysis of the frequency of idiomorph types in Mexican populations [J]. Aust Plant Pathol, 39: 217-225.

Cordeiro Z J M C, Matos A P M, Silva S O, et al, 2004. Impact and management of black sigatoka in Brasil [J]. XVI Reunião Internacional ACORBAT: 63-69.

Cruz-Cruz C, Ramirez-Tec G, Garcia-Sosa K, et al, 2010. Phytoanticipins from banana (*Musa acuminata* cv. Grande Naine) plants with antifungal activity against *Mycosphaerella fijiensis*, the causal agent of black-sigatoka [J]. Eur J Plant Pathol, 126: 459-463.

Daval S, Lebreton L, Gazengel K, et al, 2010. Genetic evidence for differentiation of *Gaeumannomyces graminis* var. *tritici* into two major groups [J]. Plant Pathol, 59: 165-178.

Diaz-Trujillo C, Kobayashi A, Zwiers L H, et al, 2009. Developing tools for *Mycosphaerella fijiensis* studies [J]. Gewasbescherming Jaargang, 40: 3.

Donzelli B G G, Chuchill A C L, 2007. A quantitative assay using mycelial fragments to assess virulence of *Mycosphaerella fijiensis* [J]. Phytopathology, 97 (8): 916-929.

Drader T, Kleinhofs A, 2010. A synteny map and disease resistance gene comparison between barley and the model monocot *Brachypodium distachyon* [J]. Genome, 53: 406 - 417.

Dubey S C, Tripathi A, Singh S R, 2010. ITS-RFLP fingerprinting and molecular marker for detection of *Fusarium oxysporum* f. sp. *ciceris* [J]. Folia Microbiol, 55 (6): 629 - 634.

Fahleson J, Nakyanzi M, Okori M, et al, 2009. Genetic analysis of *Mycosphaerella fi jiensis* in the Ugandan Lake Victoria region [J]. Plant Pathol, 58: 888 - 897.

Ferreira C F, Van der Lee TAJ, Zapater M F, et al, 2009. A genetic linkage map of *Mycosphaerella fijiensis* using SSR and DArT markers [J]. Fungal Genet Rep, 565 Abstract. N° 513.

Garcia S A L, Van der Lee T A J, Ferreira C F, et al, 2010. Variable number of tandem repeat markers in the genome sequence of *Mycosphaerella fijiensis*, the causal agent of black leaf streak disease of banana (*Musa* spp.) [J]. Genet Mol Res, 9 (4): 2207 - 2212.

Giménez C, Palacios G, Colmenares M, 2006, *Musa* methylated DNA sequences associated with tolrance to *Mycosphaerella fijiensis* toxins [J]. Plant Mol Biol Repor, 24: 33 - 43.

Gladieux P, Zhang X G, Róldan-Ruiz I, et al, 2010. Evolution of the population structure of *Venturia inaequalis*, the applescab fungus, associated with the domestication of its host [J]. Mol Ecol, 19: 658 - 674.

Goodwin S B, Waalwijk C, Kema G H J, 2004. Genetics and genomics of *Mycosphaerella graminicola*: a model for the Dothideales [M]//Arora D K, Khachatourians G G. Applied mycology and biotechnology. New York, USA: Elsevier: 315 - 330.

González-Fernández R, Prats E, Jorrín-Novo J J, 2010. Proteomics of plant pathogenic fungi [J]. J Biomed Biotechnol Article, ID 932527.

Halkett F, Coste D, Rivas Platero G F, et al, 2010, Genetic discontinuities and disequilibria in recently established populations of the plant pathogenic fungus *Mycosphaerella fijiensis* [J]. Mol Ecol, 19: 3909 - 3923.

Johanson A, Jeger M J, 1993. Use of PCR for detection of *Mycosphaerella fijiensis* and *M. musicola*, the causal agents of sigatoka leaf spots in banana and plantain [J]. Mycol Res, 97: 670 - 674.

Johanson A, Crowhurst R N, Rikkerink E H A, et al, 1994. The use of species-specifi cDNA probes for the identification of *Mycosphaerella fijiensis* and *M. musicola*, the causal agents of sigatoka disease of banana [J]. Plant Pathol, 43: 701 - 707.

Jones D R, 2003. The distribution and importance of the *Mycosphaerella* leaf spot diseases of banana [R]//Jacome L, Lepoivre P, Marin D, et al. *Mycosphaerella* leaf spot diseases of bananas: present status and outlook. Proceedings of the Second International Workshop on *Mycosphaerella* leaf spot diseases of bananas, San José, Costa Rica: 25 - 42. INIBAP, France.

Kang S T, Rouf Mian M A, 2010. Genetic map of the powdery mildew resistance gene in soybean PI 243540 [J]. Genome, 53: 400 - 405.

Kema G H J, 2009. Sequencing the major *Mycosphaerella* pathogens of wheat and banana

[J]. ActaHort，828：147－152.

Kema G H J，Goodwin S G，Hamza S，et al，2002. A combined amplified fragment length polymorphism and randomly amplified polymorphim DNA genetic linkage map of *Mycosphaerella graminicola*，the *Septoria Tritii* leach blotch pathogen of wheat [J]. Genetics，161：1497－1505.

Kuhn M L，Gout L，Howlett B J，et al，2006. Genetic linkage map and genomic organization in *Lettospheria maculans* [J]. Eur J Plant Pathol，114：17－31.

Ma A，Michailides T J，2005. Advances in understanding molecular mechanisms of fungicide resistance and molecular detection of resistant genotypes in phytopathogenic fungi [J]. Crop Protec，24：853－863.

Martins N F，Togawa Rc，Souza Jr M T，et al，2008. Identifi caçãoe caracterização functional de genes em estudos comparatives dos genomas de *Mycosphaerella graminicola* e *Mycosphaerella fijiensis* [J]. Bol Pesq Desenvol，237.

Manzo-Sánchez G，Zapater M F，Luna-Martínez F，et al，2008. Construction of a genetic linkage map of the fungal pathogen of banana *Mycosphaerella fijiensis*，causal agent of black leaf streak disease [J]. Curr Genet，53：299－311.

Meng X Q，Shoemaker R C，Yang X B，1999. Analysis of pathogenicity and genetic variation among *Phytophthora sojae* isolates using RAPD [J]. Mycol Res，103 (2)：173－176.

Michlemore R W，Hulbert S H，1987. Molecular markers for genetic analysis of pathogenic fungi [J]. Annu Rev Phytopathol，25：383－404.

Mobambo K N，Gauhl F，Vuylsteke D，et al，1993. Yield loss in plantain from black Sigatoka leaf spot and field performance of resistant hybrids [J]. Field Crop Res，35：35－42.

Mourichon X，Fullerton R A，1990. Geographical distribution of two types of *Mycosphaerella musicola* Leach (*Cercospora musae*) and *Micosphaerella fijiensis* (*Cercospora fijiensis*)，respectively，agents of sigatoka disease and black leaf streak disease in bananas and plantains [J]. Fruits，445：213－218.

Muller R，Pasberg-Gauhl C，Gauhl F，et al，1997. Oligonucleotide fingerprinting detects genetic variability at different levels in Nigerian *Mycosphaerella fijiensis* [J]. J Phytopathol，154：25－30.

Nakayashiki H，Nguyen Q B，2008. RNA interference：roles in fungal biology [J]. Curr Opin Microbiol，11 (6)：1－9.

Neu C，Kaemmer D，Kahl G，et al，1999. Polymorphic microsatellite markers for the banana pathogen *Mycosphaerella fijiensis* [J]. Mol Ecol，8：523－525.

Njambere E N，Vandemark G，Chen W，2010. Development and characterization of microsatellite markers of the fungal plant pathogen *Sclerotinia trifoliorum* [J]. Genome，53：494－500.

Rivas G G，Zapater M F，Abadie C，et al，2004. Founder effects and stochastic dispersal at the continental scale of the fungal pathogen of bananas *Mycosphaerella fijiensis* [J]. Mol Ecol，13：471－482.

Rodrigues-Garcia C M，Raigosa-Flores N，Conde-Ferraz L，et al，2006. Variation in eletrophoretic karyotype among Mexican isolates of *Mycosphaerella fijiensis* [J]. Can J Plant Pathol，28：236 - 241.

Pariaud B，Ravigné V，Halkett F，et al，2009. Aggressiveness and its role in the adaptation of plant pathogens [J]. Plant Pathology，58 (3)：409 - 424.

Portal O，Yovanny I，De Vleesschauwer D，et al，2011. Analysis of expressed sequence tags derived from a compatible *Mycosphaerella fijiensis*-banana interaction [J]. Plant Cell Rep，DOI：10.1007/s00299 - 011 - 1008 - z.

Stergiopoulos I，Van den Burg H A，Ökmen B，et al，2010. Tomato Cf resistance proteins mediate recognition of cognate homologous effectors from fungi pathogenic on dicots and moncots [J]. Proc Natl Acad Sci USA，107：7610 - 7615.

Stover R H，Simmonds N W，1987. Bananas [M]. 3rd ed. England，UK：Longman Scientific and Technical Essex.

Ubayasena L，Bett K，Tar' an B，et al，2010. Genetic control and QTL analysis of cotyledon bleaching resistance in green field pea (*Pisum sativum* L.) [J]. Genome，53：346 - 359.

Weld R J，Plummer K M，Carpenter M A，et al，2006. Approaches to functional genomics in filamentous fungi [J]. Cell Res，16 (1)：31 - 44.

Xu J R，Peng Y L，Dickman M B，et al，2006. The dawn of fungal pathogen genomics [J]. Annu Review Phytopathol，44：337 - 366.

Xu X，Roberts T，Barbara D，et al，2009. A genetic linkage map of *Venturia inaequalis*，the causal agent of apple scab [J]. BMC Res Notes，2：163.

Yang B J，Zhong B，2008. Fourteen polymorphic microsatellite markers for the fungal banana pathogen *Mycospharella fijiensis* [J]. Mol Ecol Res，8：910 - 912.

Zandjanakou-Tachin M，Vroh-Bi I，Ojiambo P S，et al，2009. Identification and genetic diversity of *Mycosphaerella* species on banana and plantain in Nigeria [J]. Plant Pathol，58：536 - 546.

Zapater M F，Rakotonantoandro S，Cohen S，et al，2004. Polymerase chain reaction—restriction fragment length polymorphism markers for the fungal banana pathogen *Mycospharella fijiensis* [J]. Mol Ecol Notes，4：80 - 82.

Zapater M F，Duchemin M，Dussart J F，et al，2008. Microsatellite markers for the fungal banana pathogens *Mycosphaerella fijiensis*，*Mycosphaerella musicola* and *Mycosphaerella eumusae* [J]. Mol Ecol Resour，8：1121 - 1125.

Zhong S，Steffenson B J，Martinez J P，et al，2002. A molecular map and electrophoretic karyotype of the plant pathogenic fungus *Cochliobolus sativus* [J]. Mol Plant-Microbe Interact，15 (5)：481 - 92.

第 13 章　香蕉枯萎病尖孢镰孢古巴专化型的基因组学

Raman Thangavelu①,a,*　Ganga Devi Perumal①,b　Mohammed Mustaffa①,c
Subramaniam Sreeramanan②　Xavier Rathinam③

摘　要：尖孢镰孢古巴专化型（*Foc*）引发的巴拿马枯萎病是对香蕉最具破坏性的疾病之一，造成香蕉严重减产，特别是对全球的香蕉经济品种卡文迪什。感染该病原体会导致植株枯萎和死亡。这种真菌在土壤中可存活很长时间。基于对宿主感染的鉴别，将 *Foc* 分为 4 个生理小种。其中 4 号生理小种（VCG 01213/01216）不仅是重要的菌株还会产生严重后果，因为危及对香蕉出口行业十分重要的卡文迪什香蕉。尽管可运用多种方法对 *Foc* 的表征进行描述，但是常用的方法是营养体亲和型（VCG）和 DNA 指纹识别。到目前为止，全球已经确定有 21 个 VCG，而仅亚洲就报道有 15 个 VCG。基于从世界各地采集的 *Foc* 的分子特性描述，将所有 VCG 分为两组。第一组包含 4 号生理小种的 VCG，第二组包含 1 号和 2 号生理小种的 VCG。每一 VCG 内的分离种群产生了非常相似的带型并且它们密切相关。各组内分离种群之间的密切关系表明每组有共同的祖先。相同生理小种的不同 VCG 菌株之间出现交叉反应。田间测试中也报道了 1 号和 2 号生理小种的 VCG 之间出现了类似的交叉反应。为了能在植株和土壤的感染早期就鉴定和识别出 *Foc* 病原体，研究人员开发出了 SCAR 及 rDNA-ITS / IGS 区域靶向分子标记，这些标记不仅有利于检疫目标的实现，也有助于实施更有效的管理。

① National Research Centre for Banana，Thiruchirapalli 620102，Tamil Nadu，India.

② School of Biological Sciences，Universiti Sains Malaysia（USM），Georgetown，11800，Penang，Malaysia；e-mail：sreeramanan@usm. my.

③ Department of Biotechnology，AIMST University（AIMST），Semelling，08000，Kedah，Malaysia；e-mail：rxavier77@yahoo. com.

a. e-mail：bananathanga@rediffmail. com，rtbanana@gmail. com.

b. e-mail：ganga _ gns@yahoo. com.

c. e-mail：directornrcb@gmail. com.

* 通信作者。

生物技术利用抗真菌基因，打破物种壁垒，为抗真菌香蕉品种的培育提供了另一种科学方法。各种重要的农艺基因可引入优质香蕉种质。为提高对真菌病原体特别是*Foc*的抗性而进行的香蕉基因工程取得了巨大进步，而这表明了研究人员对香蕉种群的高度关注。

13.1 引言

尖孢镰孢古巴专化型（*Foc*）在香蕉上引发的巴拿马枯萎病是世界上对香蕉最具破坏性的疾病之一（Ploetz and Pegg，1997）。该疾病于20世纪50年代摧毁了中美洲/加勒比地区的主要出口品种Gros Michel（AAA基因组）（Stover，1962）。而种植者之后将Gros Michel换成卡文迪什品种（AAA基因组），因其耐*Foc*的1号生理小种。遗憾的是，该品种也在枯萎病的袭击下溃败，因为出现了一种致病性强的*Foc*类型（VCG 01213/16），而该类型现在正对数十亿美元的香蕉出口业构成严重威胁，也威胁着小规模香蕉种植者的生计（Ploetz，2005）。

真菌侵染香蕉植株的根系，定殖于球茎和假茎的维管束系统，一般在种植5～6个月后诱导典型的萎蔫症状（Blomme et al.，2011）。症状表达既呈现在外部（叶子变黄和萎蔫，假茎出现纵向分裂，等等），又反映在内部（微观上组织变色，呈红褐色）（Wardlaw，1961；Stover，1962）。感染的植株一般不产生果串，严重患病时，整株死亡。

真菌以厚垣孢子在土壤中和植株残体上生存，并通过根毛、根尖和天然伤口沿着侧根基部侵入宿主。它还可以通过农具、害虫和寄生线虫所造成的伤口入侵。小分生孢子和菌丝也能够感染宿主。侵入后，真菌在皮质定殖并进入维管束系统中的木质部（Rishbeth，1955）。一旦真菌进入到维管束系统内，*Foc*可从根系移动到球茎和假茎中（Rishbeth，1955）。定殖后，该病原体将阻碍植株的维管束系统，从而导致植株萎蔫，最终死亡（Ploetz and Pegg，2000）。

病原体的传播受到多种因素影响，例如被感染的球茎或吸根的转移，与被感染的种植材料或农具/搬运工具相接的土壤（Ploetz，1994），以及地表水和根（Moore et al.，1995）。通过风扬起灰尘和垃圾，均能传播*Foc*。

13.2 *Foc*的分类

传统上，科学家基于*Foc*对少量田间不同的香蕉宿主品种的致病性，将

Foc 的菌株分为 4 个生理小种（Moore et al.，1995）。这种分类属于人工分组，并不能反映它们严格的遗传关系，因为科学家对这一病害系统的毒力和抗性的遗传学了解很少。不过，这一分类也提供了许多有用信息。*Foc* 小种的特点如下。

1 号生理小种：遍及世界各地，几乎摧毁了以 Gros Michel 占主导的热带地区的香蕉出口业。1 号生理小种导致的香蕉枯萎病困扰美洲热带香蕉产区超过 60 年。直到 20 世纪 60 年代将易感的 Gros Michel 替换成有抗性的卡文迪什香蕉品种后，病情才得到控制（Stover，1962）。此外，它还攻击其他品种，如 Silk（AAB）、Pome（AAB）、Abacá、Maqueño（AAB）、Pisang Awak（ABB）和 I. C. 2'（AAAA）（Ploetz et al.，1990；Bentley et al.，1995）。最近，印度国家香蕉研究中心（NRCB）发现，1 号生理小种的某些 VCG 会攻击 2 号生理小种的感病体，反之亦然。此外，在印度，发现 1 号生理小种的 VCG 0124 会攻击 4 号生理小种的特感品种 Grand Naine（Thangavelu and Mustaffa，2010）。

2 号生理小种：也广泛分布于所有的香蕉种植区。它会使 Bluggoe（ABB）、Monthan（ABB）和其他亲缘相近的煮食蕉（ABB）致病。它还会影响象腿蕉（*Ensete ventricosum*）和一些培育的四倍体，如 Bodles Altafort。

3 号生理小种：20 世纪中期，Waite（1963）在中南美洲研究蝎尾蕉属（*Heliconia*）的几个物种中首次发现枯萎病，并将致病的菌株命名为尖孢镰孢古巴专化型的 3 号生理小种。澳大利亚垂花粉鸟蝎尾蕉（*H. chartacea*）也被报道暴发枯萎病（Bentley et al.，1998；Moore et al.，2001）。洪都拉斯和哥斯达黎加蝎尾蕉（*Heliconia* spp.）暴发的枯萎病也被记载。这种 *Heloconia* 的致病菌株 *Foc* 对芭蕉属植物不致病或弱致病（Ploetz，1990）。

4 号生理小种：发生在大部分香蕉种植地区，如加那利群岛、越南、中国台湾、印度尼西亚、中国、澳大利亚、菲律宾、南非、柬埔寨、马来西亚、巴西等（Ploetz，1990）。此外，它还发生在苏门答腊岛、爪哇岛、苏拉威西岛、哈马黑拉岛和与巴布亚新几内亚边界附近的巴布亚省（Shivas and Philemon，1996；Davis et al.，2000）。然而，印度还未报道过发现 4 号生理小种。这个小种是最具破坏性的，因为它既会侵入对 1 号和 2 号生理小种易感的无性系，还会影响卡文迪什和 Pisang Mas（AA）。

4 号生理小种分成两组：亚热带 4 号生理小种（SR4）（VCG 0120、0121、0129、01211 在澳大利亚发现，0120 在加那利群岛和南非发现，0122 在菲律宾发现）和热带 4 号生理小种（TR4）（VCG 01213－01216 复合体）。SR4 威胁卡文迪什香蕉，这种香蕉在亚热带的低温环境下易患病。然而在热带环境下，TR4 会更猛烈地攻击卡文迪什香蕉而无需任何发病诱因。TR4 与 SR4 的区别在于二者在遗传学上截然不同，而且 TR4 在热带地区攻击卡文迪什香蕉

(Ploetz, 2004)。尽管宿主范围是相似的，无性系李林蕉（AA）仅受 TR4 影响。依据卡文迪什的萎蔫，菌株 TR4（VCG 01213）首次在中国台湾被鉴定（Su et al., 1986），并且被认为起源于马来半岛和苏门答腊岛（Buddenhagen, 2009）。尽管只在印度尼西亚（哈马黑拉岛、伊里安查亚、爪哇岛、苏拉威西岛和苏门答腊岛）、马来西亚半岛、中国南部、中国台湾（Simmonds and Shepherd, 1955; Shivas and Philemon, 1996; Ploetz and Pegg, 2000）和澳大利亚北部发现了 TR4 菌株，但是这严重影响与卡文迪什香蕉出口的相关贸易。自 1967 年以来，TR4 对中国台湾的卡文迪什香蕉产业产生严重破坏；自 20 世纪 90 年代初，它摧毁了印度尼西亚和马来西亚的新卡文迪什香蕉种植园；并且于 1997 年和 1999 年之间在澳大利亚北部引发了田间流行病。2004 年，中国南部广东省的香蕉种植园出现严重的 *Foc* 感染，这加剧了 TR4 的威胁。近年来，这一小种（TR4）在中国已毁坏了超过 20 000 hm^2 的卡文迪什香蕉，并且正威胁着菲律宾巨大的香蕉出口业（Daniells, 2009）。

而对于 SR4，该菌株 VCG 0120 - 01215 对澳大利亚亚热带地区（新南威尔士州和昆士兰州）、加那利群岛、马德拉群岛和南非（纳塔尔和德兰士瓦）的卡文迪什品种造成了严重损害。现今在印度香蕉主产区未发现有 *Foc* TR4（Thangavelu et al., 2001）。虽然拉丁美洲未发现 TR4，但是对香蕉出口地区，包括只种植卡文迪什香蕉的拉丁美洲来说仍是巨大威胁。

13.3 病原菌多样性

对 *Foc* 种内遗传变异的分析促进了对长期病害管理策略的研究，如寄主植物抗性、生物控制和系统获得性抗性的激活剂。此外，对 *Foc* 的 VCG 之间的进化关系进行推断可能有助于评估新型致病型的发展可能性。此外，系统发育研究将有助于准确诊断工具的开发，这对检疫管理部门检测和快速识别 *Foc* 的引入型外来菌株至关重要。

一般来说，尖孢镰孢没有已知的有性阶段。因为没有减数分裂重组，真菌的变异被认为是突变造成的（Kistler and Miao, 1992）。*Foc* 是一种高度可变的病原体（Ploetz, 1990），因此，全面了解 *Foc* 的病原体多样性对于选择或培育对镰孢枯萎病有持久抗性的香蕉品种是必要的，也有利于推动并实施有意义的国际和本地检疫管制，以限制该病原体的传播（Bentley et al., 1995）。此外，这些研究对于确定病原体的起源和研究真菌种群是如何在不同的地理区域和种内或种间产生变化将十分有益。这将帮助科学家了解致病性如何改变、新品种如何演变等科学问题。

当前已有几种技术用于描述 *Foc* 的表征，评估 *Foc* 内遗传变异和确定 *Foc*

的菌株之间的系统发育关系。描述 *Foc* 菌株的表征可以采用以下几种方法：营养亲和性（VCG）分析（Ploetz and Correll，1988；Brake et al.，1990；Ploetz，1990；Moore et al.，1993；Moore，1994）、挥发性有机化合物的产生（Stover，1962；Moore et al.，1991）、电泳核型（Boehm et al.，1994；Miao，1990）、随机扩增多态性 DNA（RAPD）分析（Bentley et al.，1995）、限制性片段长度多态性（RFLP）分析（Koenig et al.，1997）、DNA 扩增指纹（DAF）分析（Bentley and Bassam，1996；Bentley et al.，1998）和核基因与线粒体基因的 DNA 测序（O'Donnell et al.，1998）。这些方法对于测定致病型十分有用（Pegg et al.，1995），因为现在还没有测定小型植物致病性的测试。尽管上述方法中的一些已经成熟，但是对 *Foc* 多样性的鉴定还是广泛使用培养物中挥发性化合物的产生、营养亲和性（VCG）分析和分子鉴定技术等方法。

13.3.1　挥发物产生

Foc 病原体产生的挥发性化合物可用来区分 *Foc* 病原体菌株的一致、可重复性状（Stover，1962）。当 *Foc* 的菌株在无菌的、含水稻源淀粉底物上培养，它们会在培养物的顶部空间产生有气味的化合物醛，而这可以通过气相色谱法对其进行分析。产生醛的这些菌株被称为品种 Odoratum，而不产生醛的那些菌株被称为品种 Inodoratum。结果发现，受 2 号生理小种影响的 Bluggoe 的菌株没有产生气味；而 1 号生理小种的菌株属于 Odoratum 或 Inodoratum 型（Stover，1962）。Pegg 等（1996）描述了 VCG 的 0120、0121、0122、0126、0129、01210、01211、01213、01215、01216 和 01219 属于 Odoratum 组，而 VCG 的 0123、0124、0125、0128、01212、01214、01217、01218 和 01220 属于 Inodoratum 组。Moore 等（1991）还利用高压液相色谱法分析澳大利亚 *Foc* 菌株的挥发性化合物的产生，并发现 VCG 菌株属于 4 号生理小种的 0120、0129 和 01211、01213/16，产生典型的特征挥发物，而 VCG 菌株属于 1 号生理小种的 0123、0124、0124/0125、0125 和 01220 以及属于 2 号生理小种的 VCG 0128，没有产生挥发物。这种技术也被用于描述 *Foc* 的菌株表征（Moore et al.，1991；Pegg et al.，1993）。前面提及的澳大利亚菌株中的挥发性物质生产、VCG 和致病性之间绝对具有关联。这是一种描述 *Foc* 种群内菌株表征的有用且可靠的方法，也为田间菌株的首次表征描述提供更快速和廉价的方法（Moore et al.，1991）。

13.3.2　营养亲和性

营养亲和性已用于研究包括 *Foc* 在内的几种植物病原真菌的多样性、遗传学、生态学和种群生物学。小种内 *Foc* 的个别菌株可通过无性融合来形成稳定的异核体（Puhalla，1985），据称属于同一营养亲和群（VCG）（Ploetz and Correll，1988）。这一性状是由基因控制的。按照惯例，这是由彼此互补的硝酸盐营养缺陷型突变体（*nit*）利用硝酸盐的能力来确定（图 13 - 1）。如果一种真菌的两种不同菌株配对时不能形成异核体，则称它们为营养不亲和（Puhalla，1985）。从遗传学角度来说，相比营养不亲和菌株，营养亲和菌株更有可能相似。例如，相比营养不亲和的菌株（Moore et al.，1995），营养亲和的真菌菌株可能彼此在某些性状上十分相似，如菌落大小（Croft and Jinks，1977；Correll et al.，1986）、抗生素的产生（Croft and Jinks，1977）、毒力（Correll et al.，1985，1986；Gorden et al.，1986；Bosland and Williams，1987）和同工酶谱（Bosland and Williams，1987）。1985 年，Puhalla 首次报道了 *Foc* 的 VCG，而迄今为止，已在世界范围内确定了 21 个 VCG（Correll et al.，1987；Ploetz and Correll，1988；Moore，1994；Pegg et al.，1995）（表 13 - 1）。仅在亚洲，就已确定了 15 个 VCG（Ploetz，1990；Moore，1994；Pegg et al.，1995），而根据记录，数量最多的 VCG 出现在分布了大量小果野蕉的印度 - 马来西亚地区（Pegg et al.，1994，1996）。非洲和美洲只发现了少数 VCG（Ploetz，1993）。在澳大利亚，Brake 等（1990）使用了 VCG 分析法来圈定 *Foc* 的 245 个澳大利亚菌株中的 6 个 VCG。1 号生理小种的菌株存在于 VCG 的 0124 和 0125，2 号生理小种的菌株存在于 VCG 0128，

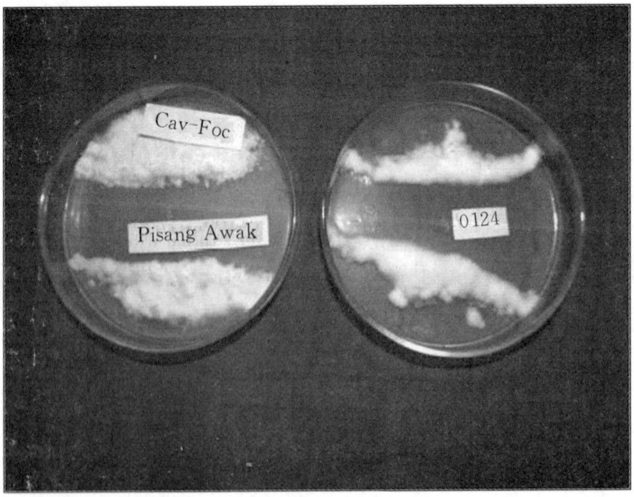

图 13 - 1　Pisang Awak 和 VCG - 0124 的 *nit* - M 检验器及来自 cv. Grand Naine（AAA，卡文迪什组）的 *Foc* 的 *nit* - *1* 之间在营养亲和性测试中形成的异核体

4 号生理小种的菌株存在于 VCG 的 0120、0129 和 01211。对 *Foc* 的澳大利亚菌株现场反应和温室致病性测试也证明了 VCG 和致病性的关联。基于这些 VCG 数据，Pegg 等（1993）提出，祖先 VCG 的 0120 和 0124－0125 分别在小果野蕉和野蕉多样性的中心独立进化；根据 Ploetz（1990）提出的突变假说，小范围分布的 VCG 有可能来自祖先 VCG。

表 13－1　尖孢镰孢古巴专化型（*Foc*）菌株的营养亲和群、小种和来源，以及鉴定它的品种

（Jones，2000；Ploetz，2005；Thangavelu，2008）

VCG	VCG 复合物	生理小种	栽培品种及基因组组群	起源地
0120	0120－01215	1，4	AA：SH－3142、SH－3362； AAA：Gros Michel、Highgate、Pisang Ambon Putih、Pisang Ambon、Dwarf Cavendish、Williams、Mons Mari、Grand Nail、Lacatan； AAB：Prata、Lady Finger、Pacovan、Hua Moa、Silk	南非、加拿大、开曼群岛、澳大利亚、巴西、洪都拉斯、哥斯达黎加、印度尼西亚、牙买加、尼日利亚、马来西亚、西班牙、葡萄牙、中国台湾、佛罗里达州（美国）、法国
0121	无	4	AAA：Gros Michel、Cavendish	印度尼西亚、中国台湾
0122	无	4?	AAA：Cavendish； ABB：Saba	菲律宾
0123	无	1	AAA：Gros Michel、Grand Nail； AAB：Silk、Latundan、Pisang Keling； ABB：Pisang Awak、Kluai Namwa	马来西亚、菲律宾、中国台湾、泰国
0124	0124－0125 0128－01220	1，2	AAA：Williams、Grand Naine； AAB：Silk； AB：Ney Poovan； ABB：Pisang Awak、Monthan	澳大利亚、巴西、布隆迪、中国、古巴、刚果民主共和国、海地、洪都拉斯、印度、牙买加、马拉维、马来西亚、墨西哥、卢旺达、尼加拉瓜、泰国、坦桑尼亚、乌干达、佛罗里达州（美国）
0125	无	1	AAAA：Jamaica 1242； AAB：Silk	澳大利亚、巴西、布隆迪、中国、古巴
0126	无	1	AA：Pisang Berlin； AAA：Highgate； AAB：Maqueno、Pisang Manrung	洪都拉斯、印度尼西亚、巴布亚新几内亚、菲律宾

（续）

VCG	VCG 复合物	生理小种	栽培品种及基因组组群	起源地
0128	0124－0125－0128－01220	1，2	AAB：Lady Finger、Maca、Manzano、Maqueno、Silk； ABB：Pisang Awak、Monthan	澳大利亚、巴西、布隆迪、中国、古巴、刚果民主共和国、海地、洪都拉斯、印度、牙买加、马拉维、马来西亚、墨西哥、卢旺达、尼加拉瓜、坦桑尼亚、泰国、乌干达、佛罗里达州（美国）
0129	无	4	AAA：Mons Mari； AAB：Lady Finger	澳大利亚
01210	无	1	AAA：Gros Michel； AAB：Manzano	开曼群岛、古巴、佛罗里达州（美国）
01211	无	4	AA：SH－3142； ABB：Monthan、Pisang Awak	澳大利亚、印度
01212	无	？4	AB：Ney Poovan； AAB：Silk、Kisubi； ABB：Pisang Awak、Bluggoe	坦桑尼亚、印度
01213	01213－01216	T4	AA：Pisang Lilin、Pisang Mas	澳大利亚、印度尼西亚、马来西亚、中国台湾
01214	无	2	ABB：Harare、Mbufu	马拉维
01215	0120－01215	1，4	AA：SH－3142、SH－3362； AAA：Gros Michel、Highgate、Pisang Ambon Putih、Pisang Ambon、Dwarf Cavendish、Williams、Mons Mari、Grand Naine、Lacatan； AAB：Prata、Lady Finger、Pacovan、Hua Moa、Silk	开曼群岛、澳大利亚、巴西、哥斯达黎加、洪都拉斯、印度尼西亚、牙买加、马来西亚、尼日利亚、葡萄牙、西班牙、中国台湾、佛罗里达州（美国）
01216	01213－01216	T4	AAA：Pisang Ambon、Valery、Williams、Grand Naine、Novaria、Red、Pisang Udang、Pisang Susu、Pisang Nangka、Pisang Barangan； AAB：Pisang Raja Serah、Pisang Rastali、Pisang Rajah、Relong； ABB：Pisang Awak Legor、Saba、Pisang Kepok、Pisang Caputu、Pisang Kosta； Unknown：Pisang Batan	澳大利亚、印度尼西亚、中国台湾、马来西亚

（续）

VCG	VCG 复合物	生理小种	栽培品种及基因组组群	起源地
01217	无	?	ABB：Harare、Mbufu	马来西亚
01218	无	?	AAB：Pisang Raja Serah、Silk；ABB：Pisang Awak、Kluai Namwa、Pisang Kepok、Pisang Siam	印度尼西亚、马来西亚、泰国、印度
01219	无	?	AAA：Pisang Ambon、Pisang Ambon Putih；Unknown：PIsang Raja Garing	印度尼西亚
01220	0124－01216 0128－01220	4?	ABB：Pisang Awak、Ducasses、Kayinga、Zambia、Kluai Namwa、Bluggoe、Harare、Kholobowa、Dwarf Bluggoe、Mbufu、Burrro Criolla、Pelipita、Icecream；AAB：Silk	澳大利亚、巴西、布隆迪、中国、古巴、刚果民主共和国、海地、洪都拉斯、印度、牙买加、马拉维、马来西亚、墨西哥、尼加拉瓜、坦桑尼亚、卢旺达、泰国、乌干达、佛罗里达州（美国）
01221	无	?	ABB：Kluai Namwa	泰国

13.3.3　分子鉴定

虽然 VCG 提供了有效手段，将 *Foc* 分成多个在遗传上分离的亚种群，但是并没有指示属于不同 VCG 的分离种群之间的遗传关联。营养亲和的分离种群被认为在每个 vic 位点上有相同的等位基因（Correll，1991）。然而，每个 vic 位点上的一次突变能导致亲缘的分离种群变得营养不亲和（Bentley et al.，1995，1998）。此外，一些 VCG 通过在分离的 VCG 之间形成异核体来产生 VCG 复合体（Ploetz，1990；Bentley et al.，1998）。在这些情况下，分子表征技术可以用来确定每个 VCG 内分离种群之间的遗传相似性和 VCG 之间的遗传亲缘关系（Pegg et al.，1996）。Bentley 等（1995）称 RAPD－PCR 的带型普遍是 VCG 特有的，因而它有可能确定不同的 VCG 之间的遗传亲缘关系，以及每一 VCG 内分离种群的遗传亲缘关系。对代表不同 VCG 和小种的尖孢镰孢古巴专化型的分离种群进行全球采集，通过在视觉上比较它们的 RAPD－PCR 带型和进行带型分析，将 *Foc* 的分离种群分为两大组。1 组包括属于 VCG 0120、0121、0122、0126、0129、01210、01211、01213、01215、01216

和 01219 的分离种群，而 2 组包括属于 VCG 0123、0124、0125、0128、01212、01214、01217、01218 和 01220 的分离种群。每个 VCG 内的分离种群产生相似的带型，并且亲缘关系很近（Bentley et al.，1995）。通过 RAPD-PCR 分析而划分的两组与之前基于产生的挥发物和电泳核型分析的分类一致。*Foc* 的澳大利亚分离种群中，DNA 指纹组和小种之间有直接的关联。同样，利用引物 OPB－07 对印度 *Foc* 分离种群的鉴定表明 *Foc* 中存在广泛变异，分为 23 个不同的组。然而，没有基于宿主来源/种群/地理来源的关联（图 13－2）。Sorensen 等（1993）通过 RAPD-PCR 技术，分析了代表澳大利亚的 3 个小种及 7 个 VCG 的 *Foc* 的遗传变异，发现两个不同集群：一个集群包含 4 号生理小种（VCG 0120、0129 和 01211），而另一个集群包含 1 号和 2 号生理小种（2 号生理小种的 VCG 0128，1 号生理小种的 VCG 0124、01245 和 0125）。对于亚洲分离种群，所有 4 号生理小种的分离种群的 RAPD-PCR 带型与来自澳大利亚、加那利群岛、南非的 4 号生理小种分离种群的 RAPD-PCR 带型相似，也与来自洪都拉斯的推定 4 号生理小种分离种群的 RAPD-PCR 带型相似（Pegg et al.，1994）。

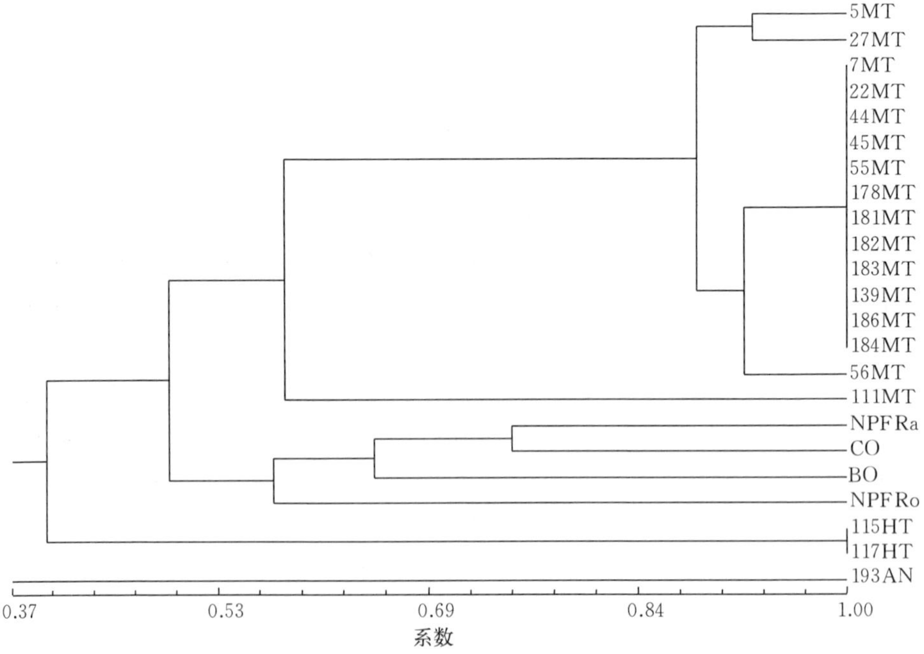

图 13－2 展示来自 cv. Monthan（ABB）的 *Foc* 菌株变异的树状图

Bentley 等（1998）采用修饰的 DNA 指纹分析研究了 *Foc* 的 350 个分离群，并进一步对其划分谱系。*Foc* 的两大组之间的遗传相似度为 60%。谱系内

VCG 之间的平均遗传相似性为 80%。由于 *Foc* 的这些谱系来自尖孢镰孢的其他专化型，这可能意味着 *Foc* 的个别谱系有独立的起源，因此这些谱系彼此不同。Koenig 等（1997）利用基因组 DNA 的限制性片段长度多态性（RFLP）和 19 个匿名的单拷贝 DNA 探针，研究了 13 个 VCG 或 VCG 复合体中的 165 个菌株的系统发育，并确定了这 13 个 VCG 或 VCG 复合体中 165 个菌株中的 10 个谱系。谱系Ⅰ和Ⅱ这两个最大的谱系分别包含了 65 个和 43 个菌株，并且分别对应于 VCG 0124/5/8/12/20 和 0120/15 复合体，而这些复合体与尖孢镰孢西瓜专化型这种西瓜病原体的遗传相似性大于这些复合体之间的相似性。而疏远的亲缘关系为 *Foc* 的部分种群独立进化的假说提供了进一步支持。但是，也有科学家认为，这一病原体是和可食用香蕉及它们东南亚的野生二倍体祖先（小果野蕉和野蕉）共同进化，并且跟随受感染的香蕉球茎和附着的土壤被传播至其他国家（Stover，1962）。各组内菌株之间的密切关系还表明每组都有共同的祖先。此外，在巴布亚新几内亚和南太平洋岛国等地区种植着高度易感的品种，但是没有出现尖孢镰孢古巴专化型，这表明致病性菌株不易从尖孢镰孢的其他种群进化而来（Pegg et al.，1993）。针对东南亚菌株之间的遗传多样性而进行的 RAPD-PCR 分析，结果也支持了共同进化假说（Pegg et al.，1993）。共同进化假说对于抗性基因型的选择具有重要影响，因为这些基因型最有可能出现在宿主和病原体存在极大多样性的地区（Vavilow，1922）。

最近，Groenewald（2006）也采用了 AFLP（扩增片段长度多态性）技术来研究 *Foc* 菌株的自然种群内和自然种群之间的基因组关系，还探讨了 AFLP 标记是否能辨别 *Foc* 的 VCG。对 AFLP 数据的距离分析清楚地将 *Foc* 划分为两个主要进化分支。聚类分析树上，进化支 1 包括 VCG 0120、0120/15、0121、0122、0126、0129、01213、01213/16、01216 和 01219，自展值为 100%，而这些 VCG 都采集自世界各地，包括来自南非的菌株。74%自举支持的进化支 2 包括 VCG 0123、0124、0125、01217 和 01218，其中包括来自澳大利亚的菌株。来自进化支 2 的 *Foc* 菌株与其他专化型的尖孢镰孢的关系比它与进化支 1 的 *Foc* 菌株的关系更近。此外，AFLP 分析将 *Foc* 菌株分为 7 组基因型。第一组基因型包含 VCG 0120 和 VCG 复合体 0120/15；第二组基因型包含 VCG 0126、0129、0122 和 01219；第三组基因型包括热带 4 号小种的菌株 VCG 01213、01216 和 01213/01216；第四组基因型包括 VCG 0121；第五组基因型包括 VCG 0123 和 01217；第六组基因型包括 VCG 01218；第七组基因型包括 VCG 0124 和 0125。最后得出的结论是每个 VCG 内的菌株普遍产生了相似的带型，因为关系紧密，不依赖于地理来源或宿主来源。VCG 0120 菌株是同质的，尽管它们地理来源不同（亚洲、澳大利亚、南非、中美洲和南美洲)、宿主不同（卡文迪什，High Gate 和 Lady Finger 香蕉品种）。这对未来的疾病管理计划十分重要（Groenewald，2006）。

13.4 小种和 VCG 之间的关系

尖孢镰孢古巴专化型的小种和 VCG 之间的关系还未明确定义（Bentley et al.，1995）。Ploetz（1990）发现世界范围内采集的菌株中，多个 VCG 会发生在一个小种内，而单个 VCG 可能有来自不止一个小种的代表。在印度，通过构建硝酸盐营养缺陷型突变体，很多分离自对 1 号生理小种易感品种的 *Foc* 菌株与分离自对 2 号生理小种易感品种的 *Foc* 菌株发生了交叉反应，反之亦然。来自对 2 号生理小种易感的品种 Monthan（ABB，44MT、45MT、181MT）的 *Foc* 菌株 *nit-1* 突变体通过 *nit*-M 测试表明与来自对 1 号生理小种易感的品种 Rasthali（19RT）的 *Foc* 菌株形成异核体（Thangavelu，2008）。同样，来自 1 号生理小种易感的品种 Rasthali（Silk，AAB）的 *Foc* 菌株 *nit-1* 突变体通过 *nit*-M 测试器表明与来自对 2 号生理小种易感的品种 Monthan 的 *Foc* 菌株形成异核体（Thangavelu，2008）。某些情况下，例如，来自 cv. Hill 香蕉（Pome，AAB）的 *Foc* 菌株的 *nit-1* 突变体通过 *nit*-M 测试器与来自 1 号和 2 号生理小种的 *Foc* 菌株形成了异核体。用来验证以上结果的盆栽研究也证实了 1 号生理小种和 2 号生理小种的各种 *Foc* 菌株之间的交叉反应。*Foc* 菌株 5MT（2 号生理小种）引发了对 1 号生理小种易感的 cv. Karpuravalli 的枯萎病。同样，1 号生理小种的 *Foc* 菌株 19RT 引发了对 2 号生理小种易感的 cv. Monthan 的枯萎病。与此相反的是，尖孢镰孢古巴专化型的澳大利亚种群内，小种和 VCG 之间的相关性较好，发现只有一个小种出现在每个 VCG 内（Brake et al.，1990；Moore et al.，1993）。

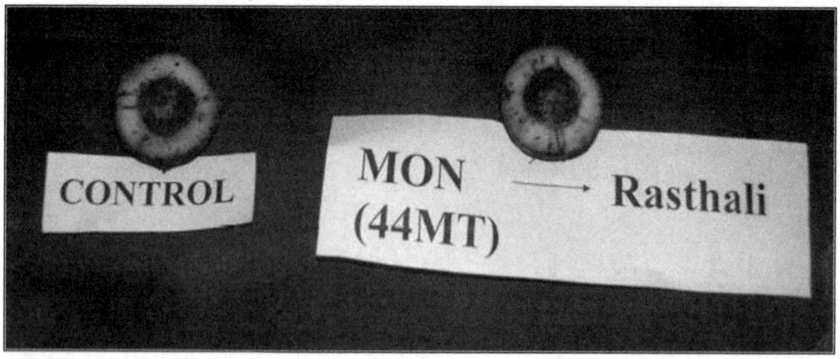

图 13-3　盆栽条件下，来自 cv. Monthan（ABB）的 *Foc* 2 号生理小种在对 1 号生理小种易感的 cv. Rasthali（Silk，AAB）上的交叉反应

（对照组：cv. Monthan 接种了来自相同品种的 *Foc* 菌株）

13.5　诊断 *Foc* 的标记

若在疾病发展的早期阶段采取控制措施，能对植物疾病进行最高效的管理。传统方法进行的疾病诊断和病原体鉴定涉及分离病原体和利用接种测试进行鉴定，既费时又费力（Alves-Santos et al.，2002）。分子生物学和生物工程的技术进步推动了开发快速、专业、灵敏的工具来检测植物病原体（Miller and Martin，1988）。

以形态特征为基础对 *Foc* 病原体的鉴定除了耗费大量时间，还需要关于镰孢分类的专门知识（Jurado et al.，2006）。此外，利用显微镜检术来鉴定镰孢则效率不高，并且难以在植物组织中检测，而植物组织中又可能包括各种疾病复合体（Yergeau et al.，2005）。在这种形势下，一种快速诊断 *Foc* 的试验法应运而生，这种方法不仅能阻止该病原体通过种植材料进入和传播，还能帮助种植者在种植前就知道田地里是否存在 *Foc* 病原体。利用诊断工具在感染早期鉴定出生长植物的这种疾病将能帮助控制该疾病。诊断工具的发展将对 *Foc* 的流行病学和真菌种群遗传学的基础研究帮助很大（Schilling et al.，1996）。

基于特异性 DNA 序列的聚合酶链式反应（PCR）的诊断方法已成功应用到检测和鉴定各种重要的真菌植物病原体，并成为常规检测方法。Sharon（2003，未发表）对澳大利亚 *Foc* 的 9 种代表性菌株的核糖体 DNA（rDNA）间隔（IGS）区进行测序。对 *Foc* 菌株的 IGS 区的序列分析显示，VCG 0124 的一个代表性菌株中有 196 bp 的插入。这一插入序列显示与称为 *Foxy* 的短分散核因子（SINE）的一个新家族同源，*Foxy* 在尖孢镰孢番茄专化型（*Fol*）中鉴定出。这一插入序列被称为 *Foxy-Foc*。基于这些核苷酸多态性，研究人员设计利用寡核苷酸引物序列来特定扩增 *Foc* 的澳大利亚 1 号生理小种（R1）、亚热带 4 号生理小种（SR1）和热带 4 号生理小种（TR4）的菌株。带有一系列 *Foc*、尖孢镰孢和镰孢菌株的这些引物的筛选显示，设计用来特定扩增 1 号生理小种、亚热带 4 号生理小种和热带 4 号生理小种的引物仅扩增了 1 号生理小种、亚热带 4 号生理小种和热带 4 号生理小种的澳大利亚 *Foc* 菌株。不幸的是，这些引物也扩增与 *Foc* 亲近的外来菌株，其原因可能是 IGS 区域缺乏序列变化性（Sharon，2003，未发表）。同样，研究人员利用来自印度的几个 *Foc* 菌株的 rDNA-ITS 区域的变异设计了序列特征性扩增区域（SCAR），以及核糖体 DNA 内转录间隔（rDNA-ITS）区靶向引物（图 13-4）。这些标记只能扩增出印度的致病性 *Foc*。这些引物也通过田间生长的植株得以验证（Thangavelu，2010，未发表）。除了特异性 DNA 序列如 rDNA-IGS/ITS 区的扩增，结合随机扩增多态性 DNA（RAPD）的 DNA 指纹分析法也是鉴定真菌病原体的强大工具（Koike et al.，1995；Mes et al.，1999；Fungaro et al.，2004；Lin et al.，2009）。利用这一 RAPD 技术，Lin 等（2008）在中国台湾开发出了一项检测 *Foc* 的 4 号生理小种的分子方法。通过 PCR 扩增，源自随机引物 OP－A02 扩增片段的引物集 *Foc－1/Foc－2* 产生了 242 bp 大小的 DNA 片段，而该片段特定于 *Foc* 的 4 号生理小种。因为有最优化的 PCR 参数，这一分子方法十分灵敏，可以高效地检测 50～2 000 ng的宿主基因组 DNA 中含量低至 10 pg 的 *Foc* DNA。这些参数还显示出，通过使用带有 *Foc－1/Foc－2* 引物集的 PCR 测试法，很容易将 *Foc* 的 4 号生理小种与 *Foc* 的 1 号、2 号生理小种区分开，并且分离尖孢镰孢的其他专化型。这种 PCR 诊断方法可用来筛选自然感染 *Foc* 的香蕉样本，也能迅速测试出香蕉的培育材料中是否存在 *Foc* 的 4 号生理小种，这样能帮助抵抗 *Foc* 4 号生理小种的攻击。

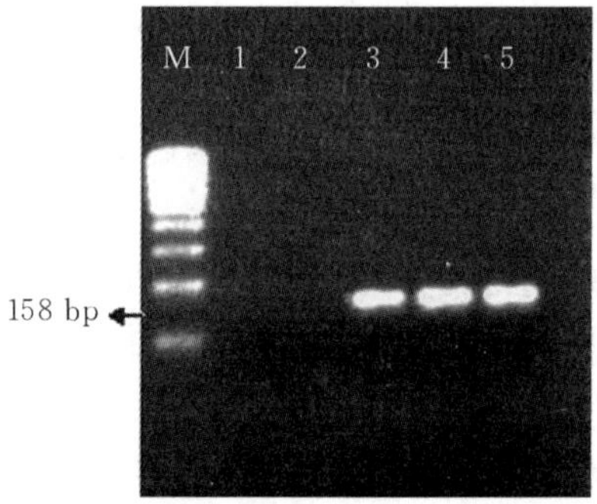

图 13－4　通过 SCAR 标记对感染了 *Foc* 的土壤和感染了 *Foc* 的香蕉球茎和假茎中的 *Foc* 病原体进行 PCR 扩增

M. 标记（100 bp）　1. 控制（未感染）　2. 负控制　3. 感染了 *Foc* 的土壤　4. 感染了 *Foc* 的球茎　5. 感染了 *Foc* 的假茎

13.6　香蕉对真菌病原体抗性的生成策略

真菌病原体导致了严重的香蕉减产。通过轮作、环境卫生、种植对疾病有抗性的植物和使用杀真菌剂可以控制真菌病害。杀真菌剂仍然是控制疾病的主要手段。但是使用杀真菌剂有几大问题：①价格昂贵，发展中国家和地区的小规模种植者负担不起；②它们对环境有潜在的或直接的危害；③杀真菌剂的扩大应用会加速新的恶性真菌品种的出现。杀真菌剂无法控制维管枯萎病和土传疾病，如侵染香蕉的镰孢枯萎病。对于许多真菌性疾病，无法通过植物育种来解决，因为根本没有天然来源的抗性材料可供育种者使用。生物技术提供了另一种科学手段，打破物种壁垒，利用抗真菌基因来产生抗真菌的植物新品种。同时，其他有重要农艺学意义的基因也可引入优质的香蕉种质中。

香蕉在整个生命周期中会受到许多病原体的攻击，而它的细胞壁无疑是第一道防线，在抵抗非宿主病原体上十分有效。但在宿主专一性的相互作用下，病原体可以躲开植物的防御机制，并在香蕉宿主植物内部进行全身扩散（亲和互作）或被阻止传染植物（不亲和互作）。亲和互作常常导致传染位置上的组织坏死，称为过敏反应（HR）（Govrin and Levine，2000；Jabs and Slusarenko，2000）。

操纵过敏反应能为植物的生产创造条件，尤其是对于能抵抗广泛病原体的香蕉。除了过敏反应，植物进化出了一系列抗病原体的防卫机制，包括细胞壁木质化增加（Lamb and Dixon，1997），合成小抗生素分子或植物抗毒素（Lamb and Dixon，1997；Punja，2001），以及生产活性氧簇（Govrin and Levine，2000；Liang et al.，2001）。然而，这些机制的生物化学复杂性导致它们很难应用在对于真菌抗性的分子育种上。植物通过识别病原体而激活的一组大量基因是编码病程相关（PR）蛋白的基因（Dixon et al.，1996；Punja，2001）。PR 蛋白被定义为宿主植物编码的蛋白质，但是只在疾病状况下或相关胁迫条件下诱发。迄今为止，研究人员已对 PR 蛋白的 10 多个家族进行了分类，包括 β-1,3-内切葡聚糖酶（PR2 蛋白）、内切几丁质酶（PR3 蛋白）、蛋白酶抑制剂（PR6 蛋白）、蛋白酶（PR7 蛋白）和细胞外氧化酶（PR9 蛋白）。

受到病原体攻击后，许多植物在感染部位产生抗性，甚至在之前未感染的组织上保持长时间的抗性，而这种抗性不仅抵抗首次触发一个或更多防卫机制的病原体，还广谱抗其他病原体。这种诱导的抗性或系统获得性抗性（SAR）最容易出现在精选的植物上，并伴随着大型基因集合的表达，尤其是那些编码 PR 蛋白的基因（Verberne et al.，2000；Donofrio and Delaney，2001；Lee et al.，2001）。当前科学家对这一 SAR 反应下潜藏的机制进行了大量研究，旨在弄清这一机制，因为这种 SAR 反应有可能成为促进基因转化的潜在基因的重要来源，而这种基因转化是帮助香蕉获得对真菌病原体的抗性，特别是针对尖孢镰孢古巴专化型（*Foc*）的抗性。而科学家的努力主要集中于发现导致 SAR 的信号，然后弄清大量防卫基因的诱导。其他分子中，水杨酸、茉莉酮酸甲酯和乙烯似乎涉及导致 SAR 的生物化学反应的级联的不同步骤（Dong，1998；Dempsey et al.，1999）。基于抗性植物的信号转导，科学家未来将有可能开发出一种可靠方法，利用分子育种手段使得香蕉植株产生对广谱的病原体的“免疫”。

13.7 控制真菌疾病的转基因香蕉

让香蕉获得疾病抗性的基因工程的重要步骤之一是测试抗目标真菌病原体，特别是 *Foc* 的新蛋白质的效力。“抗真菌蛋白”策略依赖基因编码蛋白质的组成或病原体诱导表达，而这一表达展现出杀真菌毒性或抑制真菌作用（Datta et al.，1999）。这一策略利用的是单个或多个基因，因为相比更复杂的防卫机制如 SAR，单个或多个基因更适于抗性基因工程。对这些基因的鉴定使得之后可以通过基因工程技术开发出的转基因植物评估它们的特定作用和在疾病反应途径中的重要性（Pink and Puddephat，1999；Melchers and Stuiver，2000；Echeverria et al.，2008；Sreeramanan et al.，2010）。

科学界普遍接受的观点是大多数 PR 蛋白在植物对真菌感染的防卫反应中起到重要作用，因为 PR 蛋白要么单独显示出体外的抗真菌活性，要么结合其他类型的 PR 蛋白显示出体外的抗真菌活性。PR 蛋白已成为开发抗真菌作物，尤其是香蕉的良好候选基因（Datta et al.，2000；Sagi，2000；Punja，2001；Chakrabarti et al.，2003；Pei et al.，2005；Maziah et al.，2007；Sreeramanan et al.，2008；Sreeramanan et al.，2010）。因此，抗真菌蛋白的细胞内或细胞间靶向可能产生植物抗性，而这也意味着抗真菌蛋白策略针对的广谱抗性可能需要靶向植物组织中不同位置的多个抗真菌蛋白的共表达。例如，通过除去短羧基末端前肽而从细胞外靶向液泡 PR 蛋白确实是可行（Melchers et al.，1993）。有趣的是，涉及大豆植物抗毒素诱导子释放因子的产生，并且表达大豆 β-1,3-葡聚糖酶基因的转基因烟草植株显示出对两种真菌——*Phytophtora parasitica* cv. *nicotiane* 和 *Alternaria longipes* 的高度抗性（Yoshikawa et al.，1993）。由于 β-1，3-聚糖是真菌细胞壁的主要成分，并且广泛分布于植物病原真菌中，因而 Yoshikawa 等（1993）认为这一系统会导致广谱抗性。然而，这可能需要同时发生细胞内和细胞外表达。关于这一方法，Sreeramanan 等（2008）使用微小的分生组织和 β-1，3-葡聚糖酶开发出了一种有效的、农杆菌介导的对香蕉品种 Rasthali（AAB）的转化方法，而结果显示对 *Foc*（1 号生理小种；VCG 01217）的抗性增强。

在番茄中，仅表达几丁质酶或 β-1，3-葡聚糖酶转基因的转基因植物易于感染尖孢镰孢，但是表达这两种基因的植株的抗性比只表达几丁质酶或 β-1，3-葡聚糖酶的植株的抗性高很多（Jongedijk et al.，1995）。同样，Jach 等（1995）认为，表达大麦 β-1，3-葡聚糖酶和几丁质酶基因的烟草植株比只表达这两种酶之一的烟草植株表现出更高的抗性。表达编码水稻几丁质酶基因的烟草植株和其他表达苜蓿酸性葡聚糖酶的烟草植株是分开培育的。将两个转基因株系杂交生成杂交植物。对每一个转基因杂合的杂交种和纯合的自交子代的评估显示，两个转基因在保护植物免受 *Cercospora nicotiane* 攻击上的联合效果比单个转基因的好。Sreeramanan 等（2010）利用粒子轰击法，成功地将几丁质酶和 β－1,3 葡聚糖酶基因导入 Silk Banana（芭蕉属物种 AAB 组）cv. Rasthali。因为 *Foc* 的细胞壁具有几丁质、β－1,3－葡聚糖，所以几丁质酶和 β－1,3－葡聚糖酶基因的组合表达显示出它们协同作用将耐受性提高到更高水平。最后，得出的结论是，抗真菌基因的联合表达是提高对镰孢枯萎病的耐受性的一种有效方法（Sreeramanan et al.，2010）。这些结果表明，PR2（β－1,3－葡聚糖酶）和 PR3（几丁质酶）的联合表达能抵抗真菌感染，对植物产生有效保护。

科学家要特别注意的是 PR 蛋白对菌根真菌可能产生的影响，因为菌根真菌十分重要，能刺激包括香蕉在内的许多植物的生长（Declerck et al.，1995）。例如，在表达 class 11（细胞外）PR2 蛋白的烟草植株中观察到泡囊-丛枝菌根真菌 *Glomus mosseae* 产生的定殖延迟，而当其他 PR 蛋白（PR1a、PR3、PR4 和 PR5）表达时，没有观察到负效应（Vierheilig et al.，1995）。尽管这些有益真菌已发展出了一套机制，避免对寄主的抗真菌蛋白产生伤害，但是测试转基因植物中表达的，抗已知菌根真菌的每个抗真菌蛋白的活性还是十分有价值的。

除了 PR 蛋白，植物还表达大量的肽（长度小于 100 个氨基酸），具有抗微生物特性，包括硫堇、富含半胱氨酸的防卫素，其在结构上与昆虫防御素相关并在超过 20 多种植物的几种组织中发现，非特异性脂转移蛋白、hevein 型肽和 knottin 型肽（Cammue et al.，1995；Broekaert et al.，1997；Sagi，2000）。香蕉中，基因枪法可以产生数百个转基因香蕉系，表达单独编码或联合编码防卫素型抗微生物肽（AMP）和 nsLTP 的各种基因（Sagi et al.，1995）。Chakrabarti 等（2003）称，表达非洲爪蟾蜍抗菌肽的合成替代类似物的转基因香蕉能抗两种真菌病原体，即尖孢镰孢古巴专化型和斐济球腔菌，而非洲爪蟾蜍抗菌肽是一种来自非洲爪蟾蜍（*Xenopus laevis*）皮肤分泌物的蛋白质。

来自植物的核糖体灭活蛋白（RIP）会干扰真菌代谢（Kim et al.，2003）。RIP 通过裂解 28S rRNA 亚基中的特定腺嘌呤的 *N*－糖苷键，抑制真核靶细胞的翻译（Stirpe et al.，1992）。这些靶细胞被分成两组：含有两个亚基的单链Ⅰ型和Ⅱ型蛋白，其中一链（A 外源凝集素）能黏合在真菌细胞表面。非植物水解酶（细菌几丁质酶和真菌几丁质酶，溶菌酶）和抗微生物肽（昆虫杀菌肽）的表达是另一种构建对植物病原真菌抗性的策略（Sagi，1999；Punja，2001；Emani et al.，2003）。此外，Pei 等（2005）开发出表达人溶菌酶的转基因香蕉植株，该植株对 *Foc* 的 4 号生理小种表现出一定程度的抗性。

抗性基因产物充当病原体 Avr 因子的受体来间接识别 Avr 因子，而这可能会触发一个或多个信号转导途径，这反过来又会激活植物的顺从反应，从而预防病原体感染（Staskawicz et al.，1995；Punja，2001）。转基因植物中抗性（R）基因的表达有好几个例子。有报告称，植物 R 基因通过细胞质核苷酸结合位点和富含亮氨酸重复（NBS－LRR）结构域来编码蛋白质，而 NBS－LRR 结构域能赋予植物对多种病原体，特别是 *Foc* 4 号生理小种的抗性（Dang and Jones，2001；Echeverria et al.，2008）。Echeverria 等（2008）报道了来自香蕉的核苷酸结合位点（NBS）型的候选抗病基因的表征和转录多态性与对 *Foc* 4 号生理小种的抗性的关联。他们使用简并引物来扩增抗 *Foc* 4 号生理小种的野生香蕉（*Musa acuminata* ssp. *malaccensis*）的基因组 NBS－LRR。随后，Echeverria 等（2008）建议，可使用 RNA 干扰（RNAi）技术来测试多个香蕉抗性基因候补（RGC）的功能，方法是将它们在香蕉抗病基因型中的对应目标沉默，这样就能识别特定的潜在 R 基因。

科学家推测香蕉植物—病原体的相互作用会生成信号，激活参与植物防御反应的核基因，从而诱导出对胁迫进行反应的化合物和各种酶，例如过氧化氢、酚类化合物、苯丙氨酸解氨酶、过氧化物酶和多酚氧化酶（De Ascensao and Dubery，2000；Sreeramanan et al.，2005）。过氧化氢的积累已被确定为发生在宿主病原体识别的最早事件之一，并且推测它在植物防御中起着重要作用（Baker and Orlandi，1995；Punja，2001；Sreeramanan et al.，2005）。科学家构建出了防卫反应的另一种活化剂，该活化剂是编码葡萄糖氧化酶的基因通过表达产生的过氧化氢（Punja，2001）。研究人员观察到马铃薯植株中通过 *Aspergillus niger* 的葡萄糖氧化酶基因转化的过氧化氢浓度升高，并且导致对晚疫病真菌 *Phytophthora infestants* 的抗性提升（Wu et al.，1995）。有趣的是，研究人员还发现在通过抗真菌基因转化的香蕉中进行孢子接种之后，过氧化氢和苯丙氨酸解氨酶的活性对 *Foc*（1 号生理小种，VCG 01217）的抗性最敏感、反应最强烈（Sreeramanan et al.，2005）。然而，过氧化氢浓度的降低似乎也会增加抗病性，因为来自热带豆科牧草 *Stylosanthes humili* 的病程相关过氧化物酶（Shpx6a）在转基因烟草和油菜系中的组成型表达分别导致了对 *Phytophthora parasitica* var. *nicotiane* 和 *Leptospheria maculans* 的少量但重要的抗性（Kazan et al.，1998）。香蕉中，相比接种 *Foc* 1 号生理小种的未转化的小植株，包含几丁质酶和 β－1,3－葡聚糖酶的基因的转基因植物 Rasthali 小植株显示出更高的几丁质酶、β－1,3－葡聚糖酶、过氧化氢酶、苯丙氨酸解氨酶、过氧化物酶和多酚氧化酶活性（Sreeramanan et al.，2005）。

旨在提高对真菌病原体，特别是对 *Foc* 抗性的香蕉基因工程上取了巨大进步，表明了真菌病原体在芭蕉属作物中的重要性。尽管科学家在香蕉转化上取得了更大进展，但是在识别潜在抗真菌基因并将其导入香蕉以降低 *Foc* 产生的损失方面，科学家仍面临着巨大挑战。

缩略语

AFLP：扩增片段长度多态性
AMP：抗微生物肽
Avr factor：无毒因子
DAF：DNA 扩增指纹
Foc：尖孢镰孢古巴专化型
Fol：尖孢镰孢番茄专化型
HR：过敏反应
IGS：间隔区
ITS：内部转录间隔区
NBS-LRR：核苷酸结合位点-富含亮氨酸重复
nit：硝酸盐营养缺陷型突变体
nsLTP：非特异性脂质转移蛋白
PCR：聚合酶链式反应

参考文献

Alves-Santos F M, García-Sanchez R B, Díaz-Mínguez J M, 2002. A DNA-based procedure for in planta detection of *Fusarium oxysporum* f. sp. *phaseoli* [J]. Phytopathology, 92: 237-244.

Baker C J, Orlandi E W, 1995. Active oxygen in plant pathogenesis [J]. Annu Rev Phytopathol, 33: 299-322.

Bentley S, Bassam B J, 1996. A robust DNA amplification fingerprinting system applied to analysis of genetic variation within *Fusarium oxysporum* f. sp. *cubense* [J]. J Phytopathol, 144: 207-213.

Bentley S, Pegg K G, Dale J L, 1995. Genetic variation among a world-wide collection of isolates of *Fusarium oxysporum* f. sp. *cubense* analyzed by RAPD-PCR fingerprinting [J]. Mycol Res, 99: 1378-1384.

Bentley S, Pegg K G, Moore N Y, et al, 1998. Genetic variation among vegetative compatibility groups of *Fusarium oxysporum* f. sp. *cubense* analyzed by DNA fingerprinting [J]. Phytopathology, 88: 1283-1293.

Blomme G, Eden-Green S, Mustaffa M, et al, 2011. Major diseases of banana [M]//Pillay

M, Tenkouano A. Banana breeding: progress and challenges. Boca Raton, FL, USA: CRC Press: 85 - 119.

Boehm E W A, Ploetz R C, Kistler H C, 1994. Statistical analysis of electrophoretic karyotype variation among vegetative compatibility groups of *Fusarium oxysporum* f. sp. *cubense* [J]. Mol Plant-Microbe Interact, 7: 196 - 207.

Bosland P W, Williams P H, 1987. An evaluation of *Fusarium oxysporum* from crucifers based on pathogenicity, isozyme polymorphism, vegetative compatibility and geographical origin [J]. Can J Bot, 65: 2067 - 2073.

Brake V M, Pegg K G, Irwin J A G, et al, 1990. Vegetative compatibility groups within Australian populations of *Fusarium oxysporum* f. sp. *cubense*, the cause of Fusarium wilt of bananas [J]. Aust J Agric Res, 41: 863 - 870.

Broekaert W F, Cammue B P A, De Bolle M F C, et al, 1997. Antimicrobial peptides of plants [J]. Crit Rev Plant Sci, 16: 297 - 323.

Buddenhaggen I, 2009. Understanding strain diversity in *Fusarium oxysporum* f. sp. *cubense* and history of introduction of 'Tropical Race 4' to better manage banana production [J]. Acta Hort, 828: 193 - 204.

Cammue B P, Thevissen A K, Hendriks M, et al, 1995. A potent antimicrobial protein from onion (*Allium cepa* L.) seeds showing strong sequence homology to plant lipid transfer protein [J]. Plant Physiol, 109: 445 - 455.

Chakrabarti A, Ganapathi T R, Mukherjee P K, et al, 2003. MSI - 99, a magainin analogue, imparts enhanced disease resistance in transgenic tobacco and banana [J]. Planta, 216: 587 - 596.

Correll J C, 1991. The relationship between formae speciales, races, and vegetative compatibility groups in *Fusarium oxysporum* [J]. Phytopathology, 81: 1061 - 1064.

Correll J C, Puhalla J E, Schneider R W, et al, 1985. Differentiating races of *Fusarium oxysporum* f. sp. *pisi* based on vegetative compatibility (Abstr.) [J]. Phytopathology, 75: 1347.

Correll J C, Puhalla J E, Schneider R W, 1986. Identification of *Fusarium oxysporum* f. sp. *apii* on the basis of colony size, virulence and vegetative compatibility [J]. Phytopathology, 76: 396 - 400.

Correll J C, Klittich C J R, Leslie J F, 1987. Nitrate non-utilizing mutants of *Fusarium oxysporum* and their use in vegetative compatibility tests [J]. Phytopathology, 77: 1640 - 1646.

Croft J H, Jinks J L, 1977. Aspects of the population genetics of *Aspergillus nidulens* [M]// Smith J E, Pateman J A. Genetics and physiology of *Aspergillus*. New York, USA: Academic Press: 339 - 360.

Daniells J W, 2009. Global banana disease management - getting serious with sustainability and food security [J]. Acta Hort, 828: 411 - 416.

Daniells J, Davis D, Peterson R, et al, 1995. Gold finger: not as resistant to Sigatoka/

yellow Sigatoka as first thought [J]. Infomusa, 4-6.

Dang I L, Jones J, 2001. Plant pathogens and integrated defence responses to infection [J]. Nature, 411: 826-833.

Datta K, Velazhahan R, Oliva N, et al, 1999. Over expression of the cloned rice thaumatin-like protein (PR-5) gene in transgenic rice plants enhances environmental friendly resistance to *Rhizoctonia solani* causing sheath blight disease [J]. Theor Appl Genet, 98: 1138-1145.

Datta K, Koukoliková-Nicola Z, Baisakh N, et al, 2000. *Agrobacterium*-mediated engineering for sheath blight resistance of indica rice cultivars from different ecosystems [J]. Theor Appl Genet, 100: 832-839.

Davis R I, Moore N Y, Bentley S, et al, 2000. Further records of *Fusarium oxysporum* f. sp. *cubense* from New Guinea [J]. Aust Plant Pathol, 29: 224.

De Ascensao A R D C F, Dubery I A, 2000. Panama Disease: Cell wall reinforcement in response to elicitors from *Fusarium oxysporum* f. sp. *cubense* race four [J]. Amer Phytopathol Soc, 90: 1173-1180.

Declerck S, Plenchette C, Strullu D G, 1995. Mycorrhizal dependency of banana (*Musa acuminate*, AAA group) cultivar [J]. Plant Soil, 7: 183-187.

Dempsey D A, Shah J, Klessig D F, 1999. Salicylic acid and disease resistance in plants [J]. Crit Rev Plant Sci, 18: 547-575.

Dixon R A, Lamb C J, Masoud S, et al, 1996. Metabolic engineering: prospects for crop improvement through the genetic manipulation of phenylpropanoid biosynthesis and defense responses-a review [J]. Gene, 179: 61-71.

Dong X, 1998. SA, JA, ethylene and disease resistance in plants [J]. Curr Opin Plant Biol, 1: 316-323.

Donofrio N M, Delaney T P, 2001. Abnormal callose response phenotype and hypersu sceptibility to *Peronospora parasitica* in defense-compromised *Arabidopsis nim 1-1* and salicylate hydroxylase-expressing plants [J]. Mol Plant-Microbe Interact, 14: 439-450.

Echeverria S P, Dale J L, Harding R M, et al, 2008. Characterization of disease resistance gene candidates of the nucleotide binding site (NBS) type from banana and correlation of a transcriptional polymorphism with resistance to *Fusarium oxysporum* f. sp. *cubense* race 4 [J]. Mol Breed, 22: 565-579.

Emani C, Garcia J M, Lopata-Finch E, et al, 2003. Enhanced fungal resistance in transgenic cotton expressing an endochitinase gene from *Trichoderma virens* [J]. Plant Biotechnol J, 1: 321-326.

Fungaro M H P, Vissotto P C, Sartori D, et al, 2004. A molecular method for detection of *Aspergillus carbonarius* in coffee beans [J]. Curr Microbiol, 49: 123-127.

Gorden T R, Correll J C, McCain A H, 1986. Host specificity and vegetative compatibility in *Verticillium alboatrum* (Abstr.) [J]. Phytopathology, 76: 1111.

Govrin E M, Levine A, 2000. The hypersensitive response facilitates plant infection by the necrotrophic pathogen *Botrytis cinerea* [J]. Curr Biol, 10: 751-757.

Groenewald S, Van den berg N, Marasas W F O, et al, 2006. The application of hight-hroughput AFLP's in assessing genetic diversity in *Fusarium oxysporum* f. sp. *cubense* [J]. Mycol Res, 110: 297-305.

Jabs T, Slusarenko A J, 2000. The hypersensitive response [M]//Slusarenko A J, Fraser RSS, Van Loon L C. Mechanisms of resistance to plant diseases. Dordrecht, The Netherlands: Kluwer Academic Publishers: 279-323.

Jach G, Görnhardt B, Mundy J, et al, 1995. Enhanced quantitative resistance against fungal disease by combinatorial expression of different barley antifungal proteins in transgenic tobacco [J]. Plant J, 8: 97-109.

Jones D R, 1999. Introduction to banana, abaca, and enset [M]//Jones D R. Diseases of banana, abaca, and enset. Wallingford, UK: CABI Publishing: 1-36.

Jones D R, 2000. Diseases of banana, abacá and enset [M]. Wallingford, UK: CABI Publishing.

Jongedijk E, Tigelaar H, Van Roekel J S C, et al, 1995. Synergistic activity of chitinases and β-1, 3-glucanases enhances fungal resistance in transgenic tomato plants [J]. Euphytica, 85: 173-180.

Jurado M, Vázquez C, Marín S, et al, 2006. PCR-based strategy to detect contamination with mycotoxigenic *Fusarium* species in maize [J]. Syst Appl Microbiol, 29: 681-689.

Kazan K, Goulter K C, Way H M, et al, 1998. Expression of a pathogenesis-related peroxidase of *Stylosanthes humilis* in transgenic tobacco and canola and its effect on disease development [J]. Plant Sci, 136: 207-217.

Kim J K, Jang I C, Wu R, et al, 2003. Coexpression of a modified maize ribosome-inactivating protein and a basic chitinase gene in transgenic rice plants confers enhanced resistance to sheath blight [J]. Transgen Res, 12: 475-484.

Kistler H C, Miao V, 1992. New modes of genetic change in filamentous fungi [J]. Annu Rev Phytopathol, 30: 131-152.

Koenig R L, Ploetz R C, Kistler H C, 1997. *Fusarium oxysporum* f. sp. *cubense* consists of a small number of divergent and globally distributed clonal lineages [J]. Amer Phytopathol Soc, 87: 915-923.

Koike M, Watanabe M, Nagao H, et al, 1995. Molecular analysis of Japanese isolates of *Verticillium dahliae* and *V. alboatrum* [J]. Lett Appl Microbiol, 21: 75-78.

Lamb R, Dixon R A, 1997. The oxidative burst in disease resistance [J]. Annu Rev Plant Physiol & Plant Mol Biol, 48: 251-275.

Lee M W, Qia M, Yang Y, 2001. A novel jasmonic acid inducible rice *myb* gene associates with fungal infection and host cell death [J]. Mol Plant-Microbe Interact, 14: 527-535.

Liang H, Maynard C A, Allen R D, et al, 2001. Increased *Septoria musiva* resistance in transgenic hybrid poplar leaves expressing a wheat oxalate oxidase gene [J]. Plant Mol Biol，45：619 - 629.

Lin Y H，Chang J Y，Liu E T，et al，2009. Development of a molecular marker for specific detection of *Fusarium oxysporum* f. sp. *cubense* race 4 [J]. Eur J Plant Pathol，123：353 - 365.

Maziah M，Sreeramanan S，Puad A，et al，2007. Production of transgenic banana cultivar，Rasthali (AAB) via *Agrobacterium*-mediated transformation with a rice chitinase gene [J]. J Plant Sci，2：504 - 517.

Melchers L S，Stuiver M H，2000. Novel genes for disease resistance breeding [J]. Curr Opin Plant Biol，3：147 - 152.

Melchers L S，Sela-Buurlage M B，Vloemans S A，et al，1993. Extracellular targeting of the vacuolar tobacco proteins AP24，chitinase and β - 1,3 - glucanase in transgenic plants [J]. Plant Mol Bio，21：583 - 593.

Mes J J，Weststeijn E A，Herlaar F，et al，1999. Biological and molecular characterization of *Fusarium oxysporum* f. sp. *lycopersici* divides race 1 isolates into separate virulence groups [J]. Phytopathology，89：156 - 160.

Miao V P W，1990. Using karyotype variability to investigate the origins and relatedness of isolates of *Fusarium oxysporum* f. sp. *cubense* [M]//Ploetz R C. *Fusarium* Wilt of Banana. St Paul，MN，USA：American Phytopathological Society：55 - 62.

Miller S A，Martin R R，1988. Molecular diagnosis of plant disease [J]. Annu Rev Phytopathol，26：409 - 432.

Moore N Y，1994. *Fusarium* wilt of banana：pathogen variability and host pathogen interaction [D]. Australia：University of Queensland.

Moore N Y，Hargreaves P A，Pegg K G，et al，1991. Characterization of strains of *Fusarium oxysporum* f. sp *cubense* by production of volatiles [J]. Aust J Bot，39：161 - 166.

Moore N Y，Pegg K G，Allen R N，et al，1993. Vegetative compatibility and distribution of *Fusarium oxysporum* f. sp. *cubense* in Australia [J]. Aust J Exp Agric，33：797 - 802.

Moore N Y，Bentley S，Pegg K G，et al，1995. *Musa* disease fact sheet：*Fusarium* wilt of banan [Z]. INBAP，Montpellier，France，5：4.

Moore N，Pegg K G，Buddenhagen I W et al，2001. *Fusarium* wilt of banana：a diverse clonal pathogen of a domesticated clonal host [M]//Summerell B A，Leslie J F，Backhouse D，et al，Paul E Nelson memorial symposium：*Fusarium*. St. Paul，MN，USA：The American Phytopathological Society：212 - 224.

O' Donnell K，Kistler H C，Cigelnik E，et al，1998. Multiple evolutionary origins of the fungus causing Panama disease of banana：concordant evidence from nuclear and mitochondrial gene genealogies [J]. Proc Nat Acad Sci USA，95：2044 - 2049.

Pegg K G，Moore N Y，Sorensen S，1993. Fusarium wilt in the Asian Pacific region [C]. Valmayor R V，Hwang S C，Ploetz R，et al. Proceedings of international symposium on recent developments in banana cultivation technology. INIBAP/ASPNET：Los Bafi os，

Laguna, the Philippines: 255 - 269.

Pegg K G, Moore N Y, Sorenson S, 1994. Variability in populations of *Fusarium oxysporum* f. sp. *cubense* from the Asia/Pacific region [C]. Jones D R. The improvement and testing of *Musa*: a global partnership. Proceeding of the First Global Conference of the International Musa Testing Program, FHIA, Honduras. INIBAP, Montpellier, France: 70 - 82.

Pegg K G, Shivas R G, Moore N Y, et al, 1995. Characterization of a unique population of *Fusarium oxysporum* f. sp. *cubense* causing Fusarium wilt in Cavendish bananas at Carnarvon, Western Australia [J]. Aust J Agric Res, 46: 167 - 178.

Pegg K G, Moore N Y, Bentley S, 1996. *Fusarium* wilt of banana in Australia: a review [J]. Aust J Agric Res, 47: 637 - 650.

Pei X, Chen S, Wen R, et al, 2005. Creation of transgenic bananas expressing human lysozyme gene for panama wilt resistance [J]. J Int Plant Biol, 47: 971 - 977.

Pink D, Puddephat I, 1999. Development of disease resistance genes by plant transformation—a 'mix and match' approach [J]. Trends Plant Sci, 4: 71 - 75.

Ploetz R C, 1990. Fusarium wilt of banana [M]. St. Paul, MN, USA: American Phytopathological Society.

Ploetz R C, 1993. Fusarium wilt (panama disease) in Africa: current status and outlook for smallholder agriculture [M] // Gold C S, Gemmill B. Biological and integrated control of highland banana and plantain pests and diseases. Ibadan, Nigeria: IITA: 312 - 323.

Ploetz R C, 1994. Panama disease: return of the first banana menace [J]. Int J Pest Manag, 40: 326 - 336.

Ploetz R C, 2000. Panama disease: a classic and destructive disease of banana [J]. Online. Plant Health Progress.

Ploetz R, 2004. Diseases and pests: a review of their importance and management [J]. Infomusa, 13: 11 - 16.

Ploetz R C, 2005. Panama disease, an old enemy rears its ugly head: parts 1 and 2 [M]// Plant Health Progress. APSnet, Online doi: 10. 1094/PHP - 2005 - 1221 - 01 - RV.

Ploetz R C, Correll J C, 1988. Vegetative compatibility among races of *Fusarium oxysporum* f. sp. *cubense* [J]. Plant Dis, 72: 325 - 328.

Ploetz R C, Pegg K G, 1997. Fusarium wilt of banana and Wallace's line: was the disease originally restricted to his Indo-Malayan region [J]? Australasian Plant Pathol, 26: 239 - 249.

Ploetz R C, Pegg K G, 2000. Fusarium wilt [M] // Jones D R. Diseases of banana, abacá and enset. Wallingford, UK: CABI Publishing: 143 - 159.

Ploetz R C, Herbert J, Sebasigari K, et al, 1990. Importance of Fusarium wilt in different banana-growing regions [M]//Ploetz R C. Fusarium wilt of banana. St Paul, MN, USA: American Phytopathol Soc: 9 - 26.

Puhalla J E, 1985. Classifi cation of strains of *Fusarium oxysporum* on the basis of vegetative compatibility [J]. Can J Bot, 63: 179 - 183.

Punja Z K，2001. Genetic engineering of plants to enhance resistance to fungal pathogens—a review of progress and future prospects [J]. Can J Plant Pathol，23：216 - 235.

Rishbeth J，1955. *Fusarium* wilt of bananas in Jamaica：some observations on the epidemiology of the disease [J]. Ann Bot，19：293 - 329.

Sagi L，2000. Genetic engineering on banana-future possibilities [M] // Jones D. Diseases of banana，plantain，abaca and enset. Wallingford，UK：CABI：465 - 515.

Sagi L，Panis B，Remy S，et al，1995. Genetic transformation of banana and plantain（*Musa* spp.）via particle bombardment [J]. Bio/Technology，13：481 - 485.

Schilling A G，Möller E M，Geiger H H，1996. Polymerase chain reaction-based assays for species-specific detection of *Fusarium culmorum*，*F. graminearum* and *F. avenaceum* [J]. Phytopathology，86：515 - 523.

Shivas R G，Phileman E，1996. First record of *Fusarium oxyporum* f. sp. *cubense* on banana in Papua New Guinea [J]. Australia Plant Pathol，25：260.

Simmonds N W，Shepherd K，1955. Taxonomy and origins of cultivated bananas [J]. J Linn Soc Bot（London），55：302 - 312.

Sorensen S，Pegg K G，Dale J L，1993. RAPD-PCR analysis of genetic variation within Australian populations of *Fusarium oxysporum* f. sp. *cubense* [M] // Valmayor R V，Hwang S C，Ploetz R，et al. Proc. int symp recent dev banana cultivation technol. Los Banos，Laguna，the Philippines：INIBAPASPNET：285 - 295.

Sreeramanan S，Maziah M，Sariah M，et al，2005. Bioassay method for testing Fusarium wilt disease tolerance in transgenic banana [J]. Sci Hort，108：378 - 389.

Sreeramanan S，Maziah M，Xavier R，2008. Transfer of β - 1,3 - glucanase gene into banana for tolerance to *Fusarium* wilt race 1 disease using *Agrobacterium*-mediated transformation system [J]. Transgen Plant J，2：176 - 185.

Sreeramanan S，Maziah M，Sariah M，et al，2010. Genetic engineering for tolerance to *Fusarium* wilt race 1 in *Musa sapientum* cv. Rastali（AAB）using biolistic gun transformation system [M] // Tripathi L. Tree and forestry science and biotechnology，global science books. Japan：65 - 75.

Staskawicz B J，Ausubel F M，Baker B J，et al，1995. Molecular genetics of plantdisease resistance [J]. Science，268：661 - 667.

Stirpe F，Barbieri L，Battelli L G，et al，1992. Ribosome-inactivating proteins from plants：present status and future prospects [J]. Bio/Technology，10：405 - 412.

Stover R H，1962. Fusarial wilt（panama disease）of bananas and other *Musa* species [M]. Kew，England：Commonwealth Mycological Institute.

Su H J，Hwang S C，Ko W H，1986. Fusarial wilt of Cavendish bananas in Taiwan [J]. Plant Dis，70：814 - 818.

Thangavelu R，2008. Banana wilt. final technical report of network project on wilt of crops with special reference to cultural morphological molecular characterization and pathogenic variability of isolates in India [M]. Kanpur, India: Indian Institute of Pulses Research: 266 - 285.

Thangavelu R, Mustaffa M M, 2010. First report on the occurrence of a virulent strain of *Fusarium* wilt pathogen (race 1) infecting Cavendish (AAA) group of bananas in India [J]. Plant Dis, 94: 1379.

Thangavelu R, Sundararaju P, Sathiamoorthy S, et al, 2001. Status of Fusarium wilt of banana in India [M] // Molina A B, Nik Masdek N H, Liew K W. Banana Fusarium wilt management: towards sustainable cultivation. Los Banos, Laguna, the Philippines: INIBAP-ASPNET: 58 - 63.

Vavilow N I, 1922. The law of holmologous series in variation [J]. J Genet, 12: 47.

Verberne M C, Verpoorte R, Bol J F, et al, 2000. Overproduction of salicylic acid in plants by bacterial transgenes enhances pathogen resistance [J]. Nat Biotechnol, 18: 779 - 783.

Vierheilig H, Alt M, Lange J, et al, 1995. Colonization of transgenic tobacco constitutively expressing pathogenesis-related proteins by the vesicular arbuscular mycorrhizal fungus *Glomus mosseae* [J]. Appl Environ Microbiol, 61: 3031 - 3034.

Waite B H, 1963. Wilt of *Heliconia* spp. caused by *Fusarium oxysporum* f. sp. *cubense* race 3 [J]. Trop Agric (Trinidad), 40: 299 - 305.

Wardlaw C W, 1961. Banana diseases, including plantains and abaca [J]. London, UK. Longmans, Green and Co. Ltd.

Wu G, Shortt B J, Lawrence E B, Levine E B, et al, 1995. Disease resistance conferred by expression of a gene encoding H_2O_2-generating glucose oxidase in transgenic potato plants [J]. Plant Cell, 7: 1357 - 1368.

Yergeau E, Filion M, Vujanovic V, et al, 2005. A PCR-denaturing gradient gel electrophoresis approach to assess *Fusarium* diversity in asparagus [J]. J Microbiol Meth, 60: 143 - 154.

Lin Y H, Chen K S, Liou T D, et al, 2008. Development of a molecular method for rapid differentiation of watermelon lines resistant to *Fusarium oxysporum* f. sp. *niveum* [J]. Botanical Studies, 50: 273 - 280.

Yoshikawa M, Tsuda M, Takeuchi Y, 1993. Resistance to fungal diseases in transgenic tobacco plants expressing the phytoalexin elicitor-releasing factor, β-1,3-endoglucanase, from soybean [J]. Naturwissenschaften, 80: 417 - 420.

第 14 章　香蕉的抗病相关基因：以 *NBS*、*Pto*、*NPR1* 基因为例

Santy Peraza-Echeverria[①,a,*]

Virginia Aurora Herrera-Valencia[①,b]　Andrew James[①,c]

摘　要： 科学家将图位克隆、实验生理学和基因工程的方法用于模式植物、易处理的植物，如拟南芥和番茄，从而了解抗不同生物胁迫，如病毒、细菌、真菌和卵菌引发的植物病害的分子和生理基础。对于难处理的多年生和部分不育的重要作物，如香蕉，这些知识使得科学家们可对影响它们抗病性的关键调节剂的同系物进行鉴定和克隆。本章，将概述香蕉中两种抗病性（R）基因同系物：R 基因的 *NBS*（核苷酸结合位点）和 *Pto*（丝氨酸或苏氨酸激酶）的克隆和分子表征。此外，还将描述与病程相关基因非表达子 1（*NPR1*）同源的 3 个香蕉基因最近的克隆和表征描述，而 *NPR1* 是系统获得性抗性（SAR）的主转录调节子。最后，将描述这些基因对于培育香蕉抗病性的潜在用途。

14.1　引言

香蕉在许多发展中国家是主食，并且对热带地区的众多农民来说是重要的出口商品。而那些威胁世界香蕉生产，每年造成香蕉严重减产的几种病害中，真菌性病害是科学家们关注的焦点。香蕉黑叶斑病属叶面真菌病，其病原体是斐济球腔菌（*Mycosphaerella fijiensis*），该病原菌主要通过空气传播，该病是危害香蕉产业经济的最重要叶部病害（Churchill，2010）。例如，根据估计，

---

① Unidad de Biotecnología，Centro de Investigación Científi ca de Yucatán，Calle 43 No. 130 Colonia Chuburná de Hidalgo，Mérida，Yucatán，México. C. P. 97200.

a. e-mail：santype@cicy. mx.

b. e-mail：vicky@cicy. mx.

c. e-mail：andyj007@cicy. mx.

* 通信作者。

在墨西哥南部和中美洲各地，用来控制斐济球腔菌的农药喷雾剂费用高达生产成本的30%（Agrios，2005）。对许多农民来说，密集使用杀真菌剂不仅价格昂贵，也会给种植园工人带来严重的健康风险，并且威胁环境。另一个破坏性真菌病是由经土壤传播的尖孢镰孢古巴专化型（*Foc*）引发的巴拿马枯萎病。20世纪50年代，*Foc*的1号生理小种对以Gros Michel品种为基础的香蕉产业造成毁灭性打击。现在，这种病原体的一个新的小种*Foc* 4号生理小种一跃成为当前卡文迪什亚组商用香蕉品种的严重威胁，因为该香蕉品种抗*Foc* 1号生理小种，但易感*Foc* 4号生理小种（Ploetz and Pegg，2000；Ploetz，2005）。科学家发现部分野生香蕉对这两种真菌病害表现出抗性，但是这一抗性的分子性质目前仍未知。因此，认识与香蕉抗病性有关的基因对于香蕉的遗传改良和可持续生产至关重要。

为了感知大批潜在的有害微生物，并将这种感知转化为有效的免疫应答，植物进化出复杂的策略。初级免疫应答会识别微生物病原体的共同特征，如鞭毛蛋白、壳多糖、糖蛋白和脂多糖（Chisholm et al.，2006；Göhre and Robatzek，2008）。这些微生物大分子物质被视为病原体相关分子模式（PAMP），被宿主植物的跨膜模式识别受体（PRR）识别出。PR蛋白在细胞表面感知到微生物时会启动PAMP触发免疫（PTI），这是一种有效的基底防御反应，通常会中止病原体攻击，产生非宿主抗性。然而，许多病原体可通过传递毒力因子（效应子）攻破PTI防线，因为毒力因子能抑制基底防御信号进入植物细胞。作为反PTI抑制的防御策略，植物部署了抗性（R）蛋白，这种蛋白能直接或间接地识别给定的效应子，并激活效应子触发免疫（ETI），从而产生疾病抗性（Chisholm et al.，2006）。在攻击位点，ETI常常以程序性细胞死亡的方式告终，这限制了病原体的进一步蔓延。这种类型的细胞自杀也被称为过敏反应（HR）。这种局部反应常常会触发植物全身的非特异性抗性，这种现象称为系统获得性抗性（SAR）（Durrant and Dong，2004）。而SAR一经触发，就能提供对一系列病原体的抗性。SAR的关键调节器是编码转录调节子的*NPR1*（病程相关基因非表达子1）基因（Pieterse and Van Loon，2004）。*NPR1*基因的超量表达增强了对多种病原体的抗性水平（Cao et al.，1998；Friedrich et al.，2001；Lin et al.，2004；Makandar et al.，2006）。重要的是，该抗性通常不与任何不利的植物表型相关联。

在实际病害控制中，ETI介导的抗性有几个吸引人的特点。当协同反应被及时诱导出，它能有效地阻止病原体发展，而不对植物产生附带伤害。此外，农民不需要任何投入，也不会对环境产生不利影响。因此，ETI介导的抗性是植物育种和遗传改良计划的首要任务之一。为了有效利用ETI基因，对其进行克隆和表征描述是十分重要的。研究人员采用图位克隆的方法，从模式植物拟南芥和许多作物中克隆出涉及 ETI 反应的几个抗病相关基因（Hammond-Kosack and Parker，2003；Gurr and Rushton，2005；Van Ooijen et al.，2007）。然而，由于香蕉染色体易位和分离群体小的问题，迄今为止还很难对香蕉实施图位克隆的方法。最近，科学家有可能基于序列同源性来分离类 R 基因或类 *NPR1* 基因，并且进行简并 PCR（Pei et al.，2007；Peraza-Echeverria et al.，2007，2008，2009；Azhar et al.，2008；Endah et al.，2008；Miller et al.，2008；Zhao et al.，2009）。这些序列的可用性为进一步的功能分析和潜在应用铺平了道路。关于最后一点，分子生物技术利用遗传转化将单个或多个抗病相关基因导入当前的香蕉品种，为构建香蕉的抗病性提供了广泛的策略。总体而言，相比传统育种策略，通过分子生物技术对植物进行的遗传改良具有几个优势，如克服物种屏障，解决无性繁殖植物（如香蕉）不育的问题，剔除不需要的遗传连锁性状（遗传累赘）的能力，还能将基因迅速导入优良的经济品种。本章介绍了在克隆香蕉的抗病相关基因和运用不同分子生物技术策略来培育香蕉抗病性上的最新进展。

14.2　*NBS* 抗性候选基因

科学家预测大多数鉴定的植物 R 基因会编码具有细胞质核苷酸结合位点和富含亮氨酸重复（NBS-LRR）结构域的蛋白质，而该结构域会赋予植物对各种病原体和害虫，包括病毒、细菌、真菌、线虫和昆虫的抗性（Dangl and-Jones，2001）。植物基因组中富含 NBS-LRR 基因，从拟南芥和水稻中分别分离出 149 个和 480 个 NBS-LRR 基因（Meyers et al.，2003；Zhou et al.，2004），大多分布在聚簇中（Hulbert et al.，2001）。科学家认为 NBS-LRR 蛋白识别病原体并以激活信号转导路径的方式反应，从而产生抗病性（Belkhadir et al.，2004）。NBS-LRR 种类可分成两个子类：TIR 和非 TIR，而这取决于与果蝇 Toll 和哺乳动物白细胞介素 1 受体（TIR）同源的 N 末端上是否存在结构域（Meyers et al.，1999）。非 TIR-NBS-LRR 基因存在于单子叶植物和双子叶植物中，而 TIR-NBS-LRR 基因似乎仅限于双子叶植物中（Meyers et al.，2003；Zhou et al.，2004）。NBS 结构域包括 3 个基序和几个功能未知的保守基序，而科学家预测这 3 个基序结合 ATP 或 GTP。该区域与动物细胞中细胞凋亡的两种激活剂——APAF－1 和 CED 同源。科学家通过类推程序性细胞死亡的这些鉴定充分的调节剂，认为 NBS-LRR 蛋白中对应的结构域可能发挥着分子内信号转导器的作用（Van der Biezen andJones，1998）。生化证据揭示了番茄 Mi－1 和 I2 的非 TIR-NBS-LRR 抗性蛋白的 NBS 结构域能够体外结合和水解 ATP（Tameling et al.，2002），而这再次佐证了 NBS 结构域发挥着信号转导作用的观点。NBS-LRR 基因的 TIR 和非 TIR 子类也可通过 NBS 结构域中发现的基序来区分，或通过 NBS 激酶 2 基序的最后部分的单个氨基酸残基来区分，而大部分情况下该残基若是天冬氨酸，则证明是 TIR 子类，若是色氨酸，则证明是非 TIR 子类（Meyers et al.，1999）。带有靶向 R 基因 NBS 结构域的保守 GVGKTT（P-loop）、GSRRIIITTRD 或 GLPLA 基序的简并引物的 PCR 导致了大量包含 NBS 的基因从多种植物物种包括香蕉中分离（Pei et al.，2007；Azhar et al.，2008；Miller et al.，2008；Peraza-Echeverria et al.，2008）。这种 PCR 方法分离的 R 基因同系物被称为抗性候补基因（RGC）。

对于香蕉，科学家已从野生和栽培蕉中分离出几个 NBS-RGC 基因。香蕉的 NBS-RGC 序列的几个特点表明它们属于抗病基因的非 TIR-NBS-LRR 子类。例如，已知抗性基因的 NBS 结构域的特征基序（Meyers et al.，1999；Pan et al.，2000）存在于每个香蕉 RGC 中，并且在相似的位置上（图 14－1）。

这些基序中的一个，即高度保守的 P-loop 显示出结合来自番茄的 NBS-LRR 抗性蛋白 I2 和 Mi 中的 ATP（Tameling et al.，2002），而这表明香蕉 NBS-RGC 蛋白有可能也结合 ATP。科学家在香蕉 NBS-RGC 序列的 N 末端区域发现了非 TIR（nT）基序（Bai et al.，2002），该基序只与 NBS 序列的非 TIR 子类相关联（Peraza-Echeverria et al.，2008），而在香蕉 NBS-RGC 蛋白的对应区域没有发现与 TIR 子类相关的基序。此外，香蕉的 NBS-RGC 基因 *RGC1*、*RGC2* 和 *RGC5* 表明非 TIR 结构域中存在推定卷曲螺旋（CC）结构，而这是该区域的另一个共同特征（Peraza-Echeverria et al.，2008）。例如，在水稻基因组的非 TIR 结构域中，535 个 NBS 序列中的 174 个含有 CC 基序（Zhou et al.，2004）。香蕉中发现的非 TIR-NBS-LRR 基因与分离自其他没有 TIR 结构域的单子叶植物的 R 基因的结构一致（Meyers et al.，1999；Pan et al.，2000；Bai et al.，2002；Zhou et al.，2004）。非 TIR-NBS-LRR 子类在植物基因组中广泛存在，而拟南芥、葡萄、毛果杨和水稻中分别含有 55 个、196 个、233 个和 480 个非 TIR-NBS-LRR 相关序列（Meyers et al.，2003；Zhou et al.，2004；Tuskan et al.，2006；Velasco et al.，2007）。通过类推，香蕉基因组中很可能富含这种类型的序列。基于对分离自几个双子叶和单子叶植物种的 R 基因的分析，科学家发现如果 P-loop 附近存在基序 RNBS-A-non-TIR 或激酶 2 基序的末端存在色氨酸残基（W），可以预测 R 基因中缺失 TIR 结构域（Meyers et al.，1999）。这种预测也可用于目前已分离的香蕉 NBS-RGC 序列（图 14－1）（Pei et al.，2007；Azhar et al.，2008；Milleret al.，2008；Peraza-Echeverria et al.，2008）。

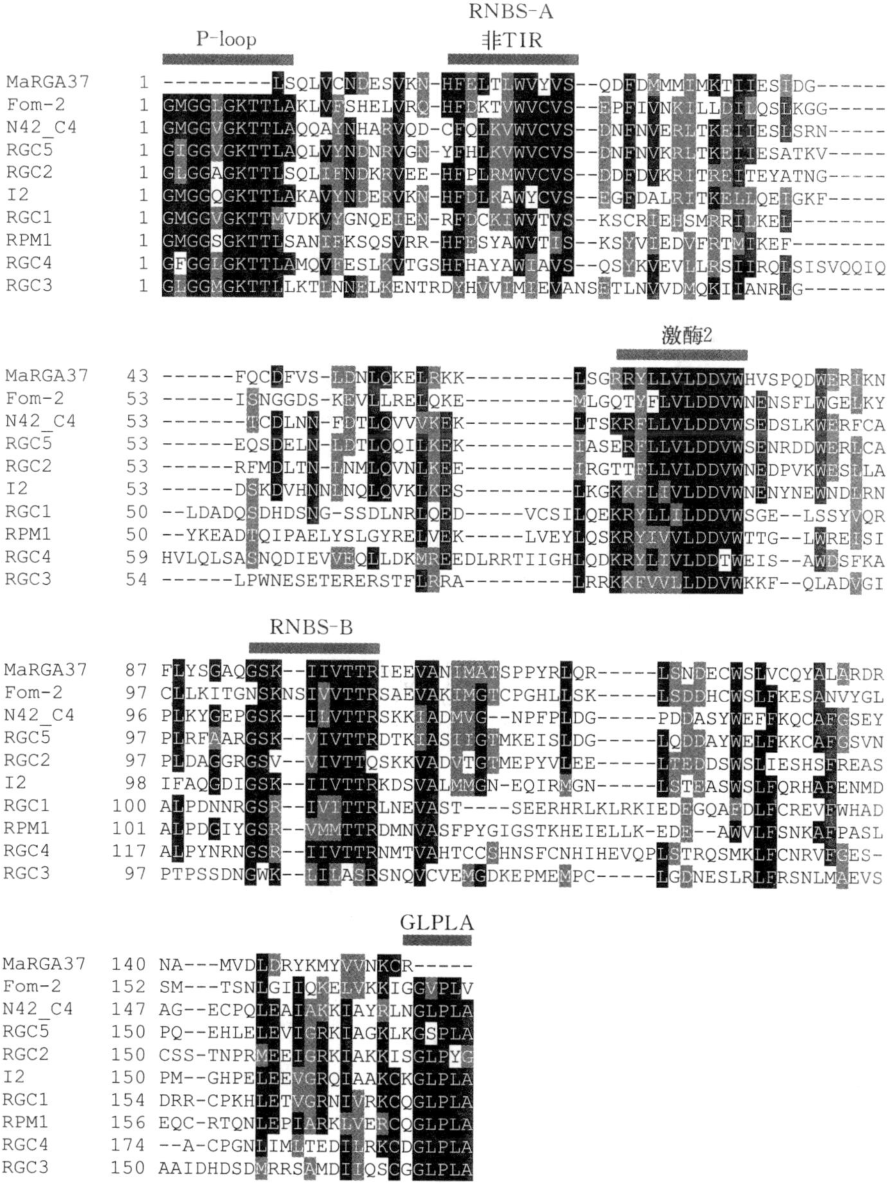

图 14 - 1　一些香蕉 NBS-RGC、RGC1 至 RGC5（EU239819 至 EU239823）（Peraza-Echeverria et al.，2008）、MaRGA37（ER936009）（Miller et al.，2008）、N42 - C4（AM931310）（Azhar and Heslop-Harrison，2008）和分别赋予对尖孢镰孢（I2 和 Fom - 2）和丁香假单胞菌具有抗性的 3 种 R 蛋白——I2（AAD27815）、Fom - 2（AAS80152）和 RPM1（A57072）的预测氨基酸序列的 ClustalX 比对

（保守的 NBS 基序标示在比对序列上面，黑色阴影标示相同的氨基酸，灰色阴影标示保守性置换）

迄今为止，单子叶植物 NBS-LRR 的 R 基因结构中未发现有 TIR 结构域，即使是在完整的水稻基因组序列中也不曾发现（Bai et al.，2002；Cannon et al.，2002；Zhou et al.，2004）。系统发育分析支持将香蕉的 NBS-RGC 序列划分到非 TIR 子类，因为它们都与非 TIR 子类的其他 NBS 序列一样聚集（Pei et al.，2007；Azhar et al.，2008；Peraza-Echeverria et al.，2008；Miller et al.，2008）。科学家猜测在单子叶植物和双子叶植物的趋异过程中，单子叶植物的 NBS-LRR 基因可能丧失了 TIR 结构域（Pan et al.，2000）。由于香蕉和其他单子叶植物有共同的进化起源，因而所有的香蕉 R 基因结构中都可能没有这一结构域。因此，之前这么多研究中只分离得到非 TIR 子类的 NBS 序列，这个不太可能是扩增过程中使用的简并 PCR 方法产生的人为假象，相反，更多的可能是由于芭蕉科家族本身不存在 TIR 结构域。事实上，现今正进行的对香蕉基因组的测序将揭开这个迷雾（www.musagenomics.org）。

NBS-RGC 序列也被用于筛选源自野生种——小果野蕉 Calcutta 4（AA）、野蕉 Pisang Klutuk Wulung（PKW）（BB）和经济三倍体品种小果野蕉 Grand Naine（AAA）的细菌人工染色体（BAC）文库（Miller et al.，2008）。在 3 种香蕉基因型中，称为 MaRGA08 的香蕉 NBS-RGC 探针作为确认的 BAC 中的单拷贝和多拷贝出现。对于小果野蕉 Calcutta 4，BAC 克隆主要包含的是单拷贝 NBS-RGC 序列，相反的是，Grand Naine 和 PKW 的 BAC 分别含有多达 9 个和 11 个拷贝，这表明了香蕉 NBS-RGC 基因的聚类模式。这一发现与其他植物的 NBS-RGC 基因的聚类模式是一致的（Hulbert et al.，2001）。

14.3 *Pto* 抗性候选基因

Pto（对 *Pseudomonas syringae* pathovar *tomato* 的抗性）是第一个克隆的植物 R 基因，它和病原体进行了基因对基因的相互作用（Martin et al.，1993）。它编码 321 个氨基酸的胞内丝氨酸/苏氨酸蛋白激酶（STK），是迄今为止最具表征的 R 基因（Oh and Martin，2010）。*Pto* 激酶直接与两种病原体Ⅲ型效应蛋白——AvrPto 和 AvrPtoB 中的之一相互作用（Martin et al.，1993；Kim et al.，2002）。与 Prf 这种 NBS-LRR 蛋白合作，*Pto* 激活 ETI，阻止病原体的生长。在强大的花椰菜花叶病毒（CaMV）35S 启动子的控制下，番茄中 *Pto* 的超量表达在未受到病原体攻击下触发了防卫反应。*Pto* 超量表达的植物显示出抗细菌病原体 *Xanthomonas campestris* pv. *vesicatoria* 和真菌病原体 *Cladosporium fulvum* 的抗性基因型（Tang et al.，1999）。因此，番茄 *Pto* 基因或来自其他植物的同系物可用于在其他作物中构建广谱病原体抗

性。研究人员利用靶向 STK 蛋白的高度保守子域的简并引物 PCR 技术，已成功将 *Pto* 抗性候选基因从豆类、葡萄、香蕉和刺梨中分离出（Vallad et al.，2001；Di Gaspero and Cipriani，2003；Peraza-Echeverria et al.，2007；Xu and Deng，2010）。

已报道了来自香蕉栽培种的 Tuu Gia 基因组 DNA 的 *Pto*-RGC 的 7 个不同种类（Peraza-Echeverria et al.，2007）。这些序列被称为 Tg－4、Tg－6、Tg－9、Tg－10、Tg－12、Tg－13 和 Tg－67，而它们的 GenBank 检索号分别为 EF492518、EF492519、EF492520、EF492521、EF492522、EF492523 和 EF492524。*Pto*－RGC 的预测氨基酸序列和番茄 *Pto* 蛋白的对应区域之间的同一性水平很高，在 58%（Tg－67）～68%（Tg－13），而 *Pto*-RGC 的氨基酸同一性水平范围是 65.3%（Tg－10 vs Tg－12）～84.1%（Tg－4 vs Tg－13）。

拟南芥（125 Mb 的基因组大小）的完整基因组序列揭示了存在 15 个 Pto－RGC（拟南芥基因组计划 2000），而水稻基因组序列草图（420 Mb 的基因组大小）揭示了数量相似的 *Pto*-RGC，有 14 个（Goff et al.，2002）。这些发现表明，这两个植物基因组的 *Pto*-RGC 数目是保守的。也表明了植物基因组中的 *Pto*-RGC 数目相比 R 基因的 NBS-LRR 类还很少，因为在拟南芥和水稻基因组中该 NBS-LRR 类含有大量的不同基因，数量分别是 149 个和 480 个（Meyers et al.，2003；Zhou et al.，2004）。*Musa acuminata* 的基因组大小估计是 600 Mb（Dolezel et al.，1994），假定植物基因组中 *Pto*-RGC 的数目不会随着基因组大小的变化而大幅增加，而后有可能香蕉中的 *Pto*-RGC 数目类似于拟南芥和水稻中的。事实上，香蕉基因组的完整序列将为这一芭蕉科物种中的 *Pto*-RGC 序列数量提供综合数据（http://www.musagenomics.org）。

将香蕉 *Pto*-RGC 和番茄 *Pto* 预测蛋白进行比对发现，*Pto* 蛋白的几个区域，如 STK 子域（Ⅰ～Ⅸ）和参与 Avr*Pto* 的特异性结合的激活域在香蕉 *Pto*-RGC 中高度保守（Frederick et al.，1998）（图 14－2A）。

此外，*Pto* 的激活域中 3 个自磷酸化位点在所有香蕉 *Pto*-RGC 的对应区域中保守（Peraza-Echeverria et al.，2007）。在香蕉 *Pto*-RGC 中保守的 3 个 *Pto* 自身磷酸化位点（Thr195、Ser198 和 Thr199），Ser198 是 Avr*Pto*-*Pto* 介导的过敏反应所必需的（Sessa et al.，2000），而它存在于大部分香蕉 *Pto*-RGC 中，Tg－12 除外，因为 Tg－12 中丝氨酸被苏氨酸取代，从而提供了一种替代的磷酸化位点。其他两个 *Pto* 残基 Thr204 和 Tyr207 不是在体外自磷酸化，它们的突变体 $pto^{Thr204Ala}$ 或 $pto^{Tyr207Ala}$ 并不与 Avr*Pto* 在酵母中相互作用（Pedley and Martin，2003）。有趣的是，在缺少类似过敏反应的 Avr*Pto* 的情况下，本氏烟草的叶组织中突变体 $pto^{Thr204Asp}$ 和 $pto^{Tyr207Asp}$ 的瞬时表达产生了细胞死亡反应（Rathjen et al.，1999）。*Pto*-RGC 的全 cDNA 序列的克隆将可能

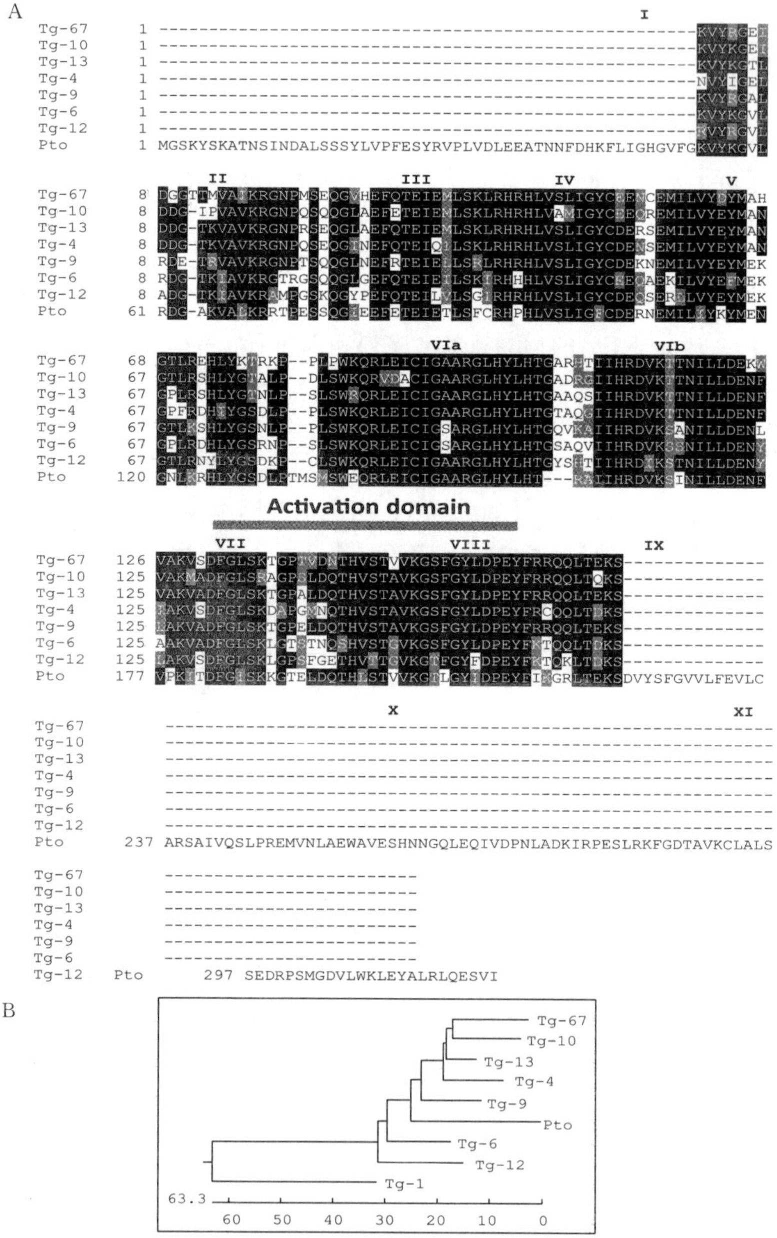

图 14－2　来自香蕉栽培种 Tuu Gia 的 Pto-RGC

A. ClustalX 比对香蕉 Pto-RGC 的预测氨基酸序列（Peraza-Echeverria et al.，2007）和番茄 Pto 抗性蛋白的全长序列（A49332）。图中标示蛋白激酶的子域和激活域（Hanks and Quinn，1991）。黑色阴影标示相同的氨基酸，灰色阴影标示保守置换

B. 香蕉 Pto-RGC、Pto R 蛋白和香蕉壁关联的类受体激酶（Tg－1）（EF492525）的预测氨基酸序列预测的系统发育树状图

通过定点诱变产生香蕉 *Pto*-RGC 的突变体。这些突变体将有助于回答一些基本问题，例如，*Pto*-RGC 编码的蛋白是否在体外自磷酸化，以及酪氨酸在 *Pto*（Tyr207）的对应位点上被天冬氨酸替换是否将导致 HR 样反应。所有香蕉 *Pto*-RGC 产物显示出了保守的丝氨酸-苏氨酸激酶子域（Hanks and Quinn，1991），这表明未覆盖的基因有可能编码活性激酶。此外，参与病原体识别和 HR 诱导的 *Pto* 激活域的大多数残基（Pedley and Martin，2003）在香蕉 *Pto*-RGC 中高度保守，这表明这些残基可能在香蕉中发挥着类似的作用。

对 *Pto* 和 *Pto*-RGC 序列的系统发育分析表明，这些序列在植物中形成了独特的一组激酶（Vallad et al.，2001；Vleeshouwers et al.，2001）。与这一发现一致的是，香蕉 *Pto*-RGC 形成了带有 *Pto* 抗疾病蛋白的一组，表明这些序列和番茄 *Pto* 蛋白有共同的进化起源，并可能在抗病性上有相似的功能（图 14-2B）(Peraza-Echeverria et al.，2007）。此外，对来自不同茄属物种的 *Pto*-RGC 的系统发育分析显示 *Pto* 直系同源基因比旁系同源基因更相似，这表明了 *Pto* 的起源可能早于茄属物种的分化（Vleeshouwers et al.，2001）。*Pto* 和 *Pto* 直系同源基因（*LhirPto*）都在本氏烟草上具有功能的事实进一步支持了 *Pto* 起源悠久的观点（Riely and Martin，2001）。Peraza-Echeverria 等（2007）支持并延伸了之前的观察结果，所有的香蕉 *Pto*-RGC 都集合成包含单子叶和双子叶植物物种的 *Pto*-RGC 的进化枝，这表明了该类型序列的起源可能早于单子叶和双子叶植物的分化。

14.4　*NPR1* 类似基因

NPR1（病程相关基因非表达子 1）是系统获得性抗性（SAR）的关键正调节物，可从远离病原体攻击位点处保护植物组织，使其免受广泛的活体营养型或死体营养型真菌、病毒或昆虫的攻击（Cao et al.，1998；Pieterse and VanLoon，2004）。植物遭受病原体攻击时，水杨酸（SA）、茉莉酸（JA）和乙烯（ET）在植物组织中积聚并充当信号发射网络。SA 集聚时，*NPR1* 将信号转导到细胞核中，而细胞核中激活病程相关（PR）基因的表达被激活，从而导致系统获得性抗性。一些 PR 蛋白显示出抗微生物特性（Durrant and Dong，2004）。*NPR1* 也调控着诱导系统抗性（ISR），至少是部分调控，而该抗性是由某些非致病根际细菌通过根部定殖触发（Pieterse et al.，1998；Spoel et al.，2003）。

科学家发现，每个基因组的 *NPR1* 同系物数量会有变化，大豆中数量是 2，拟南芥中数量是 6，虽然这 6 个中的 *AtNPR3* 和 *AtNPR4* 显示是防卫反应的负调节子（Zhang et al.，2006）。拟南芥另外两个远缘的 *NPR1* 类似基因：

AtNPR5 和 *AtNPR6* 在植物发育中发挥作用（Hepworth et al.，2005）。在非诱导的 SAR 情况下，NPR1 蛋白以无活性的低聚体形式存在于胞质溶胶中，科学家发现它低水平地组成型表达（Henanff et al.，2009）。随着 SA 的积聚，氧化还原电位发生变化，变成更加还原的状态，而 NPR1 蛋白还原为活性的单体状态并转移到细胞核。这是通过一个双向核定位序列和通过还原 NPR1 蛋白的高度保守的半胱氨酸残基之间的分子间二硫键来介导（Mou et al.，2003）。此外，为止检测到的所有功能性 NPR1 蛋白推导的氨基酸谱中，研究人员已发现 4～6 个锚蛋白重复序列（由两个环隔开的 α-螺旋组成的 33 个残基基序）作为中央结构域和 N 末端为 BTB/POZ，这两种结构域都促进蛋白质-蛋白质的结合。

一旦转运到细胞核中，单体 *NPR1* 与被称为 TGA 的转录因子家族相互作用，而 TGA 是转录因子的碱性亮氨酸拉链（bZIP）家族的子类。TGA 显示出调节 SA 响应的 PR 基因 *PR1*、*PR2*（β-1,3-葡聚糖酶）和奇异果甜蛋白类 *PR5*，这些 PR 基因都与 SAR 的活化相关（Despres et al.，2000，2003；Zhang et al.，1999；Zhou et al.，2000）。此外，科学家发现，SA 的活化抑制 JA 发送信号，并且 SA 活化的 NPR1 对 JA 基因表达负调节。*NPR1* 的核定位对于此功能不是必需的（Spoel et al.，2003）。其他基因整合来自不同途径的信号，包括被证明是 PR 基因活化剂的拟南芥 WRKY70 转录因子，以及被证明是受到 NPR1 调节的、由 JA 诱导的基因表达的阻抑剂（Li et al.，2004）。Cao 等（1998）首次报道了拟南芥中的 cDNA *NPR1* 基因序列的超表达，方法是利用构成的花椰菜花叶病毒（CaMV35S）启动子进行对超表达的研究。这些作者发现在转基因植物对人工接种细菌、卵菌病原体和 SAR 诱导剂 SA 与 2，6-二氯异烟酸（INA，SA 的功能类似物）的反应中，*PR1*、*PR2* 和 *PR5* 防卫基因的表达显著提升。他们也发现抗病性的增强与 *NPR1* 基因表达水平的提高有关。这一开创性工作的重要发现之一是通过观察发现在没有病原体或诱导物（SA、INA）的情况下，PR 蛋白的水平在野生型和转基因植物中相同。这些结果表明，*NPR1* 很可能是用来研究抗性基因工程手段的理想候选；转基因植物中超量表达的 *NPR1* 不会表达涉及抗病的基因，直到在病原体攻击下。为了分离 *NPR1* 基因，该组产生了拟南芥在 SAR 诱导中缺少的突变体——*Atnpr1*，它带有锚蛋白重复结构域的点突变（Zhang et al.，1999）。将此突变体用于创建遗传图谱，并利用定位克隆分离 *NPR1* 基因。SAR 缺陷型 *npr1* 已用于测试异源检测中 *NPR1* 推定直系同源物的功能；例如大豆中的 *NPR1* 同系物 *GmNPR1* 和 *GmNPR2*，当它们在拟南芥的 *Atnpr1* 突变体中表达时，它们能效仿 INA 诱导而诱导 PR1（Sandu et al.，2009）。

迄今为止，除了水稻，所有转基因 *NPR1* 的超量表达子中的 PR 蛋白被诱导为更高的水平，只有一个情况下例外，当受到病原体的挑战或在应用化学 SAR 诱导剂如 INA、苯并噻二唑（BTH，SA 的功能类似物）或 SA 之后。与这些发现相反的是，研究人员发现 *AtNPR1* 和水稻同源物 *OsNPR1*/*NH1* 在水稻中的超量表达会诱导至少 4 个防御基因的组成型表达。带有 *AtNPR1* 和水稻同源物的转基因水稻对细菌病原体 *Xanthomonas oryzae* 的抗性增强。然而，*AtNPR1* 超量表达子在对弱光和 SAR 诱导的 BTH 的响应中发展出类病变斑点，而 *OsNPR1* / *NH1* 超量表达子在开花前，于叶子上发展出类病变斑点，在弱光条件下矮化（Fitzgerald et al.，2004；Chern et al.，2005）。Chern 等（2005）推测拟南芥和水稻在防御基因诱导的调控之间有差异是因为水稻叶片存在更高水平的内源性 SA（Silverman et al.，1995）。总而言之，成功将超量表达的 *NPR1* 用在转基因植物中来提高抗病性需要考虑到不同植物种可能存在生理差异。

到目前为止，关于香蕉中 *NPR1* 基因的同系物的出版物只有两个得到同行评审（Endah et al.，2008；Zhao et al.，2009）。Endah 等（2008）利用 PCR 和 cDNA 末端快速扩增技术（RACE）分离两个全长类 *NPR1* 基因：*MNPR1A* 和 *MNPR1B*。这两个序列共享 78%相似性的氨基酸序列。这两个基因都含有保守的锚蛋白重复和 BTB/POZ 锌指结构域；然而，Endah 等（2008）发现，这两个基因在两个结构域的相对位置上和在一些氨基酸上存在差异（图 14 - 3）。

之后在两个对病原体尖孢镰孢古巴专化型（*Foc*）4 号小种的易感性不同的香蕉品种中，科学家通过定量 RT-PCR 技术来分析这两个基因同 *PR1* 和 *PR3* 的表达谱。通过 SA、JA 和接种 *Foc*，类 *NPR1* 和 PR 这两个基因被诱导超出基础水平（图 14 - 4、图 14 - 5 和图 14 - 6）。

对于 SA 的响应，两个品种的 *MNPR1A* 水平和基础水平没有显著不同，而两个品种的 *MNPR1B* 表达显著高于基础水平（图 14 - 4A、B）。SA 作用下也在两个品种中诱导了 *PR1* 和 *PR3* 表达，然而 *PR1* 在抗 *Foc* 品种中的表达比在易感品种中的表达更高（图 14 - 5A、B）。对于 JA 的响应，*MNPR1A* 和 *MNPR1B* 在两个品种中的表达都显著高于基础水平，然而在 JA 处理后的12 h 它们在抗 *Foc* 品种中的表达水平更高，随后 *MNPR1B* 在抗性品种中的表达回落至基础水平，但在易感品种中的表达仍保持在高于正常水平（图 14 - 4C、D）。关于 *PR1* 的表达，在施用 SA 和 JA 处理后品种之间的差异显著（图 14 - 5A、C）。而 *PR1* 在抗 *Foc* 品种中的表达在对 SA 和 JA 响应的 24 h 内明显更高，相反的是，*PR3* 在易感 *Foc* 品种中的表达在对 JA 响应的 48 h 内明显更高（图 14 - 5D）。对 *Foc* 接种两个品种的根的响应有显著差异：*MNPR1B* 表达在易感品种中的差异不大，但 *MNPR1A* 在接种之后的 24 h 中表达水平提高

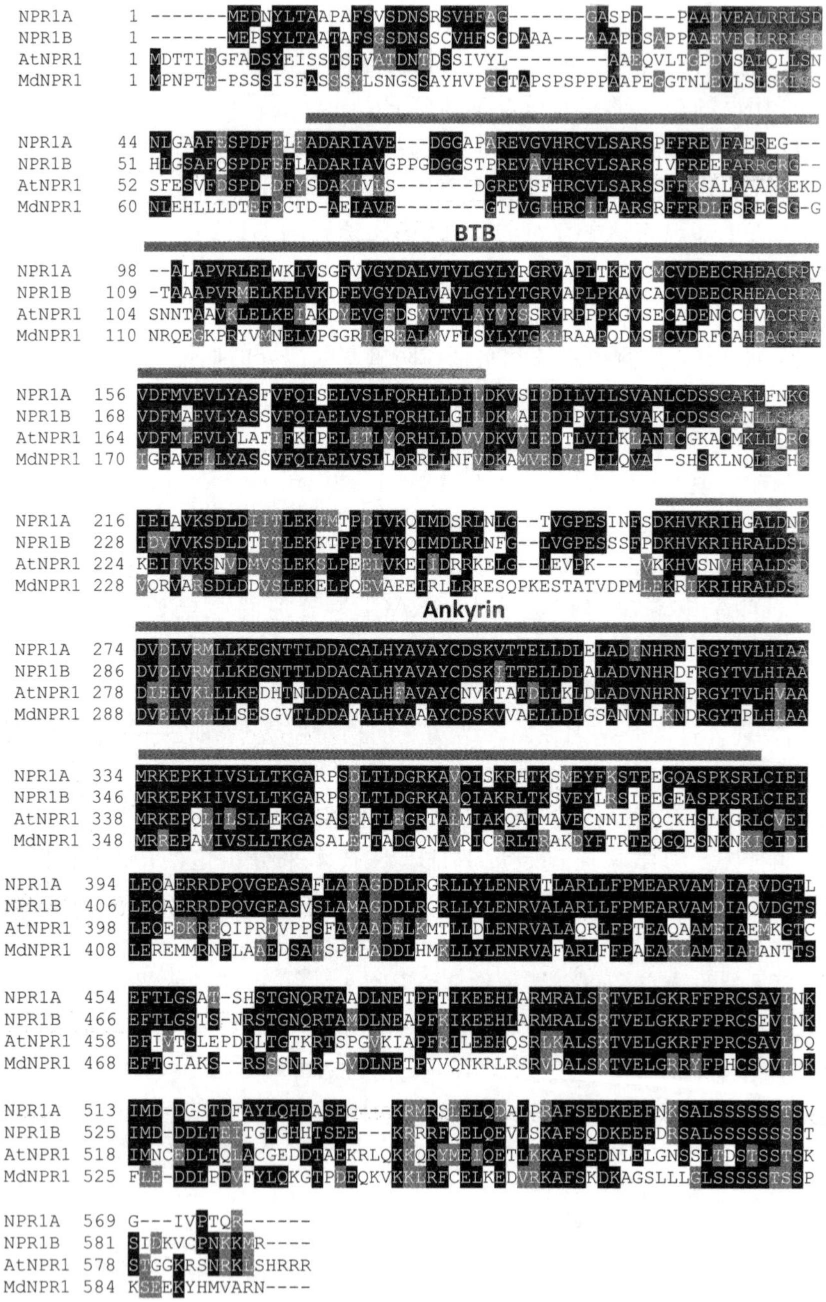

图 14－3　ClustalX 比对 3 个香蕉 *NPR1* 基因的预测氨基酸序列：NPR1A（ABI93182）、NPR1B（ABL63913）（Endah et al.，2008）、MdNPR1（ACJ04030）（Zhao et al.，2009）和拟南芥属 NPR1 蛋白（ATIG64280）

（序列上标示 BTB 和 Ankyrin 锚蛋白域。黑色阴影标示相同氨基酸，灰色阴影标示保守置换）

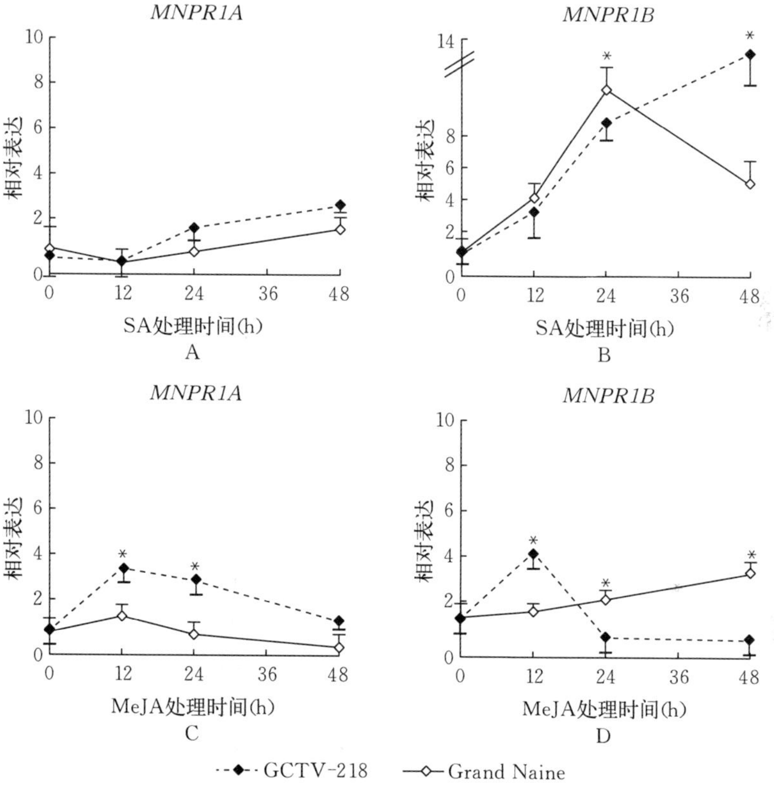

图 14-4　在经过 SA（A，B）和 JA（C，D）处理后的 48 h，类 *NPR1* 基因在易感 *Foc* 4 号小种的 Grand Naine 和抗病 GCTV-218 香蕉品种根部的表达（结果显示为 6 个植株的平均数标准误差；* 表示 $P \leqslant 0.05$ 的显著差异）

（Endah et al.，2008，经过许可）

（图 14-6A）。相比之下，接种 12 h 内 *MNPR1A* 在抗病品种中显著升高（图 14-6B）。根据观察，两品种中的 *MNPR1B* 表达不是由 *Foc* 诱导出的，因此可以不涉及 *Foc* 抗性。两个品种之间的 PR 表达差异较大，在易感品种中 *PR1* 表达并没有显著高于基础表达水平，而 *PR3* 在接种后 24 h 的表达水平显著升高（图 14-6C）。与这些结果相反的是，在抗病品种中，接种 12 h 后 *PR1* 的表达水平明显高于基础水平，但在接种 *Foc* 后的 24 h 和 48 h 之间下降到接近基础水平（图 14-6D），而 *PR3* 的表达在监测的 48 h 内并没有显著增加。

由于没有一个持续的系统能在实验室进行香蕉转化，因此 Endah 等（2008）提出了用香蕉的两个类 *NPR1* 基因来转化拟南芥，并且监测 PR 表达

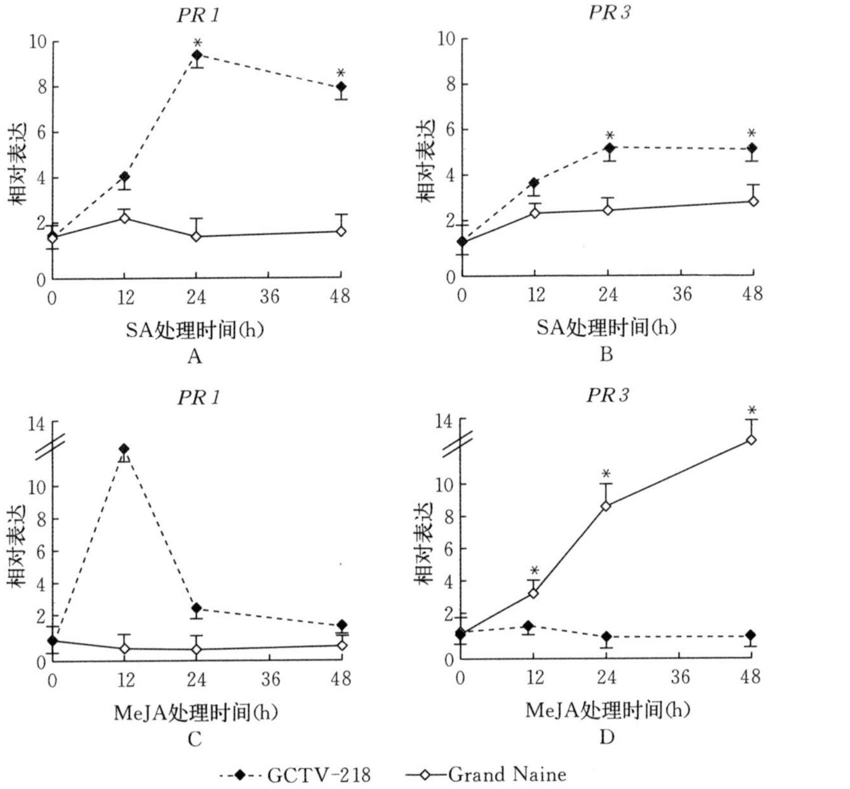

图 14-5 在经过 SA（A，B）和 JA（C，D）处理后的 48 h，*PR1* 和 *PR3* 基因在易感 *Foc* 4 号生理小种的 Grand Naine 和抗病 GCTV-218 香蕉品种根部的表达（结果显示为 6 个植株的平均数标准误差；* 表示 $P \leqslant 0.05$ 的显著差异）（Endah et al.，2008，经过许可）

和诱导抗病性的作用。关于香蕉类 *NPR1* 基因的第二份报告来自于 Zhao 等（2009）。他们从抗 *Foc* 4 号生理小种品种（cv. Dongguan Dajiao，基因组 ABB）中分离全长类 *NPR1* 基因指定的 *MdNPR1*，长度为 4 422 bp（图 14-3）。系统发育分析表明，*MdNPR1* 分组上更接近玉米 *NPR1*，而 *MNPR1A* 和 *MNPR1B* 分组上更接近来自水稻和大麦的 *NPR1*。Zhao 等（2009）认为，*MdNPR1* 可能与 Endah 等（2008）分离的类 *NPR1* 基因具有不同的功能。Zhao 等（2009）发现，易感和抗性品种中 *MdNPR1* 都呈现低水平的组成型表达。然而，科学家只在抗性品种中观察到 SA 超过基础水平的诱导，但没有在易感品种中观察到。对于 *Foc* 接种的响应，*MdNPR1* 在接种 *Foc* 的 4～8 周内，在抗病品种中的表达水平升高（是基础水平的 2～3 倍），而在易感品种中的表达水平没有这样明显的升高。

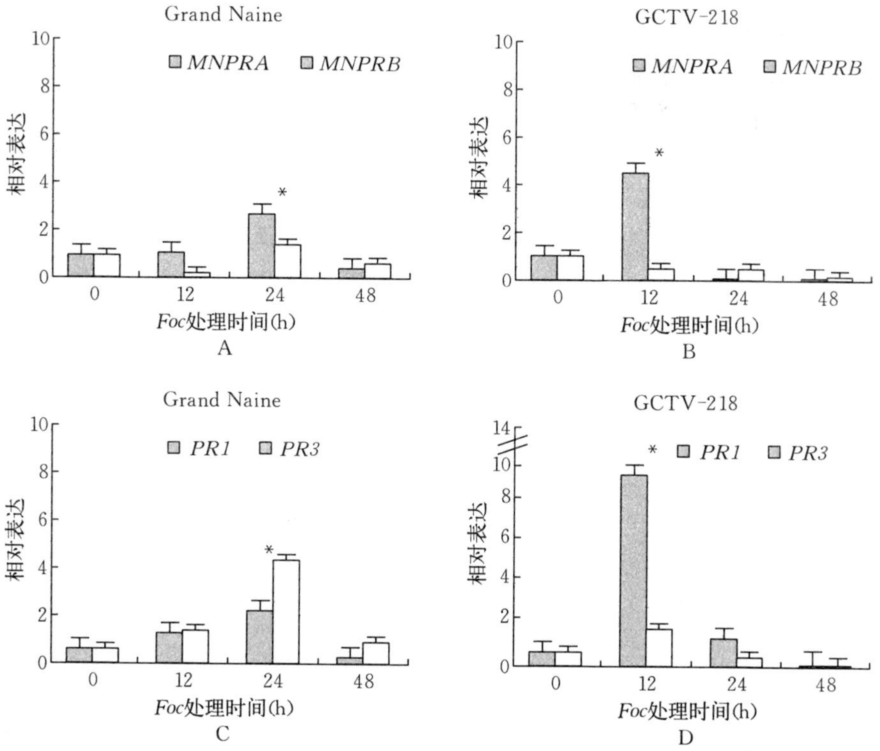

图 14－6　经过处理后的 48 h，*MNPRA* 和 *PR1*（黑条）与 *MNPRB* 和 *PR3*（白条）在感染 *Foc* 的 Grand Naine 和抗病 GCTV－218 香蕉品种根部的表达
（结果显示为 6 个植株的平均数标准误差；＊表示 $P \leqslant 0.05$ 的显著差异）
（Endah et al.，2008，经过许可）

14.5　NBS－RGC、*Pto*-RGC 和类 *NPR1* 基因在香蕉抗病性上的潜在应用

现今香蕉研究可用的 NBS-RGC、*Pto*-RGC 和类 *NPR1* 基因在帮助培育抗最具破坏性病害如香蕉黑叶斑病和巴拿马病的新香蕉品种上有着巨大潜力。在过去 20 年里，香蕉生物技术工具包显著扩大，可使香蕉引入从报告基因到抗凋亡基因的新基因（Khanna et al.，2004，2007）。因此，通过不同的分子策略和利用 NBS-RGC、*Pto*-RGC 和类 *NPR1* 基因，香蕉产生抗病性将在不久的将来成为现实。香蕉 NBS-RGC 将提供与 R 基因位点紧密联系的标记，而这些标记可作为图位克隆的工具用于高分辨率遗传作图。另一方面，后基因组时代

的技术，如 RNA 干扰（RNAi）（Waterhouse and Helliwell，2003）可以推动测试能抗最具破坏性病原体的野生香蕉植物中的多个 NBS-RGC 或 *Pto*-RGC 的功能。这些植物从抗病性转移到易感性是因为不同 NBS-RGC 或 *Pto*-RGC 的沉默会在病原体攻击后帮助鉴定一个特定的 R 基因。双元细菌人工染色体（BIBAC）库技术也能帮助对功能性 R 基因的识别，该技术可以通过根癌农杆菌将大 DNA 片段（高达 120 kb）转移到植物基因组中（He et al.，2003）。最近，科学家为品种 Tuu Gia 构建了一个 BIBAC 文库，该品种能抵抗香蕉最具破坏性的疾病，如香蕉黑叶斑病和巴拿马病（Ortiz-Vázquez et al.，2005）。这种技术结合针对香蕉研发出的高效农业转化法（Khanna et al.，2004）可利用含有作为单体或集群的 NBSRGC 的 BIBAC 克隆来转化病害易感香蕉品种（如 Grand Naine）。这种方法将导致那些准备好用于病原体抗性筛选的香蕉系的 NBS-BIBAC 集合。

对于 *Pto* 基因，之前的报道表明番茄中的超量表达激活了防御反应，从而导致了对细菌和真菌病害的抗性（Tang et al.，1999）。对香蕉使用类似的方法可能会导致对同一类型的病原体产生抗性。有趣的是，番茄 *Pto* 突变体，如 $pto^{Thr204Asp}$ 或 $pto^{Tyr207Asp}$ 的表达可以在无丁香假单胞菌的情况下，组成型激活类似 HR 的反应（Rathjen et al.，1999）。香蕉 *Pto*-RGC 的全长 cDNA 序列的克隆将使得科学家可以评估它们传递疾病抗性的潜力。关于源自拟南芥的 *NPR1* 基因，有报道显示 *NPR1* 在该种植物（Cao et al.，1998；Friedrich et al.，2001）、水稻（Chern et al.，2001）、番茄（Lin et al.，2004）、小麦（Makandar et al.，2006）、烟草（Meur et al.，2008）和胡萝卜（Wally et al.，2009）中的超量表达诱导了对细菌和真菌的抗性。同样的，水稻（Chern et al.，2005）和苹果（Malnoy et al.，2007）中内源 *NPR1* 基因的直向同源物的超量表达导致了对细菌和真菌的抗性增强。这些发现使得 *NPR1* 成为构建热带植物，如香蕉的抗病性的一个热门候选。因此，栽培品种如 Grand Naine 的香蕉类 *NPR1* 基因的超量表达在激活 SAR 反应从而产生对广泛病原体的抗性上有着一定潜力。可以预见的是，这些技术在香蕉中的应用有一个光明的未来，能揭示 NBS-RGC、*Pto*-RGC 和类 *NPR1* 基因在香蕉中的功能，以及通过 cisgenic 手段（来自相同或近缘物种的基因）来培育对病原体的抗性。

缩略语

APAF-1：凋亡蛋白酶激活因子 1

ATP：腺苷三磷酸

AVR：无毒性

BAC：细菌人工染色体

BIBAC：二元细菌人工染色体
BTB/POZ：痘病毒和锌指结构域
BTH：苯并噻二唑
CaMV35S：花椰菜花叶病毒 35S 启动子
ETI：效应子触发免疫
ET：乙烯
Foc：尖孢镰孢古巴专化型
GTP：鸟苷三磷酸
HR：过敏反应
ISR：诱导系统抗性
JA：茉莉酸
NBS：核苷酸结合位点
NBS-LRR：富含亮氨酸重复
NPR1：病程相关基因非表达子 1
PAMP：病原相关分子模式
PCR：聚合酶链式反应
PR：病程相关基因
PRR：模式识别受体
PTI：PAMP 触发免疫
R：抗性基因
RACE：cDNA 末端快速扩增技术
RGC：抗性基因候补
RNAi：RNA 干扰
SA：水杨酸
SAR：系统获得性抗性
STK：丝氨酸/苏氨酸蛋白激酶
TIR：与果蝇 Toll 和哺乳动物白细胞介素 1 受体同源

参考文献

Agrios G N, 2005. Plant Pathology [M]. San Diego, California, USA: Academic Press. Arabidopsis Genome Initiative, 2000. Analysis of the genome sequence of the flowering plant *Arabidopsis thaliana* [J]. Nature, 408: 796-815.

Azhar M, Heslop-Harrison J S, 2008. Genomes, diversity and resistance gene analogues in *Musa* species [J]. Cytogenetics, 121: 59-66.

Bai J, Pennill L A, Ning J, et al, 2002. Diversity in nucleotide binding site-leucine rich repeat genes in cereals [J]. Genome Res, 12: 1871 - 1884.

Belkhadir Y, Subramaniam R, Dangl J L, 2004. Plant disease resistance protein signalling: NBSLRR proteins and their partners [J]. Curr Opin Plant Biol, 7: 391 - 399.

Cannon S B, Zhu H, Baumgarten A M, et al, 2002. Diversity, distribution and ancient taxonomic relationships within the TIR and non-TIR NBS-LRR resistance gene subfamilies [J]. J Mol Evol, 54: 548 - 562.

Cao H, Li X, Dong X, 1998. Generation of broad-spectrum disease resistance by over-expression of an essential regulatory gene in systemic acquired resistance [J]. Proc Natl Acad Sci USA, 95: 6531 - 6536.

Chern M, Fitzgerald H A, Canlas P E, et al, 2005. Overexpression of a rice NPR1 homolog leads to constitutive activation of defense response and hypersensitivity to light [J]. Mol Plant Microbe Interact, 18: 511 - 520.

Chisholm S T, Coaker G, Day B, et al, 2006. Host-microbe interactions: shaping the evolution of the plant immune response [J]. Cell, 124: 803 - 814.

Churchill A C L, 2010. *Mycosphaerella fijiensis*, the black leaf streak pathogen of banana: progress towards understanding pathogen biology and detection, disease development, and the challenges of control [J]. Mol Plant Pathol, DOI: 10.1111/j.1364 - 3703.2010.00672.X.

Dangl L, Jones J, 2001. Plant pathogens and integrated defence responses to infection [J]. Nature, 411: 826 - 833.

Despres C, DeLong C, Glaze S, et al, 2000. The *Arabidopsis* NPR1/NIM1 protein enhances the DNA binding activity of a subgroup of the TGA family of bZIP transcription factors [J]. Plant Cell, 12: 279 - 290.

Despres C, Chubak C, Rochon A, et al, 2003. The *Arabidopsis* NPR1 disease resistance protein is a novel cofactor that confers redox regulation of DNA binding activity to the basic domain/leucine zipper transcription factor TGA1 [J]. Plant Cell, 15: 2181 - 2191.

Di Gaspero G, Cipriani G, 2003. Nucleotide binding site/leucine-rich repeats, Pto-like and receptor-like kinases related to disease resistance in grapevine [J]. Mol Gen Genom, 269: 612 - 623.

Dolezel J, Dolezelova M, Novak F J, 1994. Flow cytometric estimation of nuclear DNA amount in diploid bananas (*Musa acuminata* and *M. balbisiana*) [J]. Biol Plant, 36: 351 - 357.

Durrant W E, Dong X, 2004. Systemic acquired resistance [J]. Ann Rev Phytopath, 42: 185 - 209.

Endah R, Beyene G, Kiggundu A, et al, 2008. Elicitor and *Fusarium*-induced expression of NPR1-like genes in banana [J]. Plant Physiol Biochem, 46: 1007 - 1014.

Fitzgerald H A, Chen M S, Navarre R, et al, 2004. Overexpression of (At) NPR1 in rice leads to a BTH- and environment-induced lesion-mimic/cell death phenotype [J]. Mol Plant

Microbe Interact，17：140 - 151.

Frederick R D，Thilmony R L，Sessa G，et al，1998. Recognition specificity for the bacteri-alavirulence protein AvrPto is determined by Thr - 204 in the activation loop of the tomato Pto kinase [J]. Mol Cell，2：241 - 245.

Friedrich L，Lawton K，Dietrich R，et al，2001. *NIM1* overexpression in *Arabidopsis* potentiates plant disease resistance and results in enhanced effectiveness of fungicides [J]. Mol Plant Microbe Interact，14：1114 - 1124.

Hammond-Kosack K，Parker J E，2003. Deciphering plant-pathogen communication：fresh perspectives for molecular resistance breeding [J]. Curr Opin Biotech，14：177 - 193.

Goff S A，Ricke D，Lan T H，et al，2002. A draft sequence of the rice genome (*Oryza sativa* L. ssp. *japonica*) [J]. Science，296：92 - 100.

Göehre V，Robatzek S，2008. Breaking the barriers：microbial effector molecules subvert plant immunity [J]. Annu Rev Phytopathol，46：189 - 215.

Gurr S J ，Rushton P J，2005. Engineering plants with increased disease resistance：what are going to express [J]? Trends Biotechnol，23：275 - 282.

Hanks S K，Quinn A M，1991. Protein kinase catalytic domain sequence database：identification of conserved features of primary structure and classification of family members [J]. Method Enzymol，200：38 - 62.

He R F，Wang Y，Shi Z，et al，2003. Construction of a genomic library of wild rice and *Agrobacterium*-mediated transformation of large insert DNA linked to *BPH* resistance locus [J]. Gene，321：113 - 121.

Henanff G L，Heitz T，Mestre P，et al，2009. Characterization of *Vitis vinifera* NPR1 homologs involved in the regulation of *Pathogenesis-Related* gene expression [J]. BMC Plant Biol，9：54.

Hepworth S R，Zhang Y，McKim S，et al，2005. BLADE - ON PETIOLE-dependent signaling controls leaf and floral patterning in *Arabidopsis* [J]. Plant Cell，17：1434 - 1448.

Hulbert S H，Webb C A，Smith M，et al，2001. Resistance gene complexes：evolution and utilization [J]. Annu Rev Phytopathol，39：285 - 312.

Khanna H K，Becker D，Kleidon J，et al，2004. Centrifugation assisted *Agrobacterium*-mediated transformation (CAAT) of embryogenic cell suspensions of banana (*Musa* spp. Cavendish AAA and Lady fi nger AAB) [J]. Mol Breed，14：239 - 252.

Khanna H K，Paul J Y，Harding R M，et al，2007. Inhibition of *Agrobacterium*-induced cell death by antiapoptotic gene expression leads to very high transformation efficiency of banana [J]. Mol Plant Pathol Interact，20：1048 - 1054.

Kim Y J，Lin N C，Martin G B，2002. Two distinct *Pseudomonas* effector proteins interact with the Pto kinase and activate plant immunity [J]. Cell，109：589 - 598.

Li J，Brader G，Palva E T，2004. The WRKY70 transcription factor：a node of convergence for jasmonate-mediated and salicylate-mediated signals in plant defense [J]. Plant Cell，16：

319 - 331.

Lin W C, Lu C F, Wu J W, et al, 2004. Transgenic tomato lines expressing the *Arabidopsis NPR1* gene display enhanced resistance to a spectrum of fungal and bacterial diseases [J]. Transgen Res, 13: 567 - 581.

Makandar R, Essig J S, Schapaugh M A, et al, 2006. Genetically engineered resistance to Fusarium head blight in wheat by expression of *Arabidopsis NPR1* [J]. Mol Plant Microbe Interact, 19: 123 - 129.

Malnoy M, Jin Q, Borejsza-Wysocka E E, et al, 2007. Overexpression of the apple *MpNPR1* gene confers increased disease resistance in *Malus* × *domestica* [J]. Mol Plant Microbe Interact, 20: 1568 - 1580.

Martin G B, Brommonschenkel S H, Chunwongse J, et al, 1993. Map-based cloning of a protein kinase gene conferring disease resistance in tomato [J]. Science, 262: 1432 - 1435.

Meur G, Budatha M, Srinivasan T, et al, 2008. Constitutive expression of *Arabidopsis* NPR1 confers enhanced resistance to the early instars of *Spodoptera litura* in transgenic tobacco [J]. Physiol Plant, 133: 765 - 775.

Meyers B, Dickerman A, Michelmore R, et al, 1999. Plant disease resistance genes encode members of an ancient and diverse protein family within the nucleotide-binding superfamily [J]. Plant J, 20: 317 - 332.

Meyers B C, Kozik A, Griego A, et al, 2003. Genome-wide analysis of NBS-LRR-encoding genes in *Arabidopsis* [J]. Plant Cell, 15: 809 - 834.

Miller R N, Bertioli D J, Baurens F C, et al, 2008. Analysis of non-TIR NBS-LRR resistance gene analogs in *Musa acuminata* Colla: isolation, RFLP marker development, and physical mapping [J]. BMC Plant Biol, 8: 15.

Mou Z, Fan W, Dong X, 2003. Inducers of plant systemic acquired resistance regulate NPR1 function through redox changes [J]. Cell, 113: 935 - 944.

Oh C S, Martin G B, 2010. Effector-triggered immunity mediated by the Pto kinase [J]. Trends Plant Sci, 16: 132 - 140.

Ortiz-Vázquez, E, Kaemmer D, Zhang H B, et al, 2005. Construction and characterization of a plant transformation-competent BIBAC library of the black sigatoka resistant banana *Musa acuminata* cv. Tuu Gia (AA) [J]. Theor Appl Genet, 110: 706 - 713.

Pan Q, Wendel J, Fluhr R, 2000. Divergent evolution of plant NBS-LRR resistance gene homologues in dicot and cereal genomes [J]. J Mol Evol, 50: 203 - 213.

Pedley K F, Martin G B, 2003. Molecular basis of *Pto*-mediated resistance to bacterial speck disease in tomato [J]. Annu Rev Phytopathol, 41: 215 - 243.

Pei X, Li S, Jiang Y, et al, 2007. Isolation, characterization and phylogenetic analysis of the resistance gene analogues (RGAs) in banana (*Musa* spp.) [J]. Plant Sci, 172: 1166 - 1174.

Peraza-Echeverria S, James-Kay A, Canto-Canché B, et al, 2007. Structural and phylogenetic analysis of Pto-type disease resistance gene candidates in banana [J]. Mol Genet Genomics，278：443－453.

Peraza-Echeverria S，Dale J L，Harding R M，et al，2008. Characterization of disease resistance gene candidates of the nucleotide binding site（NBS）type from banana and correlation of a transcriptional polymorphism with resistance to *Fusarium oxysporum* f. sp. *cubense* race 4［J］. Mol Breeding，22：565－579.

Peraza-Echeverria S，Dale J L，Harding R M，et al，2009. Molecular cloning and *in silico* analysis of potential *Fusarium* resistance genes in banana［J］. Mol Breeding，23：431－443.

Pieterse C M J，Van Loon L C，2004. NPR1：the spider in the web of induced resistance signaling pathways［J］. Curr Opin Plant Biol，7：456－464.

Pieterse C M J，Van Wees S C M，Van Pelt J A，et al，1998. A novel signaling pathway controlling induced systemic resistance in *Arabidopsis*［J］. Plant Cell，10：1571－1580.

Ploetz R C，2005. Panama disease，an old nemesis rears its ugly head：part 1，the beginnings of the banana export trades［J］. Plant Health Progr，doi：10.1094/PHP－2005－1221－01－RV.

Ploetz R C，Pegg K G，2000. Fungal diseases of the root，corm and pseudostem［M］// Jones D R. Diseases of banana，abaca and enset. Wallingford，UK：CAB International：143－172.

Rathjen J P，Chang J H，Staskawicz B J，et al，1999. Constitutively active *Pto* induces a Prf-dependent hypersensitive response in the absence of *avrPto*［J］. EMBO J，18：3232－3240.

Riely B K，Marti G B，2001. Ancient origin of pathogen recognition specifi city conferred by the tomato disease resistance gene *Pto*［J］. Proc Natl Acad Sci USA，98：2059－2064.

Sandu D，Tasma I M，Frasch R，et al，2009. Systemic acquired resistance in soybean is regulated by two proteins，orthologous to *Arabidopsis* NPR1［J］. BMC Plant Biol，9：105 doi：10.1186/1471－2229－9－105.

Sessa G，D' Ascenzo M，Martin G B，2000. Thr38 and Ser198 are Pto autophosphorylation sites required for the AvrPto-Pto-mediated hypersensitive response［J］. EMBO J，19：2257－2269.

Silverman P，Seskar M，Kanter D，et al，1995. Salicylic acid in rice（biosynthesis，conjugation，and possible role）. Plant Physiol，108：633－639.

Spoel S H，Koornneef A，Claessens S M C，et al，2003. NPR1 modulates cross-talk between salicylateand Jasmonate -dependent defense pathways through a novel function in the cytosol［J］. Plant Cell，15：760－770.

Tameling W，Elzinga S，Darmin P，et al，2002. Thetomato R gene products I2 and Mi-1 are functional ATP binding proteins with ATPase activity［J］. Plant Cell，14：2929－2939.

Tang X，Xie M，Kim Y J，et al，1999. Overexpression of *Pto* activates defense responses and confers broad resistance［J］. Plant Cell，11：15－29.

Tuskan G A，DiFazio S，Bohlmann J，et al，2006. The genome of black cottonwood *Populus trichocarpa*（Torr. and Gray）［J］. Science，313：1596－1604.

Vallad G，Rivkin M，Vallejos C，et al，2001. Cloning and homology modelling of a Ptolike protein kinase family of common bean (*Phaseolus vulgaris* L.) [J]. Theor Appl Genet, 103: 1046 - 1058.

Van der Biezen E, Jones J D G, 1998. Plant disease-resistance proteins and the gene-forgene concept [J]. Trends Biochem Sci, 23: 454 - 456.

Van Ooijen G, Van den Burg H A, Cornelissenn B J C, et al, 2007. Structure and function of resistance proteins in Solanaceous plants [J]. Annu Rev Phytopathol, 45: 43 - 72.

Velasco R, Zharkikh A, Troggio M, et al, 2007. A high quality draft consensus sequence of the genome of a heterozygous grapevine variety [J]. PLoS One, 12: 1 - 18.

Vleeshouwers V G A A, Martens A, Van Dooijeweert W, et al, 2001. Ancient diversification of the *Pto* kinase family preceded speciation in *Solanum* [J]. Mol Plant Microbe Interact, 14: 996 - 1005.

Wally O Jayaraj J, Punja Z K, 2009. Broad-spectrum disease resistance to necrotrophic and biotrophic pathogens in transgenic carrots (*Daucus carota* L.) expressing an *Arabidopsis NPR1* gene [J]. Planta, doi 10.1007/s00425 - 009 - 1031 - 2.

Waterhouse P M, Helliwell C A, 2003. Exploring plant genomes by RNA - induced gene silencing [J]. Nature Rev, 4: 29 - 38.

Xu Q, Deng X, 2010. Cloning and phylogenetic analysis of serine/threonine kinase class defense-related genes in a wild fruit crop 'chestnut rose' [J]. BMC Res Notes, 3: 202.

Zhang Y, Fan W, Kinkema M, et al, 1999. Interaction of NPR1with basic leucine zipper protein transcription factors that bind sequences required for salicylic acid induction of the *PR - 1* gene [J]. Proc Natl Acad Sci USA, 96: 6523 - 6528.

Zhang Y, Cheng Y T, Qu N, et al, 2006. Negative regulation of defense responsesin *Arabidopsis* by two NPR1 paralogs [J]. Plant J, 48: 647 - 656.

Zhao J T, Huang X, Chen Y P, et al, 2009. Molecular cloning and characterization of an ortholog of *NPR1* gene from Dongguan Dajiao (*Musa* spp. ABB) [J]. Plant Mol Biol Rep, 27: 243 - 249.

Zhou J M, Trifa Y, Silva H, et al, 2000. NPR1 differentially interacts with members of the TGA/OBF family of transcription factors that bind an element of the *PR - 1* gene required for induction by salicylic acid [J]. Mol Plant Microbe Interact, 13: 191 - 202.

Zhou T, Wang T, Chen J Q, et al, 2004. Genome-wide identification of NBS genes in *japonica* rice reveals significant expansion of divergent non-TIR NBS-LRR genes [J]. Mol Genet Genomics, 271: 402 - 415.

第 15 章　芭蕉属植物分子育种案例

Michael Pillay[①,*]　Kaliyaperumal Ashokkumar[②,a]
Arun Siva Kumar Shunmugam[②,b]　Sivalingam Elayabalan[③]

摘　要： 传统芭蕉属植物育种主要采用大规模的表型轮回选择，而这会受到若干因素阻碍。分子生物学技术的快速发展及其在植物育种上的应用已经导致农作物上的显著遗传增益。标记辅助育种对于像香蕉这种具有较长生命周期的作物将非常有用。目前正在寻找芭蕉中几个重要性状的 DNA 标记，包括抗病虫害的 DNA 标记。本章将讨论在香蕉抗黑叶斑病、镰孢枯萎病、香蕉束顶病毒（BBTV）、线虫和黄单胞菌枯萎病上进行分子育种所取得的成就和发展前景。此外，还描述了在提高香蕉营养上的成果。现代植物分子和数量遗传学在过去 20 年的发展使其有可能革新大多以经验为基础的经验主义植物育种。本章概述了在香蕉分子育种中现代分子工具应用的价值。

关键词： 育种挑战，分子标记，分子育种成果

15.1　引言

目前，大多生产上改良的香蕉和大蕉品种在营养上能被消费者接受，能够抗或耐生物和非生物胁迫，并能减少采后损失，这主要是由于传统育种在过去多年不断取得稳步发展，产生了大量的杂交种（Rowe，1984；Vuylsteke et al.，1995）。一方面，人口不断膨胀，对食物的需求不断增加，但另一方面耕

① Vaal University of Technology，Private Bag X021，Vanderbijlpark 1900，Gauteng，South Africa.
② Department of Plant Sciences，University of Saskatchewan，51 Campus Drive，Saskatoon，SK，Canada，S7N 5A8.
③ Crop Tech Ltd，P. O. Box 1367 Arusha，Tanzania；e-mail：balabiotech@gmail. com.
a. e-mail：biotech. ashok@gmail. com.
b. e-mail：anbeshivam481@gmail. com.
* 通信作者。

地缩减，随之产生的压力对传统植物育种有了新的要求（Pillay et al.，2011）。然而，芭蕉属植物的传统育种方法受到其不育和其他一系列因素的制约，详细描述请见 Pillay 等（2002）、Pillay 和 Tripathi（2006，2007）的文献报道。

分子生物学技术及其在植物育种应用上的快速发展导致农作物上的显著遗传增益，而其中一些农作物已经进入市场（Newell-McGloughlin，2008）。分子和生物技术工具，如标记辅助育种、组织培养、体外诱变和遗传转化有助于解决或减少传统香蕉育种的一些限制。本章探讨了芭蕉属植物分子育种的几个方面。

15.1.1 芭蕉属植物育种挑战

芭蕉属植物是多倍体，倍性范围从二倍体（$2n=2x=22$）到四倍体（$2n=4x=44$）。大多数栽培香蕉是三倍体（$2n=3x=33$），并且不育，有 1 个、2 个或 3 个 A、B、S 或 T 基因组的各种组合。培育新的香蕉品种异常困难。理想性状的选择非常耗时，培育一个新的品种可能要耗费 12 年。香蕉育种主要通过大规模的表型轮回选择。而高度杂合性导致鉴定理想的亲本材料十分困难，而选择带有良好农艺性状的个别克隆体需要特别大的种群。但实际上，由于杂交种的结实率低，几乎不可能获得那么大的种群。一般情况下，获得的种子数很少（平均 1～1.5 个），因此收集大量种子既耗时，又耗力（Ortiz and Vuylsteke，1995；Ssebuliba et al.，2006a，b，2009）。抗病虫基因是从野生二倍体渗入，但是野生物种还携带许多不良的性状，如产量低、非单性结实。为了去除这些不理想的性状，需要通过几个回交，而这会延长育种过程。某些性状的多基因特性和低遗传性也放慢了育种过程。芭蕉属植物育种受制于其种质的遗传多样性狭窄（Pillay et al.，2001；Nyine and Pillay，2011）和缺乏有用农艺性状的野生种质的信息。目前，只有少数野生二倍体得以利用，大多作为父本应用于大量的育种计划。由于栽培的香蕉是无性繁殖，因而其遗传基础狭窄，它的多样性依赖于体细胞突变。有限的遗传变异会导致作物对真菌、细菌和病毒病原体及众多害虫缺乏抗性（Miller et al.，2009）。关于芭蕉属植物重要农艺性状的遗传学知识还很少，而众所周知的精确遗传控制只能针对相对较少的性状（第 6 章和第 7 章）。

15.1.2 生产限制

大量文献描述了芭蕉属植物的生产限制（Pillay et al.，2002；Pillay and Tripathi，2006，2007；Tenkouano et al.，2011）。简而言之，全球香蕉生产受到叶面病害、线虫、病毒和害虫的联合威胁。使用抗性品种被认为是控制病虫害的最有效、最经济和最环保的方法。两个最重要的真菌性病害包括香蕉黑叶斑病（*Mycosphaerella fijiensis* Morelet）和镰孢枯萎病［*Fusarium oxysporum* Schlect. f. sp. *cubense*（E. F. Smith）］。主要害虫包括各种线虫（*Radopholus similis*、*Pratylenchus* spp.，*Helicotylenchus*）和香蕉象鼻虫（*Cosmopolites sordidus* Germar）。新的病害，如香蕉黄单胞菌枯萎病（BXW）最近在东非得以鉴定。

15.1.3　芭蕉属植物育种目标

芭蕉属植物育种的最重要目标包括：

① 增加果穗大小和产量。

② 宿主植物抗主要病原体，包括那些引发香蕉叶斑病，镰孢与黄单胞菌枯萎病的病原体和病毒。

③ 宿主植物抗线虫和害虫。

④ 果实品质性状，例如，维生素 A、铁和锌的含量提高。

⑤ 更好地适应非生物胁迫，如干旱、高温和其他可能由预测的气候变化所施加的胁迫。

高产是育种的主要目标，次要目标是培育寄主植物对影响产量的病原体和害虫的抗性。

15.2　分子育种

分子育种（MB）是用于描述几种现代育种方法的通用术语，包括：①标记辅助选择（MAS），有几个位点调节的性状的特定等位基因的选择；②标记辅助回交（MABC），将有限数量的位点从一个遗传背景转移到另一个遗传背景，包括转基因；③标记辅助轮回选择（MARS），即鉴定和选择几个基因组区域，这些区域与在单一植株内或横跨亲缘群体中“组装”最佳表现的基因型的复杂性状的表达有关；④全基因组选择（GWS），即基于标记，而无需大量测试的选择，也无需事先鉴定与性状相关的标记子集（Ribaut et al.，2010）。作为育种计划中的间接选择工具，MABC 是最受期待并且分子标记的优点也常被引用的工具之一（Semagn et al.，2006）。

现今 MAS 还未在进行的植物育种计划中得到常规使用。相比传统表型选择，MAS 的成本相对较高，因而 MAS 的推广缓慢。为了对植物育种者有价值，MAS 产生的效益必须比传统育种的更高，或者 MABC 必须节省大量的时间以证明其所涉及的额外成本是合理的（Semagn et al.，2006）。

植物中的大量性状是多基因的，因而对单基因控制的性状进行 MABC 是利用 DNA 标记的最有效方法。通过 MABC 对数量性状位点（QTL）的改进产生了不同的结果，有部分成功和/或甚至没有几个非常成功的结果（Semagn et al.，2006）。标记辅助育种对于香蕉这类生命周期相对较长的作物十分有用。利用分子标记来间接选择改良品种能加快选择过程，因为温室或田间条件下的直接筛选方法十分耗时。科学家报告了一些对于芭蕉属植物育种者而言最重要的性状有一个寡基因上位基础（Ortiz，1995）。

15.2.1 芭蕉属植物分子标记

现在普遍接受的观点是分子标记代表育种技术在过去几十年最重要的进步，并且现已成为分子生物学在植物育种上最重要的应用。似乎在育种中，分子标记技术不像转基因生物那样有阻力（Pillay et al.，2011）。科学家正在寻找与芭蕉属植物中几个重要性状如抗病虫害有关的 DNA 标记。果实质量（颜色、质地、成熟度）是 DNA 标记选择的其他候选性状。这些性状的大多数要么只在植物生命周期的后期表达，要么难以筛选。鉴定和控制与重要性状位点相关的标记将推动基因渗入和其他 MAS 应用。获取来自不同基因组的基因，包括 S（*M. schizocarpa*）和 T（*M. textilis*）基因组将对香蕉育种越来越重要。迄今为止只有很少的标记与芭蕉属植物的目标性状连锁，且仅限于抗病性和主要基因组的标记。

甲基化敏感扩增多态性（MSAP）标记及一系列参考品种和体细胞无性系变异体被用来鉴定与球腔菌毒素（黑叶斑病）抗性相关的分子标记（Gimenez et al.，2006）。这项研究确定了与球腔菌毒素抗性有关的 4 个 MSAP 标记。这些 MSAP 标记表现出与抗性基因类似物和反转录转座子序列的高度序列相似性。这些标记被用作对球腔菌毒素的耐受型和黑叶斑病抗性的分子指示器。

Lin 等（2010）开发出一个检测中国台湾的尖孢镰孢古巴专化型 4 号生理小种菌株的可靠的分子生物学方法。通过 PCR 扩增，源自随机引物 OP－A02 扩增片段序列的引物集 *Foc－1* / *Foc－2* 产生了大小 242 bp 的 DNA 片段，其特异于 *Foc* 4 号生理小种。通过优化 PCR 参数，这一分子方法十分灵敏，能高效地检测出 50～2 000 ng 宿主基因组 DNA 中含量低至 10 pg 的 *Foc* DNA。

印度的国家香蕉研究中心（NRCB）鉴定出叶斑病抗性的推定的 RAPD 标记。该标记已被克隆、测序，并已将其转换成序列特征性扩增区域（SCAR）的标记，且已使用对照亲本来验证叶斑病（香蕉生球腔菌）抗性及其后代的表达。平行研究促成了对线虫抗性的一个假定随机扩增多态性 DNA（RAPD）标记的鉴定（S. Uma，之前的评论）。研究人员在通过诱变获得的 cv. Dwarf Cavendish 的克隆系中鉴定出一个耐盐性的 RAPD 标记（Miri et al.，2009）。

命名为 *MaSERK1* 的香蕉体细胞胚类受体激酶（SERK）基因分离自 *Musa acuminata* cv. Mas（AA），与芭蕉属植物体细胞胚发生能力和抗病反应有关（Xia et al.，2010）。该基因编码含有 628 个氨基酸的蛋白质，并与椰子、水稻、玉米、拟南芥、胡萝卜和蒺藜状苜蓿中的 SERK 基因有 82%以上的一致性。*MaSERK1* 在雄花中弱表达，但在雄花源非胚性愈伤组织中却不表达。它在雄花源胚性愈伤组织和胚性细胞悬浮液（ECS）中高度表达。ECS 的体细胞胚胎发生频率与 *MaSERK1* 转录水平呈正相关。抗 *Foc* 4 号生理小种的栽培品种 Dongguan Dajiao（ABB）叶中的 *MaSERK1* 表达是由外源水杨酸（SA）处理或接种 *Foc* 4 号生理小种而诱导的。而在易感 *Foc* 4 号生理小种的 Pisang Awak（ABB）叶中的 *MaSERK1* 表达水平在经过外源水杨酸（SA）处理或接种 *Foc* 4 号生理小种后表达水平并没有改变（Xia et al.，2010）。有科学家建议，*MaSERK1* 基因表达不仅可以作为香蕉体细胞胚发生的分子标记，还可以在宿主植物对香蕉病原体的抗性反应中发挥作用。

15.2.2　分子标记辅助渗入

标记辅助育种利用的是农艺性状和遗传标记，主要是分子标记的等位基因变异之间的关联（Stam，2003）。一般情况，这些关联是标记和目标性状的基因位点之间连锁遗传的结果。这些关联也被称为连锁不平衡。连锁不平衡出现在用于连锁作图的实验群体，例如，回交世代（BC）、F_2 分离群体、重组近交系（RIL）或双单倍体（DH）（Stam，2003）。在异花授粉植物中，如芭蕉属植物，作图群体通常由不同基因型的单株之间杂交而成的一个大型的全同胞家族组成。在植物育种者利用性状和标记之间基于连锁的关联之前，必须以一定程度的准确性来评估这些关联，从而可以安全地信赖它们，因此，标记基因型才可以用作性状基因型和表型的指示器或预测器（Stam，2003）。对于基因型之间有明显性质反差的单基因性状，如基于单一基因的宿主植物对病原体的抗性，关联评估是直接的，单基因性状和标记一起作图。但是，对于多基因遗传的数量性状，对性状和标记之间关联进行可靠评估需要大规模的田间试验，以及被称为 QTL 定位的统计技术（Stam，2003）。大蕉和香蕉的育种进程受到芭蕉属栽培多倍体的复杂遗传结构和行为的限制。由于雌性和雄性高度不育，种子活性低，并且缺乏大型分离群体，芭蕉属植物作图受到阻碍。将标记辅助育种成功整合到育种计划中的关键在于鉴定标记提供与常规育种方法相比具有实际优势的应用或以新颖的方式补充它们（Semagn et al.，2006）。标记辅助育种在下列情况下有显著优势：①当表型筛选费用昂贵，难以实施或无法实施；②当性状的遗传力低（掺入了高度受环境影响的基因）；③当所选择的性状在植物发育晚期表达时，如果实和花的特征或有幼年期的物种的成年性状；④对于不容易筛选的渗入了宿主植物抗病原体或害虫的基因，因为有特殊要求才能使这一基因表达；⑤当靶基因的表达是隐性的；⑥为了累计同一品种内的一个或多个性状的多个基因而采用被称为基因聚合的处理方法（Sharma et al.，2004；Barone et al.，2005；Yang et al.，2005）。

高精度的 MAS 方法需要开发高密度连锁图。要提升 MAS 利用率，需要改进分子标记系统。对于 MAS 来说，某些因素十分重要，包括易用性、耐用性、成本和与目标性状的连锁（De Koeyer et al.，2010）。多倍体作物的理想标记系统应该是对剂量敏感，并能通过标记来区分目标基因组区域内的杂合基因型和多个单倍体（De Koeyer et al.，2010）。

目前，已经有几种分子标记方法应用于芭蕉属植物，它们在技术要求、灵敏度和可靠性方面各有不同（第 4 章）。新的可用的标记系统尚未用于芭蕉属植物。

15.2.3 基因聚合

基因聚合是指在同一品种内累计一个或多个性状的多个基因（Barone et al.，2005；Xiang et al.，2005）。基因聚合形成的遗传资源可用于育种项目。基因聚合是在栽培种中渗入控制不同农艺性状的基因的一种有效方法，而这一基因渗入可使得该栽培种同时获得若干个性状（Semagn et al.，2006）。例如，导致宿主植物对病原体或病虫害不同小种或生物型的抗性基因聚合在一起形成了对多种小种和多种生物型有抗性的品系，而这种抗性比对单一小种或单一生物型的抗性更持久（Jiang et al.，2004）。科学家发现，相比单个基因通过基因相互作用和数量互补赋予植物的抗性，基因聚合表达产生更强的抗性，或广谱的寄主抗性（Yoshimura et al.，1995；Singh et al.，2001）。基因聚合已成功应用于多个作物育种项目，也产生了许多含有多重属性的品种和品系（Porter et al.，2000；Wang et al.，2001；Samis et al.，2002；Jiang et al.，2004）。

传统观点认为性状是定量的，无法进行基因聚合，但其实如果鉴定出影响性状的主基因，就可以通过基因聚合来对性状进行改良（Ashikari and Matsuoka，2006）。不过，传统育种方法难以进行基因聚合，因为控制抗病性的基因具有显性和上位性效应（更强的抗性基因通常会屏蔽更弱的抗性基因，如果不使用前者的毒性菌株进行筛选，则不能显示出不优选的）（Semagn et al.，2006）。此外，对两个或更多小种有相似反应的基因——称为小种非特异性或部分抗性，很难通过传统方法来鉴定和转移（Singh et al.，2001），并且如果存在更强的小种特异性基因这就更加不可能鉴定和转移。

基因聚合计划也被认为费用高昂。虽然育种者使用单一供体会使工作更简便，但是在田间试验中选择理想分离种仍需要进行表型分型。无论什么基因传入一个栽培品种中，一旦基因聚合，之后育种者再进行使用，栽培品种极有可能丢失这些基因的效用。

15.2.4　标记系统和种质鉴定

芭蕉属中最频繁使用的分子标记方法仅限于种质鉴定和多样性分析。芭蕉属种质有 1 500～3 000 种，而种质中又有着多种多样的形态变异和基因组构造（Heslop-Harrison and Schwarzacher，2007）。在比利时天主教鲁汶大学（KULeuven）国际转运中心（ITC）通过组织培养保存了约 1 000 种芭蕉属栽培品种和 180 种野生种，这些种质提供了有价值的收集参考，并且大部分是公共使用的，可免费获取用于试验和育种。亚洲许多香蕉研究人员开发出了基于田间的种质资源收集，而运作良好的网上数据库正传播有关这些收集的信息（Pollefeys et al.，2004）。虽然多样性可以通过形态学和流式细胞术进行评估，但是这些分析有局限性，并且仍然存在具有单个名称的多个基因型或有多个名称的单个基因型问题（Heslop-Harrison and Schwarzacher，2007）。因此，基于 DNA 的分子多样性研究将有助于指导植物育种者用合适的种质来进行测试和选择，并将种质收集的重点集中在能代表所有倍性水平的属的多样性（Heslop-Harrison and Schwarzacher，2007）。用于评估芭蕉属植物多样性和系统发育关系的各种技术，已在第 3 章讨论过。

15.2.5　转基因育种

香蕉传统育种受到许多因素的阻碍，包括世代时间长、三倍体和大多食用栽培品种不育（Pillay et al.，2002）。在一些野生二倍体中已经确定了对许多病虫害的抗性来源。然而，大多数地方品种通常不育，无法用于育种，而涉及野生种的杂交会导致许多不想要的性状连同所需的抗性基因一起转移。此外，也有一些疾病，如香蕉束顶病毒（BBTV）的抗性来源未知（Sagi et al.，1998）。

尽管芭蕉属植物传统育种面临着种种困难，但是现今可用的转化方法可能无法解决所有这些难题。随着对芭蕉属植物在生物和非生物胁迫下的反应的认识不断加深，遗传改良可能会有新机遇。

遗传转化为单个基因或基因组合提供了新机遇，例如与宿主植物对病原体抗性相关的那些单个基因或基因组合，还有从生物源基因组中提取并直接转移到所期望品种中去的单个基因或基因组合，这种转移使得保留了品种所有原始特征的同时还添加了所需性状。此外，由于大多数香蕉品种在自然条件下不产生种子，因此其难以发生与其他品种或物种杂交的情况。这种情况下，导入的基因会一直保留在被导入的栽培品种中（Sagi et al.，1998）。

科学家在香蕉和大蕉的遗传工程上取得了一定成功，可使外源基因转移到一些栽培品种中（Sagi et al.，2007）。但有用基因的缺乏，影响转基因表达的因素，如 RNA 干扰及转基因和那些已经存在于植物的物质之间的相互作用，以及一些性状的数量性质都仍然是科学家同意遗传转化是香蕉培育唯一选择的观点前必须考虑的问题。导入基因的方案，包括组织培养中芽的高效再生和转化方法仍然是基因工程的主要瓶颈（Sharma et al.，2005）。

15.3 芭蕉属植物转基因育种的成就和前景

15.3.1 对香蕉黑叶斑病和镰孢枯萎病的抗性

为了抗真菌病害而进行的香蕉和大蕉的转化始于 20 世纪 90 年代（Sagi et al.，1995）。各种转化技术用于生产出口香蕉品种 Williams（AAA）、水果香蕉 Gros Michel、煮食蕉 Bluggoe（ABB）和大蕉 Three Hand Planty（AAB）的转基因香蕉，这些转基因香蕉带有抗真菌肽，在体外活跃地抵抗主要病原性真菌病害，如香蕉黑叶斑病和镰孢枯萎病（Remy et al.，2000）。转基因植株在实验室条件下表现出抗黑叶斑病。从不同植物种子中分离出来的稳定的、富含半胱氨酸的小分子肽的抗菌蛋白质（AMP），也可用于发展应对真菌病害的转基因学（Sagi et al.，1998）。

当对香蕉转入葡萄芪合酶基因（*StSy*）和来自 *Trichoderma harzianum* 的几丁质酶 *ThEn-42* 基因时，对叶斑病的抗性增强（Vishnevetsky et al.，2010）。在泛素启动子的控制下，来自番茄的超氧化物歧化酶基因 *Cu*，*Zn-SOD* 也被加入这个基因盒中，以促进对真菌侵袭期间产生的自由基的清除。为期 4 年的田间试验证明了几个转基因香蕉品系对叶斑病抗性的提高。由于赋予叶斑病抗性的基因可能具有大范围的抗真菌活性，因此再生的香蕉植株也接种了灰霉菌（*Botrytis cinerea*）。还发现对叶斑病表现出抗性的最佳转基因品系对灰霉菌也具有耐受性（Vishnevetsky et al.，2010）。巴西现在正通过分析来自 cDNA 文库的 EST 数据来发现基因，而这一文库是由感染了斐济球腔菌的 *M. acuminata* ssp. *burmannicoides* Calcutta 4（抗性）和 Grande Naine（AAA 基因组，易感）的叶材料中产生的（Miller et al.，2009）。

15.3.2 对香蕉束顶病毒（BBTV）的抗性

旨在开发出对 BBTV 抗性的转基因研究正在澳大利亚和美国夏威夷进行。矮缩病毒的复制起始蛋白（Rep）是病毒复制必需的唯一病毒蛋白，代表病原体衍生性抗性的理想目标。在澳大利亚，鉴定出了一种抑制 BBTV 复制的

Rep 编码蛋白（DNA－S1）。*Rep* 基因的不同结构显示出在香蕉胚性细胞悬浮液中能显著抑制 BBTV 的复制（Tsao andTsun-Hui，2008）。在澳大利亚和美国夏威夷，研究人员使用这样的结构开发出了具有对 BBTV 抗性的转基因香蕉。这些抗性品系已在夏威夷进行了田间测试。

15.3.3　对线虫的抗性

为了获得芭蕉属植物对线虫的抗性，采用了许多方法，这些方法都依赖于引入一个额外的植物基因来编码一种称为胱抑素的蛋白，这种蛋白可以抑制寄生线虫的消化。胱抑素抑制线虫的生长、产卵，和建立到危害作物种群的水平。使用胱抑素的优点是它们是人类饮食的一部分（例如，它们存在于谷物的种子或卵子中），并且对消化或健康无影响。这种方法已用于开发一种转基因卡文迪什香蕉（AAA），这种香蕉显示出对 *Radopholous similis* 这种香蕉主要线虫的抗性（Atkinson et al.，2004）。

15.3.4　对香蕉假黄单胞菌枯萎病的抗性

感染 *Xanthomonas campestris* pv. *musacearum* 所致的香蕉黄单胞菌枯萎病（BXW）或细菌性枯萎病在中、东非的大湖区已经达到了流行病程度（Biruma et al.，2007）。香蕉种质表现出的对疾病抗性的缺乏使之成为转化的理想目标。转基因技术可能是培育出抗 BXW 流行性疾病的关键。从甜椒（*Capsicum annuum*）分离出的类铁氧化还原蛋白的两亲性蛋白（pflp）和过敏反应促进蛋白（hrap）是新的蛋白质，可以强化 harpinPSS 介导的过敏性反应（Chen et al.，2000）。携带 *pflp* 基因的转基因水稻表现出对 *Xanthomonas oryzae* pv. *oryzae* 的抗性增强（Tang et al.，2001）。在转基因兰花中 *pflp* 也表现出对 *E. carotovora* 的抗性增强（Liau et al.，2003）。诱导子诱导的抗性不具体针对某特定病原体，因此它是用于开发广谱抗性的有效策略。这个策略已用于开发抗黄单胞菌枯萎病的转基因香蕉。科学家利用基于根癌农杆菌技术的方案培育出了带有 *pflp* 或 *hrap* 基因的转基因品系（Tripathi et al.，2009）。已经通过 PCR 测定法和 Southern 印迹分析法验证了各品种的这些转化系。科学家在实验室对它们进行了抗性测试，发现转基因香蕉显示出完全抗性。这些转基因品系目前正在乌干达进行田间试验（Tripathi et al.，2010）。

15.3.5　营养增强

香蕉中含有丰富的天然抗氧化剂，如维生素 C 和维生素 E（Someya et al.，2002；Amorim et al.，2009a，b）。缺乏高浓度的维生素 A 将导致严重的健康问题，尤其对于世界低收入地区的儿童，如亚洲、非洲和拉丁美洲的部分地区（Bloem et al.，2005）。微量营养素铁和锌的缺乏也会导致严重的健康问题，如脑力和身体发育迟缓，对感染的抵抗力下降和性腺机能减退（Whittaker，1998）。通过常规育种结合使用生物技术工具对香蕉的微量营养素含量进行的遗传改良（即生物强化）有可能增加香蕉新品种中的微量营养素（铁、锌）和维生素A的浓度（Amorim et al.，2011）。芭蕉属植物营养成分的改善将对数以百万计依赖香蕉为食物的人们的维生素和营养摄入产生显著影响。

澳大利亚研究人员正在进行香蕉的转化，从而增加其维生素A、维生素E和铁的含量。大量驱动原始维生素A、维生素E或铁积聚的基因的启动子被克隆到载体中，而这些启动子既表现果实特异性又是组成型。Nakinyika、Mpologoma、Nakasabira和Sukalindizi 4个品种被选入这项研究（Dale and Tushemeirewe，2008）。Fungo和Pillay（2011）进行的一项研究表明，乌干达10种东非高地香蕉中，Nakitembe、Entukura和Nakhaki的维生素A含量是最高的，这些品种可能适用于微量营养素的转化研究。

15.4 MAS的局限和前景

MAS使用的主要限制之一是高昂的费用，遗传标记的识别和验证、遗传图谱绘制等都花费巨大。经济效益是遗传改良项目中是否使用分子标记的决定因素（Dekkers and Hospital，2002）。影响标记辅助育种使用成本的其他因素包括性状的遗传力和表型评价方法。在大多数发展中国家分子育种技术的利用缓慢的主要因素包括基础设施差、能力不足和运作支持不够，在国家层面缺乏有利政策、法律和监管框架，从而影响研究机构。尽管面临重重困难，一些发展中国家在利用生物技术改良芭蕉属植物上仍取得一定进展。

15.5 结论

虽然传统育种方案有其局限性，但是随着时间的推移，它们表现出巨大的成功。虽然基因操纵和种间杂交使得芭蕉属植物育种上取得巨大的遗传进步，但是在鉴定可能产生同时具有高均值和遗传变异的子代的亲本组合上还有很长的路要走（Tenkouano，2001）。现代植物分子和数量遗传学在过去20年的发展使其有可能革新大多以经验为基础的经验主义植物育种（Ye and Smith，2008）。分子育种有望通过在靶位点选择和堆叠有利等位基因来提高作物育种的效率（Ribaut et al.，2010）。标记技术上取得的新发展和提高，功能基因组学与 QTL 定位的整合，以及更多高密度图谱的可用性是影响未来 QTL 定位和标记辅助育种效率和有效性的其他因素（Collard et al.，2005）。包含新标记类型，如单核苷酸多态性（SNP）和表达序列标签（EST）的高密度图谱的发展将为研究人员进行 QTL 定位和标记辅助育种提供更强大的工具库（Semagn et al.，2010）。数据库中可用的 EST 和基因组序列的数量正在迅速增长（特别是来自基因组测序项目），这些序列数目的增加对于未来发现 SNP 和对新标记进行数据挖掘帮助很大（Gupta et al.，2001）。应对分子育种利用的潜在遗传、经济效益和那些从现有传统育种计划所取得的或预期取得的效益进行批判性的比较。

缩略语

AMPs：抗微生物蛋白

BBTV：香蕉束顶病毒

GWS：全基因组选择

MABC：标记辅助回交

MARS：标记辅助轮回选择

MAS：标记辅助选择

MB：分子育种

MSAP：甲基化敏感扩增多态性

PCR：聚合酶链式反应

pflp：类铁氧化还原蛋白的两亲性蛋白

QTL：数量性状位点

RAPD：随机扩增多态性 DNA

参考文献

Amorim E P，Cohen K O，Amorim V B O，et al，2009a. The genetic diversity of carotenoid-rich bananas measured by diversity arrays technology（DArT）[J]. Genet Mol Biol，32：96－103.

Amorim E P，Lessa L S，Ledo C A S，et al，2009b. Caracterização agronômica e molecular de genótipos diplóides melhorados de bananeira [J]. Rev Bras Frutic，31：154－161.

Amorim E P，De Oliveira e Silva S，De Oliveira Amorim V B，et al. 2011. Quality improvement of cultivated *Musa* [M] // Pillay M，Tenkouano A. Banana breeding：progress and challenges. Boca Raton，FL，USA：CRC Press：251－267.

Ashikari M，Matsuoka M，2006. Identification，isolation and pyramiding of quantitative trait loci for rice breeding [J]. Trends Plant Sci，11：344－350.

Atkinson H J, Grimwood S, Johnston K, et al, 2004. Prototype demonstration of transgenic resistance to the nematode Radopholous simils conferred on banana by a cystatin [J]. Transgen Res, 13: 135 - 142.

Barone A, Ercolano M R, Langella R, et al, 2005. Molecular marker-assisted selection for pyramiding resistance genes in tomato [J]. Adv Hort Sci, 19: 147 - 152.

Biruma M, Pillay M, Tripathi L, et al, 2007. Banana Xanthomonas wilt: a review of the disease, management strategies and future research directions [J]. Afr J Biotechnol, 6: 953 - 962.

Bloem M W, De Pee S, Darnton-Hill I, 2005. Micronutrient deficiencies and maternal thinness: first link in the chain of nutritional and health events in economic crises [M]//Bendich A, Deckelbaum R J. Primary and secondary nutrition. 2nd ed. Totowa, N J, USA: Humana Press: 357 - 373.

Chen C H, Lin H J, Ger M J, et al, 2000. The cloning and characterization of a hypersensitive response assisting protein that may be associated with the harpin-mediated hypersensitive response [J]. Plant Mol Biol, 43: 429 - 438.

Collard B C Y, Jahufer M Z Z, Brouwer J B, et al, 2005. An introduction to markers, quantitative trait loci (QTL) mapping and marker-assisted selection for crop improvement: the basic concepts [J]. Euphytica, 142: 169 - 196.

Dale J L, Tushemeirewe W, 2008. Biofortification of banana: a grand challenge in global health [C]. Conference on Banana and Plantain in Africa: Harnessing International Partnerships to Increase Research Impact, Mombasa, Kenya.

De Koeyer D, Douglass K, Murphy A, et al, 2010. Application of high-resolution DNA melting for genotyping and variant scanning of diploid and autotetraploid potato [J]. Mol Breed, 25: 67 - 90.

Dekkers J C M, Hospital F, 2002. The use of molecular genetics in the improvement of agricultural populations [J]. Nat Rev Genet, 3: 22 - 32.

Dita M A, Rispail N, Prats E, et al, 2006. Biotechnology approaches to overcome biotic and abiotic stress constraints in legumes [J]. Euphytica, 147: 1 - 24.

Fungo R, Pillay M, 2011. β-carotene content of selected banana genotypes from Uganda [J]. Afr J Biotechnol (in press).

Gimenez C, Palacios G, Colmenares M, 2006. *Musa* methylated sequences associated with tolerance to *Mycosphaerella fijiensis* toxins [J]. Plant Mol Biol Rep, 24: 33 - 43.

Gupta P K, Roy J K, Prasad M, 2001. Single nucleotide polymorphisms: a new paradigm for molecular marker technology and DNA polymorphism detection with emphasis on their use in plants [J]. Curr Sci, 80: 524 - 535.

Heslop-Harrison J S, Schwarzacher T, 2007. Domestication, genomics and the future for banana [J]. Ann Bot, 100: 1073 - 1084.

Jiang G H, Xu C G, Tu J M, et al, 2004. Pyramiding of insect and diseaseresistance genes into an elite indica, cytoplasm. male sterile restorer line of rice, 'Minghui 63' [J]. Plant Breed, 123: 112－116.

Liau C H, Lu J C, Prasad V, et al, 2003. The sweet pepper ferredoxin-like protein (pflp) conferred resistance against soft rot disease in *Oncidium* orchid [J]. Transgen Res, 12: 329－336.

Lin Y H, Chang J Y, Liu E T, et al, 2010. Development of a molecular marker for specific detection of *Fusarium oxysporum* f. sp. *cubense* race 4 [J]. Eur J Plant Pathol, 123: 353－365.

Miller R N G, Bertioli D G, Baurens F C, et al, 2009. Understanding plant responses to biotic stress: ongoing research in *Musa* [J]. Acta Hort, 828: 255－272.

Miri S M, Mousavi A, Naghavi M R, et al, 2009. Analysis of induced mutants of salinity resistant banana (*Musa acuminata* cv. Dwarf Cavendish) using morphological and molecular markers [J]. Iran J Biotechnol, 7: 86－92.

Newell-McGloughlin M, 2008. Nutritionally Improved Agricultural Crops [J]. Plant Physiol, 147 (3): 939－953.

Nwakanma D C, Pillay M, Okoli B E, et al, 2003. PCR-RFLP of the ribosomal DNA internal transcribed spacers (ITS) provides markers for the A and B genomes in *Musa* L [J]. Theor Appl Genet, 108: 154－159.

Nyine M, Pillay M, 2011. Breeding increases diversity of bananas [J]. Acta Hort (in press).

Ortiz R, 1995. *Musa* genetics [M]//Gowen S. Bananas and plantains. London, UK: Chapman and Hall: 84－109.

Ortiz R, Vuylsteke D, 1995. Effect of the parthenocarpy gene *P1* and ploidy in bunch and fruit traits of plantain and banana hybrids [J]. Heredity, 75: 460－465.

Pillay M, Tripathi L, 2006. Banana: an overview of breeding and genomics research in *Musa* [M] // Kole C. Genome mapping and molecular breeding in plants, vol. 4: fruits and nuts. Heidelberg, Germany: Springer-Verlag: 282－301.

Pillay M, Tripathi L, 2007. Banana breeding [M]//Kang M S, Priyadarshan P M. Breeding major food staples. Boston, MA, USA: Blackwell Publishing: 393－428.

Pillay M, Nwakanma D C, Tenkouano A, 2000. Identification of RAPD markers linked To A and B genome sequences in *Musa* [J]. Genome, 43: 763－767.

Pillay M, Ogundiwin E, Nwakanma D C, et al, 2001. Analysis of genetic diversity and relationships in East African banana germplasm [J]. Theor Appl Genet, 102: 965－970.

Pillay M, Tenkouano A, Hartman J, 2002. Future challenges in *Musa* breeding [M] // Kang M S. Crop improvement: challenges in the twenty-first century. New York: Food Products Press, Inc: 223－252.

Pillay M, Ogundiwin E, Tenkouano A, et al, 2006. Ploidy and genome composition of *Musa* germplasm at the International Institute of Tropical Agriculture (IITA) [J]. Afr J Biotechnol, 5: 1224－1232.

Pillay M, Tenkouano A, Ortiz R, 2011. Molecular breeding of other vegetatively propagated crops: lessons for banana [M]//Pillay M, Tenkouano A. Banana breeding: progress and challenges. Boca Raton, FL, USA: CRC Press: 321 - 350.

Pollefeys P, Sharrock S, Arnaud E, 2004. Preliminary analysis of the literature on the distribution of wild *Musa* species using MGIS and DIVA-GIS [R]. International Network for the Improvement of Banana and Plantain, Montepellier, France.

Porter D R, Burd J D, Shufran K A, et al, 2000. Efficacy of pyramiding greenbug (Homoptera: Aphididae) resistance genes in wheat [J]. J Econ Entomol, 93: 1315 - 1318.

Remy S, Buyens A, Cammue B P A, et al, 2000. Production of transgenic banana plants expressing antifungal proteins [J]. Acta Hort, 490: 219 - 277.

Ribaut J M, De Vicente M C, Delannay X, 2010. Molecular breeding in developing countries: challenges and perspectives [J]. Curr Opinion Plant Biol, 13: 1 - 6.

Rowe P R, 1984. Breeding bananas and plantains [J]. Plant Breed Rev, 2: 135 - 155.

Samis K, Bowley S, McKersie B, 2002. Pyramiding Mn-superoxide dismutase transgenes to improve persistence and biomass production in alfalfa [J]. J Exp Bot, 53: 1343 - 1350.

Sagi L, Panis B, Remy S, et al, 1995. Genetic transformation of banana (*Musa* spp.) via particle bombardment [J]. Bio/Technology 13: 481 - 485.

Sagi L, Gregory D M, Remy S, et al, 1998. Recent developments in biotechnological research on bananas (*Musa* spp.) [J]. Biotechnol Genet Eng Rev, 15: 313 - 317.

Sagi L, Remy S, Swennen R, 2007. Transgenic and (trans) genomic research in banana (*Musa* spp.) [J]. Afr Crop Sci Conf Proc, 8: 592 - 595.

Semagn K, Bjornstad A, Ndjiondjop M N, 2006. Progress and prospects of marker assisted backcrossing as a tool in crop breeding programs [J]. Afr J Biotechnol, 5: 2588 - 2603.

Semagn K, Bjørnstad A, Xu Y, 2010. The genetic dissection of quantitative traits in crops [J]. Electronic Journal of Biotechnology, North America, 1325102010.

Sharma K K, Bhatnagar-Mathur P, Thorpe T A, 2005. Genetic transformation technology: status and problems [J]. In Vitro Cell Dev Biol, 41: 102 - 112.

Sharma P N, Torii A, Takumi S, et al, 2004. Marker assisted pyramiding of brown planthopper (*Nilaparvata lugens* Stål) resistance genes *Bph1* and *Bph2* on rice chromosome 12 [J]. Hereditas, 140: 61 - 69.

Singh S, Sidhu J S, Huang N, et al, 2001. Pyramiding three bacterial blight resistance genes (*xa5*, *xa13* and *Xa21*) using marker-assisted selection into indica rice cultivar PR106 [J]. Theor Appl Genet, 102: 1011 - 1015.

Someya S, Yumiko Y, Okubob K, 2002. Antioxidant compounds from bananas (*Musa* Cavendish) [J]. Food Chem, 79: 351 - 354.

Ssebuliba R, Magambo M, Makumbi D, et al, 2006a. Biological factors affecting seed production in East African Highland bananas [J]. J Crop Imp, 16: 67 - 77.

Ssebuliba R, Talengera D, Makumbi D, et al, 2006b. Reproductive efficiency and breeding potential of East African highland banana [J]. Field Crops Res, 95: 250 - 255.

Ssebuliba R, Makumbi D, Pillay M, 2009. Patterns of seed set in East African highland banana (*Musa* sp.) hybrids [J]. J New Seeds, 10: 160 - 170.

Stam P, 2003. Marker-assisted introgression: speed at any cost? [M]//Th. J L Van Hintum J L, Lebeda A, Pinl D, et al, Valencia, Spain: Eucarpia Leafy Vegetables: 117 - 124.

Tang K, Sun X, Hu Q, et al, 2001. Transgenicrice plants expressing the ferredoxin-like protein (AP1) from sweet pepper show enhanced resistance to *Xanthomonas oryzae* pv. *oryzae* [J]. Plant Sci, 160: 1035 - 1042.

Tenkouano A, 2001. Current issues and future directions for *Musa* genetic improvement research at the International Institute of Tropical Agriculture [J]. Advancing Banana and Plantain R & D in Asia and the Pacific, 10: 11 - 23.

Tenkouano A, Pillay M, Ortiz R, 2011. Breeding techniques [M]// Pillay M, Tenkouano A. Banana breeding: progress and challenges. Boca Raton, FL, USA: CRC Press: 181 - 202.

Tripathi L, Mwangi M, Abele S, et al, 2009. Xanthomonas wilt: a threat to banana production in East and Central Africa [J]. Plant Dis, 93: 440 - 451.

Tripathi L, Mwaka H, Tripathi J N, et al, 2010. Expression of sweet pepper *Hrap* gene in banana enhances resistance to *Xanthomonas campestris* pv. *musacearum* [J]. Mol Plant Pathol, 11: 721 - 731.

Tsao T, Tsun-Hui, 2008. Towards the development of transgenic banana bunchy top virus (BBTV) -resistant banana plants: interference with replication [D]. Queensland University of Technology, Australia.

Vishnevetsky J, White T L, Palmateer A J, et al, 2010. Improved tolerance toward fungal diseases in transgenic Cavendish banana (*Musa* spp. AAA group) cv. Grand Nain [J]. Transgen Res, 20: 61 - 72.

Vuylsteke D, Ortiz R, Ferris S, et al, 1995. 'PITA - 9': a black sigatoka-resistant hybrid from the 'False Horn' plantain gene pool [J]. HortScience, 30: 395 - 397.

Wang X Y, Chen P D, Zhang S Z, 2001. Pyramiding and marker-assisted selection for powdery mildew resistance genes in common wheat [J]. Acta Genet Sin, 28: 640 - 646.

Whittaker P, 1998. Iron and zinc interactions in humans [J]. Amer J Clinical Nutr, 68: 442S - 446S.

Xia H, Lu X Y, Zhao J T, et al, 2010. *MaSERK1* gene expression associated with somatic embryogenic competence and disease resistance response in banana (*Musa* spp.) [J]. Plant Mol Biol Rep, 28: 309 - 316.

Zeng X T, Zhang Z Y, Du L P, et al, 2005. Development of wheat germplasm with multi-resistance to powdery mildew, stripe rust and yellow dwarf virus by molecular marker assisted selection [J]. Sci Agric Sin, 38: 2380 - 2386.

Yang W C, Sacks E J, Ivey M L L, et al, 2005. Resistance in *Lycopersicum esculentum* intraspecific crosses to race T1 strains of *Xanthomonas campestris* pv. *vesicatoria* causing bacterial spot of tomato [J]. Phytopathology, 95: 519 - 527.

Yoshimura S, Yoshimura A, Nelson R J, et al, 1995. Tagging *Xa - 1*, the bacterial blight resistance gene in rice, by using RAPD markers [J]. J Breed, 45: 81 - 85.

Ye G, Smith K F, 2008. Marker-assisted gene pyramiding for inbred line development: basic principles and practical guidelines [J]. Int J Plant Breed, 1: 1 - 10.

第 16 章　结论和芭蕉属植物研究展望

Michael Pillay[①]

摘　要：尽管科学家在香蕉基因组研究上取得了一定进展，但是仍面临许多挑战。仍需提升香蕉对病虫害的遗传抗性，为保障粮食安全而提高产量和生产效率，同时还需提高产量的稳定性，更好地利用水资源，减少有毒化学物的使用，并提高香蕉质量。本章将考察芭蕉属植物遗传改良面临的一些问题，并给未来的芭蕉属植物研究提供一些建议。新数据的出现使得科学家可以利用比较基因组学来认识芭蕉属植物遗传学。关于芭蕉属植物基因组研究的新项目越来越多，芭蕉属植物基因组联盟的成立似乎已开始取得成效。本章将简要介绍分子标记、标记辅助育种、基因聚合和连锁图谱的应用。笔者认为关于芭蕉属植物在转录组学、蛋白质组学、代谢组学水平上的研究，有必要采取更全面的方法，如对基因表达产物的定量和定性分析。

本章强调了包括 RNA 干扰、TILLING、蛋白质组学和代谢组学在内的新方法将如何革新芭蕉属植物研究。笔者也总结了香蕉的保健功能，并阐释了为什么香蕉能被视为药物的重要来源并且用作生物强化的作物。关于危害香蕉的生物胁迫的研究范围仍然很广，诸如黄单胞菌枯萎病等新病害的发病为科学家开辟了新的研究道路。已培育出抗黑叶斑病、黄单胞菌枯萎病和线虫的转基因作物在田间的效力尚未得以证明。虽然关于危害香蕉的生物胁迫，如真菌病害、线虫、昆虫、病毒和细菌的研究很多，但是关于干旱、洪涝、盐碱化、金属中毒、矿物质缺乏、不良 pH、不利温度和空气污染等非生物胁迫的研究非常匮乏。本章将讨论在解决干旱胁迫、盐胁迫和寒冷胁迫上的进展。在作物基因组测序、高分辨率的遗传图谱和准确分型上的进展将加速发现与植物育种中目标性状相关的功能性等位基因和等位基因变异。

关键词：基因组学，基因功能分析，营养价值，生物强化

① Vaal University of Technology，Private Bag X021，Vanderbijlpark 1900，South Africa；e-mail：mpillay@vut. ac. za.

16.1 引言

尽管科学家在遗传改良香蕉这种无性繁殖的作物上取得了一定进展，但是仍然面临着许多挑战。香蕉生产面临着来自生物胁迫和非生物胁迫的挑战。Heslop-Harrison 和 Schwarzacher（2007）对这些挑战进行了恰如其分地总结，除了迫切需要提升香蕉对病虫害的抗性外，还需提高产量、生产效率和产量的稳定性，更好地利用水资源，减少有毒化学物的使用，提高香蕉质量，从而提升粮食安全性。本章将考察芭蕉属植物遗传改良面临的一些问题，并为芭蕉属植物将来的研究提供一些建议。

16.2 芭蕉属植物研究采用的先进工具

16.2.1 基因组学

对芭蕉属植物进行有效的育种需要了解芭蕉属植物的基因组、进化和结构（Dolezel and Bartos，2005），以及功能基因组学，因为它推动生物信息学和基因发掘。通过 DNA 的变化、杂交和多倍体能了解香蕉基因组的进化，从而能预测基因组未来将如何在进化过程中自然改变或在植物育种的加速过程中改变（Heslop-Harrison and Schwarzacher，2007）。新数据的出现使得科学家可以从比较基因组学的角度来理解芭蕉属植物遗传学。芭蕉属物种的全基因组 DNA 序列能帮助了解所有基因，而这些基因的控制序列对于基因研究和将来用于改良香蕉都十分具有价值。对遗传工具的认识和发展意味着它有光明的前景，通过涉及有性杂交、诱变育种和靶向转化方法的育种技术将可能产生超级驯化的“新世代”（Heslop-Harrison and Schwarzacher，2007）。芭蕉属植物基因组联盟的建立有力地推动了关于芭蕉属植物基因组学的联合研究。最近芭蕉属植物基因组学的出版物表明，这一举措正在取得成果。尽管如此，仍需努力加强在该领域的研究。

16.2.2 分子标记

分子标记已广泛应用于芭蕉属植物，用来检测种质中的遗传变异和遗传关系，识别田间和组织培养种质库中的重复种质，监测组织培养材料的遗传稳定性，分析辐照过的香蕉基因型和鉴定用于育种项目的性状标记。而术语“分子标记”真正意义上是用来“标记”一个特定基因的位置或特定性状的遗传，但在科学文献中，它被广泛用于表示与基因组的某一部分相关联的 DNA 序列片段，这是不准确的。虽然芭蕉属植物研究广泛使用的分子标记是指第二个意思，但是考虑到该术语的真正意义，很少有标记与农艺性状相连。尽管目前迫切需要关联芭蕉属植物的一些性状，如单性结实、顶端优势、吸芽行为或早熟的标记，但迄今为止发展出的遗传标记还很少。这可能是由于这些性状明显缺乏大型分离群体，也有可能是由于缺乏大规模的香蕉育种计划。新标记技术（参见第 4 章）包括：多样性阵列技术（DArT），无需序列信息就能同时检测多个基因组位点上的变异（Kilian et al.，2005）；高分辨率 DNA 熔解（HRM）分析，具有优于其他分型方法的几个优势（Montgomery et al.，2007；Reed et al.，2007；Erali et al.，2008）；单核苷酸多态性（SNP）（Rafalski，2002）和定向诱导基因组局部突变技术（TILLING）(Colbert et al.，2001)，这些技术应进一步在芭蕉属植物中得到验证。如果没有可用的遗传标记，标记辅助选择则无法在芭蕉属植物中实现。

16.3　分子标记辅助育种

分子标记能绕过温室和田间条件下耗时的直接筛选方法，加快作物育种中的选择过程。分子标记特别适用于靶向性状由几个基因控制的情况。对关于农艺性状的不同数量性状位点（QTL）作图的潜在可能性和确定连锁分子标记的潜在可能性使得有可能同时传送对应多个农艺性状的多个 QTL 和聚合多个 QTL 到一个改良品种中（Dita et al.，2006）。芭蕉属植物中使用的大量有关分子标记的技术已在第 4 章进行阐述。虽然采用标记辅助选择可能有助于芭蕉属植物改良，但是现今它还未进行实际应用，主要是由于缺乏资金支持和对大多数农艺性状的认识不足。

16.4　基因聚合

培育寄主植物对病虫害的持久抗性是植物育种的一个主要任务，而将不同寄主植物抗性基因聚合到一个基因型中是实现这一目标的途径之一。关于作物中有利等位基因及 QTL 的基因渗入和聚合的实例很多（Richardson et al.，2006；Bernado，2008；Semagn et al.，2006，2010）。但是，迄今还未有关于芭蕉属植物中基因聚合的报道。为了培育抵御多重胁迫的抗性品种，科学家有必要结合多个生物技术方法，如转基因或诱变和标记辅助选择来将多个抗性基因聚合到一个品种中。

16.5　连锁图谱

尽管在支撑二倍体和三倍体香蕉遗传改良的分子图谱上取得一定进展，但是关于芭蕉属植物的作图研究进展仍落后于其他作物。Faure 等（1993）开发了芭蕉属植物的第一张部分分子连锁图谱，显示了 15 个连锁群及 77 个标记，而这些标记中 36%显著偏离孟德尔分离法则（$P \leqslant 0.05$）。第二张图谱来自自交小果野蕉二倍体 M53 的 89 个个体。该图谱显示了 11 个连锁群，也显示出了畸变标记（Noyer et al.，1997）。第三张图谱包含了 14 个连锁群（Vilarhinos，2004）。第四张（高密度）图谱是科学家采用 DArT 和 SSR 标记而得（Hippolyte et al.，2010）。尽管取得如此进展，芭蕉属植物作图的发展进程仍落后于其他作物，仍有许多空白等着科学家去填补。

第 2 章已强调了细胞遗传学图谱的重要性。要识别出芭蕉属植物中的粗线期染色体和培育出粗线期核型，仍需要进一步的研究。而一旦实现，这将为芭蕉属植物基因组的详细细胞学特征鉴定铺平道路，并推动基于图谱的 DNA 序列和细胞学图谱的集成图谱出现。

16.6 基因功能分析

前文已经提到，利用来自基因表达的信息能提高 MAS 和转基因方法的效率。了解植物自身抵御胁迫所采用的机制及更全面了解其中所涉及的基因将有助于更精确地使用标记辅助育种和转基因手段。获取的序列信息，虽有价值，但不足以解答关于基因功能、调控网络和胁迫条件下生化途径的激活等问题。

目前研究人员正对芭蕉属植物基因组进行测序。即将面临的挑战是破译由测序和其他基因组项目识别的数千个基因。现在一些强大工具可用来确定基因的功能。这些工具包括反义 RNA 抑制、靶基因替换、插入诱变、基因沉默和 TILLING 方法。关于芭蕉属植物在转录组学、蛋白质组学、代谢组学水平上的研究，有必要采取更全面的方法，如对基因表达产物的定量和定性分析。

16.7 芭蕉属植物 RNA 干扰研究：研究潜力

RNA 干扰（RNAi）是由双链 RNA（dsRNA）启动的同源依赖基因沉默技术（Fire et al.，1998）。自此，针对转基因植物中合适的真核病原体靶基因的 dsRNA 表达显示出能保护植物，抵御有害线虫、食草昆虫、寄生杂草和真菌（Niu et al.，2010）。虽然没有关于芭蕉属植物的 RNA 干扰研究的报道，但是该技术已在许多领域表现出应用潜力，例如用于作物对线虫、病毒和真菌的抗性研究。

16.7.1 对线虫的抗性

RNAi 第一次在线虫 *Caenorhabditis elegans* 中证实（Fire et al.，1998）。线虫是严重危害香蕉的害虫，它破坏香蕉以及其他作物的根，估计每次造成 1 250亿美元的全球农业损失（Chitwood，2003）。在一系列不同组织和细胞类型中表达的许多基因现在可以被靶向在不同植物寄生线虫中沉默（Rosso et al.，2009）。表达 dsRNA 发夹结构的转基因植物能靶向寄生线虫必需基因，这种转基因植物已在大豆中开发出来，它显著减少了这种线虫的繁殖力，每克根组织中的卵减少 68%（Steeves et al.，2006）。这些结果表明，在寄生虫控制过程中，RNAi 是一种研究线虫基因功能及识别潜在寄生虫目标的强大技术（Niu et al.，2010）。

16.7.2 对昆虫的抗性

研究人员也已经培育出转基因植物，用于构建由 RNAi 导致的对昆虫的抗性。转基因棉花植株表现出充分的双链 RNA 水平，足以抑制棉铃虫中肠的基因表达并且阻碍其生长（Mao et al.，2007）。这种类型的研究为以香蕉象鼻虫为靶向的类似实验铺平了道路。

16.7.3 对真菌疾病的抗性

RNAi 已被用来探索广泛真菌物种和真菌类生物的基因功能（Nakayashiki and Nguyen，2008）。这些研究表明，RNAi 机制可能广泛存在，即使 RNA 的沉默途径似乎显著多样化（Nakayashiki et al.，2006）。迄今为止，还没有出版物报道过真菌中 RNAi 介导的基因抑制，而真菌中 dsRNA 分子来自人工生长培养基或植物组织。作为一个相对较新的技术，RNAi 介导的植物抗性相比传统生物工程的作物抗性既有优点，也有缺点。

（1）通过沉默保守和必需基因，可以实现对多种病原体的抗性。

（2）这一抗性有可能更加持久。RNAi 介导的抗性是基于 RNA 杂交，而不是蛋白质之间的相互作用；少数核酸突变无法抑制分子杂交（Escobar et al.，2001）。因此害虫突破抗性的可能性更小。此外，这种生物技术是一种培育抗虫作物的灵活手段。理论上，所有显示不利敲除表型的害虫基因都可被视为潜在目标。因此这种策略将不会受到缺少抗性基因的限制。基于 RNAi 的作物保护战略的一个关键挑战和必要步骤是识别正确目标。最佳的致病性相关候选基因应与寄生状态、解毒或导致致死表型的必需基因敲除相关。

RNAi 技术的主要缺点是“脱靶”效应。RNAi 技术被视为具有高度的基因特异性，但与引入的 dsRNA 序列有部分同一性的转录物的交叉杂交会诱导非预期基因的抑制，而这可能会导致除了目标基因之外的意想不到的突变表型（Jackson et al.，2003；Ma et al.，2006）。尽管该技术在几个物种中取得了成功，但是 RNA 沉默也具有几个缺点，例如之后世代的表型不稳定（Hannon，2002），需要一个可靠的植物转化系统。

16.8 TILLING

TILLING（定向诱导基因组局部突变技术）是利用传统诱变和高通量突变发现的通用反向遗传学策略，该策略已应用到多种物种上（McCallum et al.，2000；Colbert et al.，2001）。TILLING 策略能够将诱发突变的作用和扩展的序列信息结合起来，推动功能基因组学和作物改良项目的发展。要让 TILLING 策略能够有效应用在香蕉上还面临一些挑战，而科学家也正在评估几个方法。虽然香蕉不能发展出大型突变体种群被视为一个瓶颈，但是反向遗传学可为遗传研究和植物改良提供实用方法（Jain et al.，2011）。在香蕉研究中应用 TILLING 策略似乎为发现同源和部分同源的基因靶标的多态性提供了强大而准确的平台（Till et al.，2010）。

16.9 蛋白质组学

类似于基因组学，对一个生物体或一个特定类型组织的所有蛋白质种类的系统分析和注释被称为蛋白质组学（Wasinger et al.，1995；Cahill et al.，2000；Anderson et al.，2001）。蛋白质组学分析、解决的问题是关于生物体中蛋白质丰度和分布，不同组织的表达谱和目标蛋白质个体的识别和定位（Kersten et al.，2002）。这个概念意味着需要使用高通量的分析技术。最近，由于技术的进步，使得出现了许多可用于定量和定性蛋白质谱的有价值方法（Canovas et al.，2004）。这一方法对于评估胁迫反应十分重要，因为 mRNA 水平并不总是与蛋白质积累相关（Gygi et al.，1999）。已知的是，在蛋白质转换和翻译后修饰上出现巨大差异可能导致转录组学数据和蛋白质组学数据之间的巨大变化。因此，进行蛋白质研究是十分必要的，能够提供关于它们水平和活性的信息（Zivy and De Vienne，2000）。以蛋白质组学为基础的技术使得大范围的蛋白质谱成为可能，这些技术是鉴定植物中关于响应生物和非生物胁迫的蛋白质的强大工具（Gygi and Aebersold，2000）。大量研究评估了在响应胁迫时植物组织中蛋白质水平的变化（Canovas et al.，2004；Kim et al.，2003）。大多数这些研究是在模式植物中进行，例如拟南芥和水稻（Canovas et al.，2004）。

关于芭蕉属植物蛋白质组学研究的报道很少。科学家研究了用高糖溶液处理香蕉茎尖分生组织后导致的显著上调或下调的蛋白质（Carpentier et al.，2007）。Samyn 等（2007）利用衍生肽的从头测序分析，也报道了香蕉（芭蕉属）的功能蛋白质组分析。Carpentier 等（2008）阐释了蛋白质组学的优点和缺陷，以及用来研究模式植物如序列信息很少的香蕉的转录组学。结论是每种技术都有其优缺点。当一个物种的特征描述很不充分，一个 2 - DE（二维凝胶电泳）方法将产生更好的特征描述。不过，如果对低丰度蛋白质或疏水质蛋白质感兴趣，建议其采用基因表达系列分析的高级衍生（SuperSAGE）方法。

研究人员已从芭蕉属植物基因组中鉴定出各种蛋白质。例如，香蕉中编码成熟性状的基因，包括基因型 AAA（Cavendish）、AAB（Rasthali 和 Poovan）、AB（Kanthali）和 ABB（Monthan）（Choudhury et al.，2008）。

比利时天主教鲁汶大学农作物生物工程部（KULeuven,）的热带作物改良实验室和安特卫普大学的植物生物化学和生理学实验室合作构建了蛋白质组数据库。这个数据库旨在促进数据的整合、相互联系和查询，并为感兴趣的研究人员提供与此项目相关的蛋白质组研究论文所采用的全部数据。目前，许多工具正在开发，未来这些工具将有助于查询和分析相关数据。要进一步了解芭蕉属寄主植物与相应的生物、非生物胁迫之间的各种相互作用，还需要在该领域进行更多的研究。

16.10　代谢组学

转录组学和蛋白质组学数据是破译复杂生物过程的重要步骤，但是光有这些数据还不足以充分理解它们，因为大多数生物过程最终是由细胞代谢物介导的（Dita et al.，2006）。mRNA 可变剪接、蛋白质周转率和调节蛋白质活性的翻译后修饰暗示转录组或蛋白质组中的变化并非总是对应于细胞代谢组中的变化（Sumner et al.，2003）。因此，要想完全了解控制复杂植物代谢过程的基因功能和分子事件，唯一的方法是采用综合的方式并联分析转录组、蛋白质组和代谢组（Dixon，2001）。大规模代谢物组学研究对于评估各种胁迫下代谢物的状况十分重要（Dita et al.，2006）。

芭蕉属植物代谢组学在促进功能基因组学和基因功能注释上的研究有很大潜力。然而，科学家有必要推动代谢组学的进一步应用和采取有效策略控制和规范这门科学的技术流程和基础设施（DHL Chi and P. Schenk，未发表）。

代谢组学在选育更好的作物上发挥着重要作用，因为这项技术能鉴定与关键性状相联系的分子标记，这能帮助科学家重新排列代谢相互作用，以及重新设计或选择含有目标表型和基因型的作物（Hall et al.，2008）。代谢物谱还能推动基因组学在未来植物育种计划中的应用。例如，关于香蕉的苯基丙酸类合成路径的代谢物组学研究和对血皮草科植物的比较基因组研究揭示了这两个物种之间的密切关系（Kamo et al.，2000）。科学家可以从血皮草科模式植物找出生物合成苯基丙酸类路径，然后利用来自该路径的信息来修改香蕉果实中植物抗毒素的遗传基础，影响香蕉的成熟性状。关于植物代谢物的研究显著扩大了对植物蛋白质和遗传学的认识，代谢物谱出现的变化或许能解释控制代谢路径的蛋白质和基因上的相关变化。对比蛋白质组学、代谢组学，此方法呈现更高的通量和更经济的方向（Hollywood et al.，2006）。最终转录组学、蛋白质组学、代谢组学和生物信息学的融合将为芭蕉属植物遗传改良提供全面的数据。

16.11 转录因子

转录因子（TF）是大多数生化途径中在控制基因表达上起到重要作用的蛋白质（Kasuga et al.，1999；Eulgem，2005）。基因组学研究在过去几年已确定了大量的 TF（主要是拟南芥中的）并揭示了其在转录水平调节响应胁迫的基因表达有高度复杂性和重叠性（Shinozaki and Yamaguchi-Shinozaki，2000）。对 TF 作用的认识可能会为提高对胁迫的抗性或耐受性开辟新途径（Singh et al.，2002）。一个既定的 TF 可以介导对各种胁迫的响应（Eulgem，2005；Yamaguchi-Shinozaki and Shinozaki，2005）。这一特性使得 TF 在基因转化上特别受到科学家关注，因为单个的 TF 基因可导致对各种胁迫的抗性或耐受性。根据这一原则，调节 ABA 应答基因的 TF 的过度表达赋予水稻多重胁迫的耐受性（Kim et al.，2003）。然而，不同的 TF 能响应同一胁迫，其对应的响应程度不同，但具有重叠动力学效应（Onate-Sanchez and Singh，2002）。将 TF 用于作物遗传改良需要对它们的生物学功能有全面的了解。

16.12 香蕉与可食用疫苗

随着重组 DNA 技术的出现，研究人员一直试图培育出能对传染疾病免疫的转基因植物。研究已经表明，遗传改造的植物可以作为疫苗。作为疫苗的植物具有的优势是生产成本低廉，因此发展中国家可以更容易地生产。可食用植物疫苗为安全有效的口服疫苗提供了巨大的新机遇。最近，香蕉已被视为开发疫苗的理想载体，因为它易于消化、口感好，而且儿童易于接受。关于开发不同的香蕉疫苗仍在进行研究（Sala et al.，2003；Arntzen et al.，2005；Kumar et al.，2005）。

伤寒症、狂犬病和HIV的潜在疫苗蛋白质的表达基因已被导入香蕉中（Miller，2006）。这些产品正等待临床试验。植物制成疫苗的未来在于研究人员发展出抗重要疾病的疫苗，并测试这些疫苗在宿主上的效力。随着疫苗学这个新兴发展领域中不断积聚实验数据，植物表达系统相较于传统系统的优势将继续展现（Rice et al.，2005）。

16.13 体细胞无性系变异和体外诱变

组织培养产生遗传变异的能力还未在芭蕉属植物遗传改良上得到完全利用。组织培养在植物中产生了一系列可被并入植物育种项目的变异（Jain，2001）。众所周知，涉及愈伤组织培养和体细胞胚胎发生的体细胞无性系变异有可能产生遗传变异（Larkin and Scowcroft，1981）。已有报道称，芭蕉属植物中有通过器官发生和体细胞胚胎发生而产生有农用价值的体细胞无性系的可能（Jain et al.，2011）。体外诱变策略，如用甲基磺酸乙酯（EMS）、快中子辐射和插入诱变来诱发，已应用于植物育种。这些方法分别诱导点突变、缺失或插入，并且在育种中对于作物中的生物胁迫（Bhagwatand Duncan，1998；Kowalski and Cassells，1999）和非生物胁迫（Khan et al.，2001；Fuller and Eed，2003）都十分有用。

这些技术的主要难点在于需要大量的个体才能找到期望的性状。尽管如此，通过体外选择系统，这个不足可被最小化。

利用体外选择可以筛选源自常规育种和转基因方法的推定抗胁迫品系。特别引人注目的是适用于一些无法使用合适筛选方法或筛选效率很低的非生物胁迫方法。虽然近年来高通量技术凭借自己本身的优势再辅以遗传转化，逐渐成为有吸引力的手段，但是体细胞无性系变异和体外诱变，以及随后的体外选择为育种提供了可替代方法（Dita et al.，2006）。

16.14 营养和生物强化

16.14.1 香蕉营养价值和药物生产

第4章说明了香蕉是对健康有益的植物化学物质的重要来源。许多研究报道了香蕉中的有用化合物（Wall，2006；Davey et al.，2007，2009；Arora et al.，2008；Cohen et al.，2009a，b；Amorim et al.，2011；Fungo and Pillay，2011）。香蕉富含类胡萝卜素及其前体物、维生素C、总多酚、类黄酮和具有抗氧化活性的化合物（Amorim et al.，2011），因此可作为制药行业原材料的潜在来源。例如，Englberger等（2003a，b，c）鉴定出香蕉基因型，称为 Fei 香蕉（澳蕉组），它的β-胡萝卜素含量为 560～63 600 μg/kg。来自密克罗尼西亚群岛的基因型（Karat 和 Uht en Yap，*Musa troglodytarum*）的类胡萝卜素含量是卡文迪什香蕉的 275 倍，因此这一基因型在减少与维生素 A 和/或其他微量营养素缺乏有关的健康问题上有潜在价值（Amorim et al.，2011）。这些香蕉可能用于制药工业，成为β-胡萝卜素的理想来源。香蕉含有高浓度的血清素和多巴胺，这是两个神经递质，直接控制某些激素的释放并调节睡眠和食欲的昼夜节律（Adao et al.，2005）。Kanazawa 和 Sakakibara（2000）发现，香蕉中多巴胺这种天然的强抗氧化剂在卡文迪什香蕉中的含量为：皮中 800～5 600 mg/kg，果肉中 25～100 mg/kg。多巴胺容易被人体吸收，被认为对于控制帕金森氏病起到重要作用。Someya 等（2002）发现卡文迪什香蕉中抗氧化剂没食子儿茶素在皮中的含量（每 100 g 干重中含有 158 mg）高于在果肉中的含量（每 100 g 干重中含有 29.6 mg），而 González-Montelongo 等（2010）量化了香蕉中的抗氧化活性，并观察到皮中含有大量的多巴胺和 L-多巴胺这两种包含大量抗氧化活性的儿茶酚胺。这些结果表明，香蕉可被视为药品的重要来源。

16.14.2 生物强化

生物强化是将自然强化食品提供给相对偏远地区的营养不良人口的一种方法，因为在这些地区市场上不容易出现强化食品或人们负担不起强化食品（Bouis，2003）。食品本身每日可提供的微量营养素与人体每日建议摄取量的比较，是对食物鉴定的一个重要方面，特别是维生素 A、铁和锌。在撒哈拉以南非洲的一些低收入国家，饮食中缺乏这些元素将会导致一些重大健康问题（Fungo and Pillay，2011）。维生素 A 对健康有许多有益的效应，如预防癌症、心血管疾病、白内障和视网膜黄斑疾病，以及神经、炎性和免疫疾病（Arora et al.，2008）。低收入国家的大多数人主要通过植物性食物获取维生素 A（Van de Berg et al.，2000）。因此，源自植物食物的维生素 A 的前体物，尤其是β-胡萝卜素对于减轻发展中国家的维生素 A 缺乏症是非常重要的（Kidmose et al.，2007）。通过育种或基因工程提高香蕉的营养价值可对数百万依赖于香蕉作为主食的人的健康和福祉产生积极影响。香蕉的生物强化可以通过常规育种实现，此外如果已鉴定出调控微量营养素的对应基因，还可通过转基因的方法来实现生物强化。评估香蕉矿物成分的一些研究已经公布。关于转化香蕉以提高其营养状态的单个研究正在进行（第 4 章）。要让香蕉生物强化成为现实还需进行更多的研究。

16.14.3 香蕉的采后产品和膳食纤维

香蕉加工产品的发展速度相较于其他作物比较缓慢。收获之后损失的大量香蕉，特别是南美香蕉主产国家生产的无法用于出口的大量香蕉刺激了香蕉加工发展（Narayana and Pillay，2011）。尚未有不同国家采后香蕉损失的确切数据，但是在热带地区由于各种各样的腐烂，一般会损失 40%～50%的水果和蔬菜（Mejia，2003）。

用成熟和未成熟的水果生产面粉，然后把面粉制成各种创新产品，如慢消化饼干（Aparicio-Saguilan et al.，2007）、高纤维面包（Juarez-Garcia et al.，2006）和可食用薄膜（Sothornvit and Pitak，2007），这种生产方式可提高香蕉和大蕉的利用率。香蕉面粉做的烘焙食品血糖指数低，可以作为膳食助剂，满足人群对低热量的要求（Juarez-Garcia et al.，2006）。生香蕉由于其果肉的淀粉含量，高水平的纤维素、半纤维素和木质素，使其成为难消化的碳水化合物的良好来源。

生物技术的进步正被应用到食品中来促进新产品的生产。加工工业的发展促进了香蕉加工产品的生产（薯条、面粉、干果、果酱和饮料）。虽然很多创新实验室有关于香蕉产品加工的研究，但是这些产品仍未被投入工业中（Aurore et al.，2009）。香蕉可生产乙醇的事实表明，香蕉可以用于生产生物燃料。

香蕉丰富的遗传多样性为任何尝试探索芭蕉属植物各种用途的项目提供了宝贵财富。要想更好地利用香蕉和大蕉，可以调查它们对不同类型加工的适用性（Aurore et al.，2009）。香蕉对于食品和非食品加工行业都是一个潜力巨大的原料来源。

16.15　生物和非生物胁迫

16.15.1　生物胁迫

关于影响香蕉生物胁迫范围的研究仍然十分广阔，一些新的疾病，如黄单胞菌枯萎病的发病率为科学家开辟了新的研究道路。研究人员已培育出对黑叶斑病、黄单胞菌枯萎病和线虫有抗性的转基因植物，但它们在田间的有效性尚未被验证。

16.15.2　非生物胁迫

虽然科学家关于影响香蕉的生物胁迫，如真菌疾病、线虫、昆虫、病毒和细菌进行了大量研究，但是很少有研究是关于非生物胁迫的，如干旱、洪涝、盐碱化、金属毒性、矿物质缺乏、不利的 pH、不利的温度和空气污染。

16.15.3　气候变化

据预测，气候变化将导致气温和降水量在一些地区上升，而在另一些地区下降。当前文献上反映的普遍共识是，气候变化将对大多数作物造成不利影响。香蕉是热带、亚热带作物，通常情况下最适合生长于降水量高的温暖气候。香蕉对气候条件非常敏感（Robinson，1996）。这样的变化会影响土壤的温度和湿度水平，也决定了有益生物和害虫的活力。雨量是香蕉生长和生产的主要限制因素。充足的水分对于植物至关重要，特别是在植物萌发和果实发育期间。影响香蕉生产力的非生物胁迫中，干旱的危害性最大。干旱和盐碱化在许多地区普遍存在，预计到 2050 年 50%以上的耕地将严重盐碱化（Vinocur and Altman，2005）。鉴于这些因素，科学家有必要开展芭蕉属作物对干旱的反应及对气候变化导致的极端温度耐受性状的研究。

16.15.4 芭蕉属植物抗旱性研究进展

干旱是影响全球农业的最大环境制约因素之一（Boyer，1982）。因此，培育出耐旱的农作物具有巨大的经济意义。关于香蕉抗旱性的传统育种研究事实上并不存在。2004 年，已在乌干达开展耐旱香蕉品种的尝试及鉴定工作（M. Pillay，未发表）。

对芭蕉属植物耐旱性状和相关基因的鉴定将推动培育在水资源受限条件下产量稳定的品种。为了制订促进香蕉耐旱的新策略，科学家的主要研究目标之一是了解作物耐旱状态下潜在的分子机制（Umezawa et al.，2006）。干旱会触发一系列植物反应，包括基因表达的改变，代谢物的累积，如植物激素脱落酸（ABA）或渗透活性化合物，以及合成特异性蛋白质（例如，主要是亲水性蛋白，清除氧自由基的蛋白，伴侣蛋白）（Reddy et al.，2004）。耐旱性基因工程的基本策略是引入直接参与这些活动的功能基因。生物技术加深了对植物在遭遇干旱时，在分子水平和植株水平的优先应答的理解，并且有可能识别数百个干旱条件下诱导的基因，而其中一些基因已被克隆。利用 DNA 微阵列技术可以高通量分析不同的 mRNA 表达，现已有数百个胁迫诱导的基因被鉴定，并作为基因工程的候选基因（Umezawa，2006）。

从基因表达模式到转基因植物等一系列工具，都可用来更好地了解耐旱机制。新技术，如全基因组工具、蛋白质组学、稳定同位素和热成像或荧光成像，可帮助跨越基因型和表型之间的鸿沟。基因工程和分子标记技术这两个主要的生物技术方法，现在正被用来培育耐旱作物。

现代香蕉被认为是由古老的 A 基因组和 B 基因组杂交而成。有证据表明，芭蕉属植物的耐旱性与 B 基因组相关。香蕉的根系浅，叶面积指数大，并且需要大量的水才能保证高产量。对于年降水量低于 1 200 mm 的地区，干旱导致的产量损失估计为 40%（R. Swennen，KULeuven，之前的评论）。因此，有必要更好地认识芭蕉属植物的抗旱性和耐旱品种。

科学家正在研究利用不同的渗透作用来鉴定芭蕉属植物耐旱品种（体外）。初步调查显示，Cachaco（ABB）和Lep Chang Kut（ABB）耐旱，而Mbwazirume（AAA）易受干旱影响。对不同香蕉品种的水分利用效率的表型鉴定将利用高通量蛋白质组分析，验证这些结果的转录组水平，并将表型关联到基因组结构（R. Swennen，KULeuven，之前的评论）。科学家需要调查具有很强生物多样性的芭蕉属植物的耐旱性。将生理学、蛋白质组学、转录组学和基因组学这几个领域结合起来，在未来有可能阐明芭蕉属植物耐旱的复杂机制。鉴定出的耐旱品种本身就可以使用或用于杂交以培育出对易旱环境适应力更强的植物。

16.15.5　耐寒性

已有报道关于确定耐寒香蕉品种的试验（Ram et al.，2008）。之前的研究表明，田间环境的一些香蕉品种显示出耐霜冻。科学家发现甜蕉种质中的4个品种具有耐受性，6个具有中等耐受性。煮食蕉中，发现7个种质具有耐受性，而12个具有中等耐受性。研究还发现，A基因组比例更高的品种更易受霜冻害，而B基因组比例较高的种质更耐霜冻害。

16.15.6　盐胁迫

有报道指出关于芭蕉属植物盐胁迫的研究在香蕉品种中进行（Gomes et al.，2002；Willadino et al.，2011），除了在感染了丛枝菌根真菌的Pacovan品种中进行（Yano-Melo et al.，2003），还在诱导突变体中进行（Miri et al.，2009）。这些研究表明，虽然一些香蕉品种能够承受一定量的盐胁迫，但是要获得关于作物中盐胁迫的更多信息，研究人员还须进一步的研究。

16.16　芭蕉属种质保存和交换

由于气候变化，芭蕉属的野生近缘种正处于危险之中。对于提高栽培作物抵抗生物和非生物胁迫的能力所必需的那些基因来说，野生近缘种是这些基因的宝贵来源。目前香蕉育种计划已使用野生近缘种作为抵抗香蕉黑叶斑病、线虫、香蕉象鼻虫和镰孢枯萎病及大果串体积等问题的抗性材料。许多研究表明，驯化已经使香蕉的遗传变异变窄（Pillay et al.，2001；Nyine and Pillay，2011）。气候变化可能会使得对香蕉野生近缘种的依赖更强，因为气候变化可能导致香蕉生产地区的气候过热、过冷、过湿或过干。目前科学家迫切需要做的是在香蕉野生近缘种消失之前收集它们并鉴定其各项特征。目前，现有收集保存的只是野外的野生物种多样性的一部分。

16.17 双单倍体

双单倍体（DH）技术是指利用孢子或花药培养来获得单倍体胚胎。这项技术为育种者提供了快速产生纯合系的工具。这些纯合系可以倍增和作为品种发布，或与重组自交系一样用于分子作图和/或用于育种计划（Martinez et al.，2002）。高效的 DH 生产技术可以大大减少品种培育的时间和成本（Liu et al.，2002）。

16.18 结论

除了传统育种方法，现在已有多种生物技术方法用于香蕉遗传改良。过去的 20 年中生物技术已成为育种最有前景的工具，用它来克服作物中的病虫害。虽然这些技术在一些作物中取得了显著进展，但是在芭蕉属植物中的进展仍然有限。现今依靠组织培养、遗传转化和 MAS 等技术，以及强大的新“组学”技术的进步，为突破现状创造了巨大可能（Dita et al.，2006）。基因组学辅助育种方法、生物信息学技术和代谢组学资源正逐渐成为全球作物改良计划的重要组成部分。在作物基因组测序、高分辨率遗传图和准确分型上取得的进展将加快发现与植物育种目标性状相关的功能等位基因和等位基因变异（Ortiz，2011）。本章着重介绍了作物育种研究中潜在的新领域和促进芭蕉属植物遗传改良的新兴技术。

缩略语

DArT：多样性阵列技术
HRM：高分辨率熔解
RNAi：RAN 干扰
TF：转录因子
TILLING：定向诱导基因组局部突变技术

参考文献

Adao R C，Beatriz A，Glória M，2005. Bioactive amines and carbohydrate changes during ripening of ‘Prata’ banana（*Musa acuminata* × *M. balbisiana*）[J]. Food Chem，90：705 - 711.

Amorim E P，de Oliveira e Silva S，de Oliveira Amorim V B，et al，2011. Quality improvement of cultivated *Musa* [M]// Pillay M，Tenkouano A. Banana breeding：progress and challenges. Boca Raton, FL, USA: CRC Press: 251 - 267.

Anderson N G, Matheson A, Anderson N L, 2001. Back to the future: the human protein index (HPI) and the agenda for post proteomic biology [J]. Proteomics, 1: 3 - 12.

Aparicio-Saguilan A, Sayago-Ayerdi S G, Vargas-Torres A, et al, 2007. Slowly digestible cookies prepared from resistant starch-rich lintnerized banana starch [J]. J Food Compo Anal, 20: 175 - 181.

Arora A, Choudary D, Agarwal G, et al, 2008. Compositional variation in β - carotene content, carbohydrate and antioxidant enzymes in selected banana cultivars [J]. Int J Food Sci Technol, 43: 1913 - 1921.

Arntzen C, Plotkin S, Dodet B, 2005. Plant-derived vaccines and antibodies: potential and limitations [J]. Vaccine, 23: 1753 - 1756.

Aurore G, Parfait B, Fahrasmane L, 2009. Bananas, raw materials for making processed food products [J]. Trends Food Sci Technol, 20: 78 - 91.

Bernardo R, 2008. Molecular markers and selection for complex traits in plants: learning from the last 20 years [J]. Crop Sci, 48: 1649 - 1664.

Bhagwat B, Duncan E J, 1998. Mutation breeding of banana cv Highgate (*Musa* spp., AAA Group) for tolerance to *Fusarium oxysporum* f. sp. *cubense* using chemical mutagens [J]. Sci Hort, 73: 11 - 22.

Bouis H E, 2003. Micronutrient fortification of plants through plant breeding: can it improve nutrition in man at low cost [J]. Proc Nutr Soc, 62: 403 - 411.

Boyer J S, 1982. Plant productivity and environment [J]. Science, 218: 443 - 448.

Cahill D J, Nordhoff E, O'Brien J, et al, 2000. Bridging genomics and proteomics [M]// Pennington S, Dunn M. 2000 Proteomics. New York, USA: BIOS Scientific Publishers: 1 - 17.

Canovas F, Dumas-Gaudot E, Recorbet G, et al, 2004. Plant proteome analysis [J]. Proteomics, 4: 285 - 298.

Carpentier S C, Witters E, Laukens K, et al, 2007. Banana (*Musa* spp.) as a model to study the meristem proteome: acclimation to osmotic stress [J]. Proteomics, 7: 92 - 105.

Carpentier S C, Coemans B, Podevin N, et al, 2008. Functional genomics in a non-model crop: transcriptomics or proteomics [J]? Physiol Planta, 133: 117 - 130.

Chitwood D J, 2003. Research on plant-parasitic nematode biology conducted by the United States Department of Agriculture-Agricultural Research Service [J]. Pest Manag Sci, 59: 48 - 753.

Choudhury S R, Roy S, Sengupta D N, 2008. Characterization of cultivar differences in β - 1,3 - glucanase gene expression, glucanase activity and fruit pulp softening rates during fruit ripening in three naturally occurring banana cultivars [J]. Plant Cell Rep, 28: 1641 - 1653.

Cohen K O, Amorim E P, Paes N S, et al, 2009a. Teores de fl avonóides e polifenóis totais em genótipos diplóides de bananeira [J]. Comunicado Técnico. Embrapa Mandioca e Fruticultura, 1: 1 - 4.

Cohen K O, Amorim E P, Paes N S, et al, 2009b. Vitamina C, carotenóides, compostos fenólicos e atividade antioxidante em genótipos de banana [J]. Comunicado Técnico. Embrapa Mandioca e Fruticultura, 2: 1 - 4.

Colbert T, Till B J, Tompa R, et al, 2001. High-throughput screening for induced point mutations [J]. Plant Physiol, 126: 480 - 484.

Davey M W, Stals W, Ngoh-Newilah G, et al, 2007. Sampling strategies and variability in fruit pulp micronutrient contents of West and Central African bananas and plantains (*Musa* sp.) [J]. J Agric Food Chem, 55: 2633 - 2644.

Davey M W, Van den Berg I, Markham R, et al, 2009. Genetic variability in *Musa* fruit provitamin A carotenoids, lutein and mineral micronutrient contents [J]. Food Chem, 115: 806 - 813.

Dita M A, Rispail N, Prats E, et al, 2006. Biotechnology approaches to overcome biotic and abiotic stress in legumes [J]. Euphytica, 147: 1 - 24.

Dixon R A, 2001. Natural products and plant disease resistance [J]. Nature, 411: 843 - 847.

Dolezel J, Bartos J, 2005. Plant DNA flow cytometry and estimation of nuclear genome size [J]. Ann Bot, 95: 99 - 110.

Englberger L, Darnton-Hill I, Coyne T, et al, 2003a. Carotenoid-rich bananas: a potential food source for alleviating vitamin A deficiency [J]. Food Nutr Bull, 24: 303 - 318.

Englberger L, Schierle J, Marks G C, et al, 2003b. Micronesian banana, taro, and other foods: newly recognized sources of provitamin A and others carotenoids [J]. J Food Compos Anal, 16: 3 - 19.

Englberger L, Aalbersberg W, Ravi P, et al, 2003c. Further analyses on Micronesian banana, taro, breadfruit and other foods for provitamin A carotenoids and minerals [J]. J Food Compos Anal, 16: 219 - 236.

Erali M, Voelkerding K V, Wittwer C T, 2008. High resolution melting applications for clinical laboratory medicine [J]. Exp Mol Pathol, 85: 50 - 58.

Escobar M A, Civerolo E L, Summerfelt K R, et al, 2001. RNAi-mediated oncogene silencing confers resistance to crown gall tumorigenesis [J]. Proc Natl Acad Sci USA, 98: 13437 - 13442.

Eulgem T, 2005. Regulation of the *Arabidopsis* defense transcriptome [J]. Trends Plant Sci, 10: 71 - 78.

Faure S, Noyer J L, Horry J P, et al, 1993. A molecular marker-based linkage map of diploid bananas (*Musa acuminata*) [J]. Theor Appl Genet, 87: 517 - 526.

Fire A, Xu S, Montgomery M K, et al, 1998. Potent and specific genetic interference by double-stranded RNA in Caenorhabditis elegans [J]. Nature, 391: 806 - 811.

Fuller M P, Eed M H I, 2003. The development of multiple stress resistant cauliflower using mutagenesis in conjunction with a microshoot tissue culture technique [J]. Acta Hort, 618: 71 - 76.

Fungo R, Pillay M, 2011. β-carotene content of selected banana genotypes from Uganda [J]. Afr J Biotechnol (in press).

Gomes E W F, Willadino L, Martins L S S, et al, 2002. The effects of salinity on five banana genotypes (*Musa* spp) [J]. Dev Plant Soil Sci, 92: 410 - 411.

González-Montelongo R, Lobo M G, González M, 2010. Antioxidant activity in banana peel extracts: testing extraction conditions and related bioactive compounds [J]. Food Chem, 119: 1030 - 1039.

Gosal S G, Wani S H, Kang M S, 2009. Biotechnology and drought tolerance [J]. J Crop Improv, 23: 19 - 54.

Gygi S P, Aebersold R, 2000. Mass spectrometry and proteomics [J]. Curr Opin Chem Biol, 4: 489 - 494.

Gygi S P, Rochon Y, Franza B R, et al, 1999. Correlation between protein and mRNA abundance in yeast [J]. Mol Cell Biol, 19: 1720 - 1730.

Hall R D, Brouwerand I D, Fitzgerald M A, 2008. Plant metabolomics and its potential application for human nutrition [J]. Physiol Planta, 132: 162 - 175.

Hannon G J, 2002. RNA interference [J]. Nature, 418: 244 - 251.

Heslop-Harrison J S, Schwarzacher T, 2007. Domestication, genomics and the future for banana [J]. Ann Bot, 100: 1073 - 1084.

Hippolyte I, Bakry V, Seguin M, et al, 2010. A saturated SSR/DArT linkage map of *Musa acuminata* addressing genome rearrangements among bananas [J]. BMC Plant Biol, 10: 65.

Hollywood K, Brison D R, Goodacre R, 2006. Metabolomics: current technologies and future trends [J]. Proteomics, 6: 4716 - 4723.

Jackson A L, Bartz S R, Schelter J, et al, 2003. Expression profiling reveals off-target gene regulation by RNAi [J]. Nat Biotechnol, 21: 635 - 637.

Jain S M, 2001. Tissue culture-derived variation in crop improvement [J]. Euphytica, 118: 153 - 166.

Jain S M, Till J, Suprasanna P, et al, 2011. Mutations and cultivar development of banana [M] // Pillay M, Tenkouano A. Banana breeding: progress and challenges. Boca Raton, FL, USA: CRC Press: 203 - 217.

Juarez-Garcia E, Agama-Acevedo E, Sayago-Ayerdi S G, et al, 2006. Composition, digestibility and application in breadmaking of banana flour [J]. Plant Foods Human Nutr, 61: 131 - 137.

Kamo T, Hirai N, Tsuda M, et al, 2000. Changes in the content and biosynthesis of phytoalexins in banana fruits [J]. Biosci Biotechnol Biochem, 64: 2089 - 2098.

Kanazawa K, Sakakibara H, 2000. High content of dopamine, a strong antioxidant, in Cavendish banana [J]. J Agric Food Chem, 48: 844 - 848.

Kasuga M, Liu Q, Miura S, et al, 1999. Improving plant drought, salt, and freezing tolerance by gene transfer of a single stress-inducible transcription factor [J]. Nat Biotechnol, 17: 287 - 291.

Kersten B, Bürkle L, Kuhn E J, et al, 2002. Large-scale plant proteomics [J]. Plant Mol Biol, 48: 133 - 141.

Khan A J, Hassan S, Tariq M, et al, 2001. Haploidy breeding and mutagenesis for drought tolerance in wheat [J]. Euphytica, 120: 409 - 414.

Kidmose U, Christensen L P, Agili S M, et al, 2007. Effect of home preparation practices on the content of provitamin A carotenoids in coloured sweet potato varieties (*Ipomoea batatas* Lam.) from Kenya [J]. Innov Food Sci Emerg Technnol, 8: 399 - 406.

Kilian A, Huttner E, Wenzl P, et al, 2005. The fast and the cheap: SNP and DArT-based whole genome profi ling for crop improvement [C]//Proceedings of the international congress "In the wake of the double helix: from the green revolution to the gene revolution". 27 - 31 May 2003, 443 - 461. Avenue Media, Bologna, Italy.

Kim S T, Cho K S, Yu S, et al, 2003. Proteomic analysis of differentially expressed proteins induced by rice blast fungus and elicitor in suspension-cultured rice cells [J]. Proteomics, 3: 2368 - 2378.

Kowalski B, Cassells A C, 1999. Mutation breeding for yield and *Phytophthora infestans* (Mont.) de Bary foliar resistance in potato (*Solanum tuberosum* L-cv. Golden Wonder) using computerized image analysis in selection [J]. Potato Res, 42: 121 - 130.

Kumar G B, Ganapathi T R, Revathi C J, et al, 2005. Expression of hepatitis B surface antigen in transgenic banana plants [J]. Planta, 222: 484 - 493.

Larkin P J, Scowcroft W R, 1981. Somaclonal variation—a novel source of variability from cell-cultures for plant improvement [J]. Theor Appl Genet, 60: 197 - 214.

Liu J H, Xu X Y, Deng X X, 2005. Intergeneric somatic hybridization and its application to crop genetic improvement [J]. Plant Cell Tiss Org Cult, 82: 19 - 44.

Liu W G, Zheng M Y, Polle E A, et al, 2002. Highly efficient doubled-haploid production in wheat (*Triticum aestivum* L.) via induced microspore embryogenesis [J]. Crop Sci, 42: 686 - 692.

Ma Y, Creanga A, Lum L, et al, 2006. Prevalence of off-target effects in Drosophila RNA interference screens [J]. Nature, 443: 359 - 363.

Mao Y B, Cai W J, Wang J W, 2007. Silencing a cotton bollworm P450 monooxygenase gene by plant-mediated RNAi impairs larval tolerance of gossypol [J]. Nat Biotechnol, 25: 1307 - 1313.

Martinez V A, Hill W G, Knott S, 2002. On the use of double haploids for detecting QTL in outbred populations [J]. Heredity, 88: 423 - 431.

McCallum C M, Comai L, Greene E A, et al, 2000. Targeted screening for induced muta-

tions [J]. Nat Biotechnol, 18: 455 - 457.
Mejia D J, 2003. Preservation of fruits and vegetables by combined methods Technologies [Z]. International Tropical Fruits Network: 1 - 2.
Miller H I, 2006. Biotech production is blooming in the tropics [J]. Point of View, 26 (19): 1.
Miri S M, Mousavi A, Naghavi M R, 2009, Analysis of induced mutants of salinity resistant banana (*Musa acuminata* cv. Dwarf Cavendish) using morphological and molecular markers [J]. Iran J Biotechnol, 7: 86 - 92.
Montgomery J, Wittwer C T, Palais R, et al, 2007. Simultaneous mutation scanning and genotyping by high-resolution DNA melting [J]. Nat Protoc, 2: 59 - 66.
Nakayashiki H, Kadotani N, Mayama S, 2006. Evolution and diversification of RNA silencing proteins in fungi [J]. J Mol Evol, 63: 127 - 135.
Nakayashiki H, Nguyen Q B, 2008. RNA interference: roles in fungal biology [J]. Curr Opin Microbiol, 11: 494 - 502.
Narayana C K, Pillay M, 2011. Postharvest processed products from banana [M] // Pillay M, Tenkouano A. Banana breeding: progress and challenges. Boca Raton, FL, USA: CRC Press: 269 - 284.
Niu J H, Jian H, Xu J M, et al, 2010. RNAi technology extends its reach: engineering plant resistance against harmful eukaryotes [J]. Afr J Biotechnol, 9: 7573 - 7582.
Noyer J L, Dambier D, Lanaud C, et al, 1997. The saturated map of diploid banana (*Musa acuminata*) [C]. Plant and Animal Genome Conference V, January 12 - 16, San Diego, CA, USA.
Nyine M, Pillay M, 2011. Breeding increases diversity of bananas [J]. Acta Hort (in press).
Ortiz R, 2011. Revisiting the green revolution: seeking innovations for a changing world [J]. Chron Hort, 51 (1): 6 - 11.
Onate-Sanchez L, Singh K B, 2002. Identification of *Arabidopsis* ethylene-responsive element binding factors with distinct induction kinetics after pathogen infection [J]. Plant Physiol, 128: 1313 - 1322.
Pillay M, Tripathi L, 2006. Banana: an overview of breeding and genomics research in *Musa* [M] // Kole C. Genome mapping and molecular breeding in plants, vol. 4: fruits and nuts. Heidelberg, Germany: Springer-Verlag: 282 - 301.
Pillay M, Tripathi L, 2007. Banana breeding [M] // Kang M S, Priyadarshan P M. Breeding major food staples. Boston, MA, USA: Blackwell Publishing: 393 - 428.
Pillay M, Ogundiwin E, Nwakanma D C, et al, 2001. Analysis of genetic diversity and relationships in East African banana germplasm [J]. Theor Appl Genet, 102: 965 - 970.
Pillay M, Tenkouano A, Ortiz R, 2011. Molecular breeding of other vegetatively propagated crops: lessons for banana [M] // Pillay M, Tenkouano A. Banana breeding: progress and challenges. Boca Raton, FL, USA: CRC Press: 321 - 350.
Rafalski J A, 2002. Novel genetic mapping tools in plants: SNPs and LD-based approaches

[J]. Plant Sci, 162: 329 - 333.

Ram K, Rajan S, Singh A, 2008. Evaluation of banana germplasm (*Musa* spp.) for frost tolerance under field conditions in subtropics [J]. Indian J Hort, 65: 35 - 39.

Reddy A R, Chaitanya K V, Vivekanandan M, 2004. Drought-induced responses of photosynthesis and antioxidant metabolism in higher plants [J]. J Plant Physiol, 161: 1189 - 1202.

Reed G H, Kent J O, Wittwer C T, 2007. High-resolution DNA melting analysis for simple and efficient molecular diagnostics [J]. Pharmacogenomics, 8: 597 - 608.

Rice J, Ainley W M, Shewen P, 2005. Plant-made vaccines: biotechnology and immunology in animal health [J]. Anim Health Res Rev, 6: 199 - 209.

Richardson K L, Vales M I, Kling J G, et al, 2006. Pyramiding and dissecting disease resistance QTL to barley stripe rust [J]. Theor Appl Genet, 113: 485 - 495.

Robinson J C, 1996. Bananas and plantains [M]. Cambridge, UK: Institute of Tropical and Subtropical Crops, South Africa University Press.

Rosso M N, Jones J T, Abad P, 2009. RNAi and functional genomics in plant parasitic nematodes [J]. Annu Rev Phytopathol, 47: 207 - 232.

Sagi L, Remy S, Swennen R, 2007. Transgenic and (trans) genomic research in banana (*Musa* spp.) [J]. Afr Crop Sci Conf Proc, 8: 592 - 595.

Sala F, Rigano M M, Barbante A, et al, 2003. Vaccine antigen production in transgenic plants: strategies, gene constructs and perspectives [J]. Vaccine, 21: 803 - 808.

Samyn B, Sergeant K, Carpentier S, et al, 2007. Functional proteome analysis of the banana plant (*Musa* spp.) using de Novo sequence analysis of derivatized peptides [J]. J Proteome Res, 6: 70 - 80.

Semagn K, Bjornstad A, Ndjiondjop M N, 2006. Progress and prospects of marker assisted backcrossing as a tool in crop breeding programs [J]. Afr J Biotechnol, 5: 2588 - 2603.

Semagn K, Bjørnstad A, Xu Y, 2010. The genetic dissection of quantitative traits in crops [J]. Electronic Journal of Biotechnology, North America, 1325 10 2010.

Shinozaki K, Yamaguchi-Shinozaki K, 2000. Molecular responses to dehydration and low temperature: differences and cross-talk between two stress signaling pathways [J]. Curr Opin Plant Biol, 3: 217 - 223.

Someya S, Yoshiki Y, Okubo K, 2002. Antioxidant compounds from bananas (*Musa* Cavendish)[J]. Food Chem, 79: 351 - 354.

Singh K B, Foley R C, Onate-Sanchez L, 2002. Transcription factors in plant defense and stress responses [J]. Curr Opin Plant Biol, 5: 430 - 436.

Sothornvit S, Pitak N, 2007. Oxygen permeability and mechanical properties of banana films [J]. Food Res Int, 40: 365 - 370.

Steeves R M, Todd T C, Essig J S, et al, 2006. Transgenic soybeans expressing siRNAs specific to a major sperm protein gene suppress *Heterodera glycines* reproduction [J]. Funct Plant Biol, 33: 991 - 999.

Sumner L W, Mendes P, Dixon R A, 2003. Plant metabolomics: large-scale phytochemistry in the functional genomics era [J]. Phytochemistry, 62: 817 - 836.

Till B J, Jankowicz-Cieslak J, Sagi L, et al, 2010. Discovery of nucleotide polymorphisms in the *Musa* gene pool by Eco Tilling [J]. Theor Appl Genet, 121: 1381 - 1389.

Umezawa T, Fujita M, Fujita Y, et al, 2006. Engineering drought tolerance in plants: discovering and tailoring genes to unlock the future [J]. Curr Opin Biotechnol, 17: 113 - 122.

Van den Berg H, Faulks R, Fernando Granado H, et al, 2000. The potential for the improvement of carotenoid levels in foods and the likely systemic effects [J]. J Sci Food Agric, 80: 880 - 912.

Vilarhinos A D, 2004. Cartographie génétique et cytogénétique chez le bananier: caractérisation des translocations [Z]. Ecole Nationale Supérieure Agronomique, Montpellier, France.

Vinocur B, Altman A, 2005. Recent advances in engineering plant tolerance to abiotic stress: achievements and limitations [J]. Curr Opin Biotechnol, 16: 123 - 132.

Wall M M, 2006. Ascorbic acid, vitamin A, and mineral composition of banana (*Musa* sp.) and papaya (*Carica papaya*) cultivars grown in Hawaii [J]. J Food Compos Anal, 19: 434 - 445.

Willadino L, Gomes E W F, De F e Silva E F, et al, 2011. Effect of salt stress on banana tetraploid genotypes [J]. Rev Bras Eng Agríc Ambient, 15: 53 - 59.

Wasinger V C, Cordwell S J, Cerpa-Poljak A, et al, 1995. Progress with gene-product mapping of the Mollicutes: mycoplasma genitalium [J]. Electrophoresis, 16: 1090 - 1094.

Yamaguchi-Shinozaki K, Shinozaki K, 2005. Organization of cisacting regulatory elements in osmotic- and cold-stress-responsive promoters [J]. Trends Plant Sci, 10: 88 - 94.

Yano-Melo A M, Saggin O J, Maia L C, 2003. Tolerance of mycorrhized banana (*Musa* sp. cv. Pacovan) plantlets to saline stress [J]. Agric Ecosyst Environ, 95: 343 - 348.

Zivy M, De Vienne D, 2000. Proteomics: a link between genomics, genetics and physiology [J]. Plant Mol Biol, 44: 575 - 580.